Laboratory Exercises in Human Physiology

Laboratory Exercises in Human Physiology

A Clinical and EXPERIMENTAL Approach

Second Edition

William I. Lutterschmidt

DEPARTMENT OF BIOLOGICAL SCIENCES
SAM HOUSTON STATE UNIVERSITY

Deborah I. Lutterschmidt

CENTER FOR BEHAVIORAL NEUROSCIENCE
GEORGIA STATE UNIVERSITY

Boston Burr Ridge, IL Dubuque, IA New York San Francisco St. Louis
Bangkok Bogotá Caracas Kuala Lumpur Lisbon London Madrid Mexico City
Milan Montreal New Delhi Santiago Seoul Singapore Sydney Taipei Toronto

LABORATORY EXERCISES IN HUMAN PHYSIOLOGY: A CLINICAL AND EXPERIMENTAL APPROACH, SECOND EDITION

Published by McGraw-Hill, a business unit of The McGraw-Hill Companies, Inc., 1221 Avenue of the Americas, New York, NY 10020.

This book is printed on recycled, acid-free paper containing 10% postconsumer waste.

1 2 3 4 5 6 7 8 9 0 QPD/QPD 0 9 8

ISBN 978-0-07-337805-3
MHID 0-07-337805-4

Publisher: *Michelle Watnick*
Director of Deveopment: *Kristine Tibbetts*
Executive Editor: *Colin H. Wheatley*
Developmental Editor: *Fran Schreiber*
Marketing Manager: *Lynn M. Breithaupt*
Project Manager: *Joyce Watters*
Senior Production Supervisor: *Kara Kudronowicz*
Senior Freelance Design Coordinator: *Michelle D. Whitaker*
Cover Designer: *Christopher Reese*
(USE) Cover Image: © *Dr. David M. Phillips/Getty Images*
Senior Photo Research Coordinator: *John C. Leland*
Photo Research: *Pam Carley/Sound Reach*
Compositor: *Laserwords Private Limited*
Typeface: *10/12 Minion*
Printer: *Quebecor World Dubuque, IA*

Some of the laboratory experiments included in this text may be hazardous if materials are handled improperly or if procedures are conducted incorrectly. Safety precautions are necessary when you are working with chemicals, glass test tubes, hot water baths, sharp instruments, and the like, or for any procedures that generally require caution. Your school may have set regulations regarding safety procedures that your instructor will explain to you. Should you have any problems with materials or procedures, please ask your instructor for help.

www.mhhe.com

CONTENTS

Preface . . . . *vii*

LABORATORIES

1 Scientific Investigation 1
2 Homeostasis 13
3 Diffusion, Osmosis, and Tonicity 23
4 Enzyme Activity 35
5 Action Potentials 45
6 Reflexes 59
7 Sensory Physiology 71
8 Functional Anatomy of Muscle and Mechanics of Contraction 93
9 Physiology of Muscle Contraction 107
10 Endocrine Physiology 117
11 Cardiovascular Physiology 133
12 Physiology of Blood 147
13 Respiratory Physiology 159
14 Renal Physiology 177
15 Metabolic Rate 191

Appendix 1 Scientific Writing and Preparing a Scientific Research Paper for Peer Review 201
Appendix 2 Reference Tables 211
Appendix 3 SI Unit Prefixes and Symbols 213
Appendix 4 Schematic for an Osmosis Chamber 215

Credits *217*
Index *219*

MEET THE AUTHORS

Drs. William I. and Deborah I. Lutterschmidt are a sibling research team who have published many papers together and have made significant scientific contributions to the fields of physiological ecology and behavioral endocrinology. Their collaborations and overlap in interests and experience again come together here to offer a collection of laboratory exercises for instruction in human physiology. Both authors have more than a decade of experience in laboratory pedagogy and draw from their experience to create a laboratory text that introduces not only basic physiological principles but also the process of scientific investigation.

William I. Lutterschmidt received his Ph.D. in Comparative Physiology and Physiological Ecology from the University of Oklahoma in the fall of 1997. He remained at the University of Oklahoma for an additional semester, conducting research in the Melatonin Research Laboratory and teaching Human Physiology before accepting an assistant professor's position at Sam Houston State University. His research interests focus on temperature-dependent physiological processes and how the plasticity in these processes may influence macroecological patterns of ectothermic vertebrates. He has published articles in journals such as *Comparative Biochemistry and Physiology, Brain, Behavior and Evolution, Hormones and Behavior,* and *BioScience.* William has received research grants from the National Science Foundation (NSF) and the United States Department of Agriculture (USDA) and several research and teaching awards from his respective universities. Dr. Lutterschmidt was recently promoted to associate professor in the Department of Biological Sciences at Sam Houston State University, where he teaches Human Physiology, Comparative Physiology, and Biostatistics. He also holds an adjunct summer faculty position at the University of Oklahoma Biological Station, where he teaches a course in Experimental Design in Herpetology. He resides in Huntsville, Texas with his wife Kathryn and their dogs Allie, Sammy, Burt, and Ernie.

Deborah I. Lutterschmidt received an M.S. in Physiological Ecology from the University of Oklahoma in 2000 and her Ph.D. in Comparative Physiology and Behavioral Endocrinology from Oregon State University in 2006. Her research focuses on chronobiology and the environmental and hormonal factors that regulate seasonal rhythms in the behavior and physiology of ectotherms. She is currently a postdoctoral research fellow at the Center for Behavioral Neuroscience in Atlanta, Georgia, where she is investigating the role of the hormone melatonin in mediating seasonal neuroplasticity. Deborah has extensive teaching experience in Physiology and Human Anatomy and Physiology laboratories. She also has instructed lecture courses in Human Physiology at the University of Oklahoma and Human Anatomy and Physiology at Oregon State University. Both universities have recognized her efforts and dedication to teaching with several teaching awards. Her hobbies include hiking and camping with her husband, Bill, and their Border collie mix, Kodiak. They reside in Atlanta, Georgia.

PREFACE

This laboratory text has been developed to introduce important and fundamental concepts in physiology. This text and its experiments will help students to identify and understand the functional bases of several organ systems and how these systems maintain homeostasis. More importantly, this text is designed to emphasize an experimental approach to teaching physiology and is therefore designed for a particular student and curriculum. Many of the experiments will help students to develop a clinical knowledge of physiology and to gain an appreciation for the clinical techniques needed by students in nursing, physical therapy, and related health-oriented fields. Although this laboratory text may be used independently, its presentation and format will follow closely *Vander's Human Physiology*.

The Experimental Approach

We hope this laboratory text appeals to instructors who are as concerned with students understanding the process of scientific investigation as they are with students understanding basic physiological concepts. It is in this combination of clinical and experimental approaches to the pedagogy of human physiology that most laboratory texts fail and we hope to succeed. We provide a laboratory exercise on scientific investigation to introduce basic data analyses used in physiology. Each laboratory is also organized in the format of a scientific publication with Introduction, Materials and Methods, Results, and Discussion sections. An appendix is dedicated to the topic of writing a scientific research paper.

Comparative Notes

Often, students of human physiology lose sight of taxonomic diversity and the unique physiologies among animals. Many of those who teach human physiology are comparative animal physiologists and physiological ecologists. We have found that the study of physiology is most intriguing to both students and instructors when physiological processes are placed in the broader context of physiological adaptation. Each laboratory offers a *"Comparative Note"* to introduce students to a variety of physiological solutions to the unique challenges of a species' environment. These comparative notes also list representative citations from the literature entitled *"Research of Interest"* to help direct the intrigued student to additional resources on the particular physiological topic.

Why Use This Text

Many human physiology laboratory manuals are exhaustive in that they present an overwhelming amount of material that an instructor either (1) finds inappropriate for an introductory course in human physiology or (2) cannot cover due to the limited time in a typical semester course. This laboratory text is not exhaustive in its choice of exercises, but rather has 15 carefully selected laboratories that coincide nicely with a typical semester course in Human Physiology. These 15 laboratories will allow students to master fundamental principals in human physiology without overwhelming them with superfluous material.

Other Advantages of this Laboratory Text:

- This is the only human physiology laboratory text that introduces students to physiological concepts using the scientific method.
- Each laboratory is organized in the format of a scientific publication with Introduction, Materials and Methods, Results, and Discussion sections.

- Each laboratory has a thorough introduction to the topics investigated in the experiments. This thorough introduction, in combination with a series of pre-lab exercises, will provide students with clear expectations for how to prepare for each laboratory session.
- Students should complete the pre-lab exercises before each laboratory session. These pre-lab exercises may be collected by the instructor at the beginning of the laboratory session in addition to or in lieu of a pre-lab quiz. Note that all pre-lab exercises can be removed from the laboratory manual without removing any necessary procedures for conducting the laboratory experiments.
- An instructor's manual is provided online at www.mhhe.com/labcentral. Simply contact your local McGraw-Hill sales representative to obtain the appropriate user name and password.
  - The online instructor's manual provides electronic data tables for printing overheads or projection, which will aid in the collection of class data.
  - The online instructor's manual includes an Answer Key for all pre-lab exercises and laboratory reports.
  - The online instructor's manual includes a Preparation Guide with information for obtaining or making laboratory materials.
- The laboratory text is organized into weekly laboratories. The different laboratories are indexed by colored tabs on the right-hand side of the text to help index information.
- This laboratory text utilizes Physiology Interactive Lab Simulations (Ph.I.L.S) as an alternative to classical animal-based experiments. These computer simulations demonstrate the necessary physiological concepts but do not complicate the short laboratory period with animal care issues, malfunctioning equipment, limited computer equipment (which results in students crowding around the apparatus), or, most importantly, student and instructor frustration.
- **Physiology Interactive Lab Simulations (Ph.I.L.S) 3.0.** A copy of the Ph.I.L.S CD-ROM is included with your lab manual and contains 37 lab simulations that allow students to perform experiments without using expensive lab equipment or live animals. This easy-to-use software offers students the flexibility to change the parameters of every lab experiment, with no limit to the number of times a student can repeat experiments or modify variables. The power to manipulate each experiment reinforces key physiology concepts by helping students to make predictions, view outcomes, and draw conclusions.

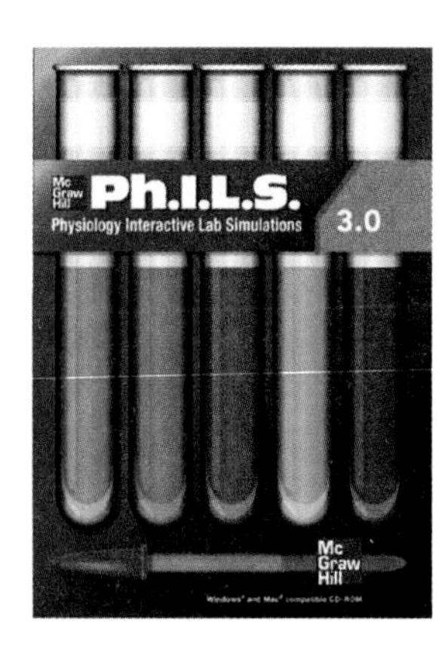

To The Student

This laboratory guide has been developed with a particular student and curriculum in mind. Laboratory exercises allow you to become familiar with fundamental concepts in physiology through the process of scientific investigation. This process includes:

- Formulating hypotheses
- Data collection
- Simple data analyses
- Evaluation of results
- Discussion of results through scientific writing

In addition to providing hands-on experience in applying physiological concepts, these laboratory exercises also promote interaction between students and the instructor while experiments are conducted and data are collected. The exercises in this laboratory guide assume that you have had formal course work or a background in:

- Introductory Biology
- Introductory Chemistry

Organization of This Laboratory Text

Both the instructor and student should note that this laboratory text will not review topics such as microscope use, cell biology, histology, or human anatomy. Many laboratory manuals are thorough in their reviews of these topics. We have deleted such activities because we believe that students should have mastered such topics in coursework prior to taking a course in human physiology. The organization of each laboratory is as follows:

- Each laboratory begins with a concise statement of **purpose** and introduces the concepts to be addressed.

- Following the purpose, **learning objectives** are presented so students know what should be mastered during each laboratory.
- **Materials** required for each experiment are listed to aid in laboratory preparation.
- An **introduction** to the laboratory presents the essential information for understanding the physiological significance of the laboratory experiments. Each introduction includes *pre-lab exercises* and *questions* that should be completed before the student attends the laboratory. Students will need to refer to their physiology textbooks to answer some of the questions presented in the pre-lab exercises.
- An easy and concise **materials and methods** section describes the procedures for each experiment in a step-by-step fashion.
- A **laboratory report** presented as *Results and Discussion* follows each set of experiments to examine a student's understanding of the physiological concepts investigated in the laboratory.

Acknowledgements

We express our sincere gratitude to colleagues who reviewed this laboratory text. Their comments and suggestions were most valuable and appreciated. We sincerely thank:

Janis Beaird
University of West Alabama

Carol A. Britson
University of Mississippi

Michael J. Buono
San Diego State University

Pat Clark
Indiana University—Purdue University, Indianapolis

Emma Coddington
Willamette University

Jonathan R. Day
California State University, Chico

Nida Schweil-Elmuti
Eastern Illinois University

Ralph E. Ferges
Palomar College

John P. Harley
Eastern Kentucky University

Allan S. Helgeson
Des Moines Area Community College

Karen L. Hinkle
Norwich University

Victor H. Hutchison
University of Oklahoma

John Arthur Knesel
The University of Louisiana at Monroe

Andrew J. Lokuta
University of Wisconsin—Madison

Virginia A. Pascoe
Mt. San Antonio College

Mark A. Shoop
Tennessee Wesleyan College

Lynnette Sievert
Emporia State University

Faith M. Vruggnik
Kellogg Community College

Robert Waldeck
University of Scranton

R. Doug Watson
University of Alabama at Birmingham

Richard Wetts
University of California, Irvine

Justin K. Williams
Sam Houston State University

References:

Fox, S. I. 2006. *Human Physiology: Concepts and Clinical Applications.* 11th Edition. McGraw-Hill Co., Dubuque, IA, p. 438.

Human Anatomy and Physiology Laboratory Manual. Department of Zoology. Oregon State University. Corvallis, Oregon.

Human Physiology Laboratory Manual. Department of Zoology. University of Oklahoma. Norman, Oklahoma.

Odenweller, C. M., C. T. Hsu, E. Sipe, J. P. Layshock, S. Varyani, R. L. Rosian, and S. E. DiCarlo. 1997. Laboratory exercise using "virtual rats" to teach endocrine physiology. Advances in Physiological Education 273:S24–S40.

Pflanzer, R. G. 1999. *Experimental and Applied Physiology.* 6th Edition. McGraw-Hill Co., Dubuque, IA. p. 550.

Randall, D., W. Burggren, and K. French. 2002. *Animal Physiology: Mechanisms and Adaptations.* New York: W. H. London: Freeman and Company, p. 736.

Sanford, G. M., W. I. Lutterschmidt, V. H. Hutchison. 2002. *The Comparative Method Revisited.* BioScience 52:830–836.

Schmidt-Nielsen, K. 1985. *Animal Physiology: Adaptation and Environment.* London: Cambridge University Press, p. 619.

Snyder, P. J., H. Peachey, P. Hannoush, J. A. Berlin, L. Loh, J. H. Holmes, A. Dlewati, J. Staley, J. Santanna, S. C. Kapoor, M. F. Attie, J. G. Haddad, Jr., and B. L. Strom. 1999. Effect of testosterone treatment on bone mineral density in men over 65 years of age. J. Clin. Endocrinol. Metab. 84:1966–1972.

Widmaier, E. P., H. Raff, and D. T. Strang. 2006. *Vander's Human Physiology: The Mechanisms of Body Function.* New York: McGraw-Hill, p. 827.

COMPARATIVE NOTES

Laboratory 1
The Comparative Phylogenetic Method: Who's More Related to Whom? 2

Laboratory 2
Body Temperature Homeostasis: Fuzzy Physiology versus Slimy Behavior 16

Laboratory 3
Switching Osmoregulatory Mechanisms: Who's Fresh and Who's Salty? 33

Laboratory 4
Some Like It Hot: Genetic Techniques from Hot Springs 40

Laboratory 5
Axon Myelination: Faster but Not Fatter 54

Laboratory 6
Diving Reflexes: Breathing Deep 64

Laboratory 7
Infrared Modality: Seeing Heat 86

Laboratory 8
Behavioral Natural History and Muscle Type: You Are What You Do 102

Laboratory 9
Muscles in Flight 115

Laboratory 10
Sex-Changing Fish: Nemo's Dad becomes Mommy! 131

Laboratory 11
Open Circulatory Systems: Spilling Blood 142

Laboratory 12
Blood or Bloods? 158

Laboratory 13
Ventilation Rate in Fishes: It's a Gas 170

Laboratory 14
Nitrogenous Waste Production: To Pee or Not to Pee? 182

Laboratory 15
Hibernate and Save Energy 196

LABORATORY 1

Scientific Investigation

PURPOSE

This laboratory will introduce you to the principles of scientific investigation and will explain how to statistically evaluate physiological data.

Learning Objectives

- Define the term *physiology.*
- Discuss the study of physiology.
- Understand some simple statistical tools for analyzing and evaluating physiological data.

Laboratory Materials

1. Scientific calculator
2. Computers with a preferred statistical software package or use of a university computer laboratory

Introduction and Pre-Lab Exercises

The first two questions that should be asked by a physiology student are; "What is physiology, and how does a scientist study physiology?" **Physiology** is the study of the functional processes of an organism and of how these functional processes allow that organism to maintain a relatively constant condition in its internal environment (*homeostasis*). How a scientist studies physiology is a much more difficult question. Our knowledge of physiology (as in all science) is based on information or data derived from experimentation.

The following exercise is designed to introduce you to a few basic but important statistical procedures used in analyzing and evaluating physiological data. Many of the concepts and statistical tests discussed here are gross simplifications that ignore many of the mathematical and theoretical concepts in statistics. Instead of focusing on statistical theory, this exercise will simply introduce statistical tools for you to use in identifying basic trends and differences in data sets.

As you work through this exercise, you should carefully consider the questions, hypotheses, and particular tests used in each of the data sets that follow. This will help you to identify which statistical tests are appropriate for particular data sets. We encourage you to continually reference this introductory exercise throughout the course. Over the next few weeks, you will become more familiar and comfortable with your newly acquired research knowledge and skills in statistics.

If this is your first introduction to statistics, don't panic. Statistics is mastered not in one laboratory session, but through repetition and practice. The experimental nature of these laboratories will help you to reinforce your newly acquired skills in statistics throughout the course. Much of your success in evaluating laboratory data and becoming a good scientist will depend upon your understanding of the scientific method and your proper use of the statistical procedures introduced in this exercise.

COMPARATIVE NOTE Laboratory 1

The Comparative Phylogenetic Method: Who's More Related to Whom?

Historically, comparative physiologists have investigated the similarities and differences within and between groups of organisms to better understand the complexity and diversity of physiological systems. Comparing and contrasting one physiological system with another often leads to new discoveries. In this lab, you will compare physiological data among subjects and groups. For example, you will compare the resting heart rates among patients with low, normal, and high blood pressure. In the Laboratory Report you will compare the blood hematocrit levels between groups of males and females. In these comparisons, however, you will investigate similarities and differences within a single species — humans (*Homo sapiens*). What if we wished to compare a human physiological trait with that of other mammals such as chimpanzees, horses, or seals? What if we wanted to compare this human trait with that of a bird, amphibian, or fish? These comparisons are much more difficult because we must consider how the physiology of these organisms is correlated with their evolutionary history or phylogeny; we must consider how phylogeny affects physiology.

In 1985, Dr. Joseph Felsenstein at the University of Washington published an important paper that changed the way comparative biologists investigate the similarities and differences among organisms. The "Comparative Phylogenetic Method" and its philosophy introduced a set of statistical procedures that allow biologists to compare traits across species. These methods infer and control for phylogeny. The phylogeny shown here indicates that a human and a chimpanzee are more closely related to each other than they are to a horse. Thus, if you are going to compare the physiology of a chimpanzee to that of a human, we would use a different set of controls or **independent contrasts** than we would when comparing the horse to a human. It is easy to see why the chimpanzee and human would have similar physiologies; they both evolved from a common ancestor. The horse does not share the most recent common ancestor, and its physiology has been shaped by a different evolutionary trajectory.

Although we have introduced the comparative phylogenetic method and some of its basic philosophical and statistical considerations within this text, we cannot begin to give this topic justice here. For the intrigued student, we offer some interesting and important readings.

RESEARCH OF INTEREST

Felsenstein, J. 1985. Phylogenies and the comparative method. American Naturalist 125:1–15.

Harvey, P.H. and M.D. Pagel. 1991. The comparative method in evolutionary biology. Oxford University Press, Oxford.

Sanford, G.M., W.I. Lutterschmidt, and V.H. Hutchison. 2002. The comparative method revisited. BioScience 52:830–836.

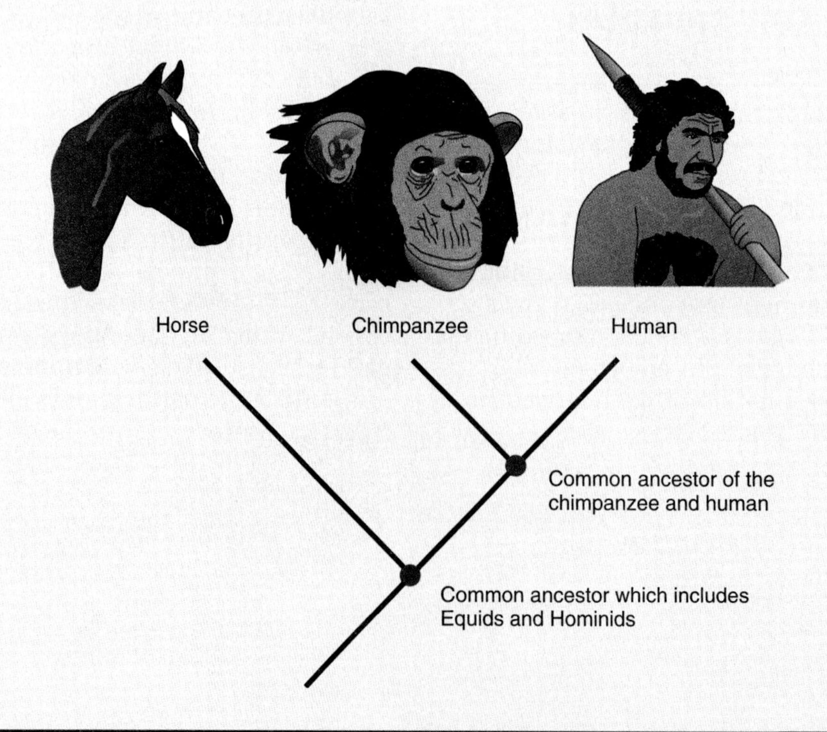

Terms and Definitions

null hypothesis (H_0)-a statement of no difference. Experiments and statistical analyses are designed to determine if there are differences between two or more groups. Statistical results provide information for determining if there is sufficient reason to reject the H_0 (the hypothesis that there are no differences between or among groups). In science we do not prove, but rather disprove. Scientific advancement is made by rejecting null hypotheses and accepting alternative hypotheses (H_A) that bring us closer to the absolute truth.

population-a defined group in space and time that includes all individuals about which inferences are made. For example, in discussions of the average or mean SAT score this year at your school, the student body at your school is the population. A population is symbolized by N.

sample-a collection of individual observations selected by a specific procedure and criteria to represent the population being sampled. A sample is a subset of the population that is used to make inferences about the population at large. For example, only 100 randomly selected SAT scores (a sample) may be used to estimate the mean SAT score of this year's student body at your school (the population). A sample is symbolized by n.

parameter-a quantity or absolute measure of central tendency to describe and/or define a characteristic of a population. Parameters are rarely reported because it is difficult to measure all individuals within a population.

statistic-an estimation of a population parameter, achieved by the use of a sample.

variable-a characteristic of an object or individual that can vary in magnitude or amount.

variate-a single score, reading, or observation of a variable. It is symbolized by $X_1, X_2, X_3 \ldots$

accuracy-the nearness of a measurement to the actual value of the variable being measured.

precision-the closeness of repeated measurements of the same quantity; repeatability of measurement.

frequency-how often a particular value of a variable occurs within a sample.

relative frequency-frequency divided by the total number in the sample; often expressed as a percentage.

mean-the sum of all variates divided by the total number of variates in the population or sample.

Population mean: $\mu = \dfrac{\Sigma x}{N}$

Sample mean: $\bar{x} = \dfrac{\Sigma x}{n}$

BOX 1.1 Example calculation of mean.

Sample: 9, 3, 5, 3

Sample size (n) = 4

$$\bar{x} = \frac{9 + 3 + 5 + 3}{4} = \frac{20}{4} = 5$$

median-the value or variate that divides the ordered sample (data set) into two equal halves. The median can be calculated by ranking all observations and using the following equations:

Odd number of observations: $X_{\frac{n+1}{2}}$

Even number of observations: $\dfrac{X_{\frac{n}{2}} + X_{\frac{n}{2}+1}}{2}$

BOX 1.2 Example calculation of median (for an even number of observations).

Sample: 9, 3, 5, 3

Sample size (n) = 4

Ranked sample:
$X_1 = 3, X_2 = 3, X_3 = 5, X_4 = 9$

$$Median = \frac{X_{\frac{4}{2}} + X_{\frac{4}{2}+1}}{2} = \frac{X_2 + X_3}{2} = \frac{3 + 5}{2} = 4$$

mode-the variate in the frequency distribution with the greatest number of observations.

BOX 1.3 Example calculation of mode.

Sample: 9, 3, 5, 3

Sample size (n) = 4

Variate with the greatest frequency = 3

range-an indication of magnitudinal difference between the smallest and largest observations.

BOX 1.4 Example calculation of range.

Sample: 9, 3, 5, 3
Sample size (n) = 4

Range = 9 − 3 = 6

variance-the degree of deviation or spread in a distribution about the mean. Variance is mathematically defined by:

$$\text{Population variance: } \sigma^2 = \frac{\Sigma(x_i - \mu)^2}{N}$$

$$\text{Sample variance: } s^2 = \frac{\Sigma(x_i - \bar{x})^2}{n - 1}$$

BOX 1.5 Example calculation of variance.

Sample: 9, 3, 5, 3
Sample size (n) = 4; Mean ($\bar{x}$) = 5

$$s^2 = \frac{(9-5)^2 + (3-5)^2 + (5-5)^2 + (3-5)^2}{4-1}$$

$$s^2 = \frac{(4)^2 + (-2)^2 + (0)^2 + (-2)^2}{3}$$

$$s^2 = \frac{16 + 4 + 0 + 4}{3} = \frac{24}{3} = 8$$

standard deviation-like variance, this describes the degree of deviation or spread in a distribution about a mean. Standard deviation is the square root of the variance and is mathematically defined by:

$$\text{Population standard deviation: } \sigma = \sqrt{\sigma^2}$$

$$\text{Sample standard deviation: } s = \sqrt{s^2}$$

BOX 1.6 Example calculation of standard deviation.

Sample: 9, 3, 5, 3
Sample size (n) = 4; Variance (s^2) = 8

$$s = \sqrt{8} = 2.83$$

standard error-also called the standard deviation of means because it standardizes sample means based upon sample size. Standard error is mathematically defined by:

$$SE = \frac{s}{\sqrt{n}} = \sqrt{\frac{s^2}{n}}$$

BOX 1.7 Example calculations of standard error.

Sample: 9, 3, 5, 3
Sample size (n) = 4; standard deviation (s) = 2.83; variance (s^2) = 8

$$SE = \frac{s}{\sqrt{n}} = \frac{2.83}{\sqrt{4}} = \frac{2.83}{2} = 1.41$$

$$SE = \sqrt{\frac{s^2}{n}} = \sqrt{\frac{8}{4}} = \sqrt{2} = 1.41$$

You have now finished reviewing the calculations for some simple but important statistics used to summarize and describe data. These statistics are **descriptive statistics** because they simply describe data. The mean, median, and mode are **measures of central tendency** because they describe how a set of observations estimate or are centered about a true population parameter. Range, variance, standard deviation, and standard error are **measures of dispersion** because they describe the amount of variation about the mean.

Statistical Exercises

EXERCISE 1.1 Descriptive Statistics

Observe the following sample of testosterone concentrations (ng/ml) measured in the blood plasma of 10 adult men. Using this data set, calculate all descriptive statistics by hand with the use of a calculator. Be sure to write out all mathematical expressions and show all work.

Sample: 4, 6, 9, 8, 10, 8, 4, 3, 5, 8

Measures of Central Tendency

Mean =

Median =

Mode =

Measures of Dispersion

Variance =

Standard deviation =

Standard error =

Now enter these data into the statistical software package provided and explained to you by your laboratory instructor. Use the "descriptive statistics function" in this software program to check your calculations; print a copy of your results. Ask your laboratory instructor for help and additional instruction as needed.

EXERCISE 1.2 The Normal Probability Distribution and The 95% Confidence Interval

You have, no doubt, taken a course in which grades assigned were based on a "normal curve." This reference is to the familiar bell-shaped normal probability distribution that is produced when some parameter of a population (e.g., height of adult men) is plotted on the abscissa (x-axis) versus the frequency of a given measurement on the ordinate (y-axis). Theoretically, the tails of the normal curve extend infinitely in both directions. If the total area under the curve is taken as 100%, approximately 34% of the population will be distributed between the mean measurement and +1.0 standard deviations of the mean. Similarly, approximately 34% of the population will lie between the mean and −1.0 standard deviations of the mean. This indicates that 68.27% of the population is distributed within the range of the mean −1.0 and +1.0 standard deviations ($\mu \pm 1\sigma = 68.27\%$) (Figure 1-1).

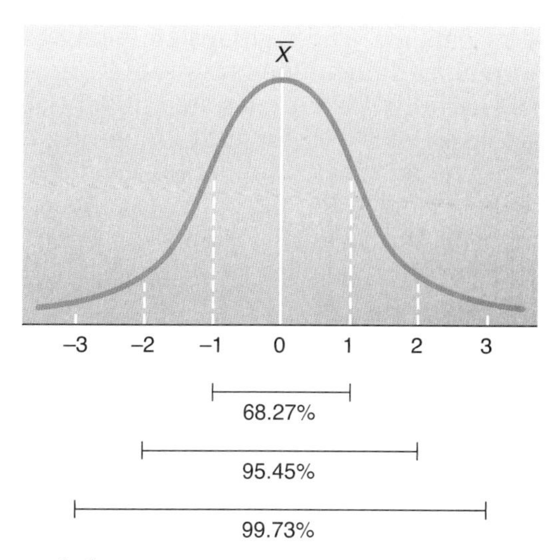

Figure 1-1 Normal distribution of a population.

For example, let us assume that the height distribution of adult men is such that the mean ($\overline{x}$) height is 70 inches and the standard deviation is 3 inches. This indicates that 68.27% of all men in the population are between 67 and 73 inches. The range $\mu \pm 2\sigma$ includes 95.45% of the population, and the range $\mu \pm 3\sigma$ includes 99.73% of the population.

If you sample a person's height, there is a 95% chance that it would fall between ±1.96 standard deviations about the mean. There is a 5% chance that it would not be included in this range. Thus, we accept or reject an observation belonging to the normal population based upon a probability of $P = 0.05$. If $P < 0.05$ we say it is statistically different and that it belongs to a different population. For example, suppose you sample an individual's height at 89 inches (7 feet 5 inches). How do you decide whether this person is one of the extremes in the normal population or if this person is part of a different population (one affected by the overproduction of growth hormone leading to gigantism)?

An easy way to determine if groups belong to different populations because they differ statistically is to calculate and plot the **95% confidence intervals** (95% CI) for each sample of a population. In theory, if the 95% CI overlap, the groups are not different from one another (i.e., they belong to the same population) as defined by the normal probability distribution. Calculations of the upper and lower limits of the 95% CI are:

Lower limit = $\overline{x} - 1.96(SE)$
Upper Limit = $\overline{x} + 1.96(SE)$

Imagine that you have just collected resting heart rate data in beats per minute (BPM) for three groups of patients. These groups are patients with low, normal, and high blood pressures. You wish to know if the resting heart rates are different among these groups. A 95% CI for each group of patients may give you an idea of similarities or differences among the resting heart rates of these patients.

Calculate the 95% CI for each of these three groups. Show all work in the space provided.

Low Blood Pressure: 73, 72, 76, 74, 71

$\bar{x}$ =

Lower limit =

Upper limit =

Normal Blood Pressure: 82, 80, 70, 75, 88

$\bar{x}$ =

Lower limit =

Upper limit =

High Blood Pressure: 92, 93, 100, 102, 98

$\bar{x}$ =

Lower limit =

Upper limit =

Now inspect Figure 1-2 and answer the questions regarding the heart rates of these three groups of patients.

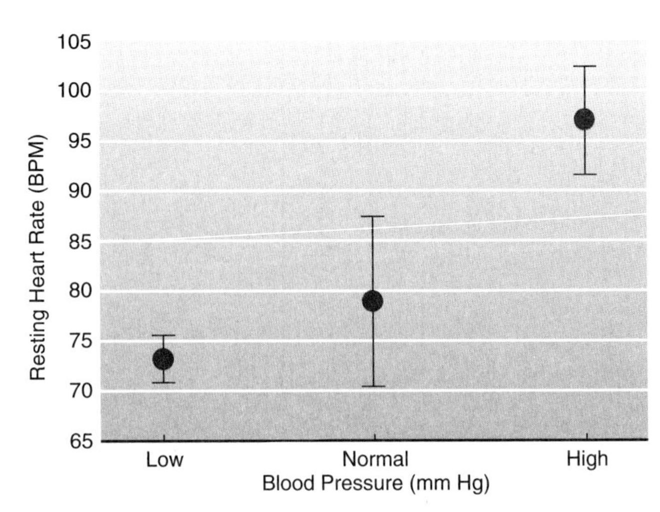

Figure 1-2 Example plot of group means and their 95% confidence intervals.

QUESTIONS

1. Compare your calculations with Figure 1-2. Do your calculations agree with the graph?

2. Is the heart rate of patients with low blood pressure statistically different from that of patients with normal blood pressure? Why or why not?

3. Of these three groups, which has the statistically different heart rate?

4. Which group has the most variable heart rate? How did you come to this conclusion?

EXERCISE 1.3 Student's *t*-test

A *t*-test is a statistical test used to compare the means of two populations or distributions. If we were interested in comparing the heart rates of patients with low and normal blood pressures, we could use a *t*-test because we are only comparing two groups. Enter the heart rate data for patients with low and normal blood pressures into your statistics program and obtain the descriptive statistics, perform a *t*-test, and report your results. Your lab instructor will help you to interpret the results of the test.

The statistical report will show a **probability** or ***P*-value.** This value will tell you if the groups are significantly different statistically. If $P > 0.05$ the individuals are not different. If $P < 0.05$, the individuals are significantly different from each other with respect to heart rate. Were your conclusions from using the 95% CI in Exercise 1.2 accurate with respect to the results of the *t*-test?

The results that should be reported for a *t*-test are the *t*-statistic, degrees of freedom, and the *P*-value following a statement of your findings.

BOX 1.8 Example result statement for a *t*-test.

The heart rates of patients with low and normal blood pressures are not significantly different from one another ($t = -1.82$; $df = 8$; $P > 0.05$).

To understand how these results were obtained, use the descriptive statistics and calculate the *t*-statistic by hand. The equation for the *t*-statistic is:

$$t = \frac{\bar{x}_1 - \bar{x}_2}{\sqrt{\frac{s_1^2}{n_1} + \frac{s_2^2}{n_2}}}; \; df = (n_1 + n_2) - 2$$

Show your work here.

EXERCISE 1.4 The Analysis of Variance

The analysis of variance (ANOVA) is somewhat like a *t*-test but is used to statistically compare more than two groups. You cannot use a *t*-test to compare several groups as pairs; an ANOVA must be used. For example, let's imagine that you now wish to determine if heart rate is statistically different among the three groups of patients with low, normal, and high blood pressures. Because you are now comparing more than two groups in this example, the ANOVA is used. Enter the data into your statistics program, calculate the descriptive statistics, perform an ANOVA, and report your results. Once again, your lab instructor will help you with interpretation of the results.

The results that should be reported are the *F*-statistic, degrees of freedom, and the *P*-value following a statement of your findings.

BOX 1.9 Example result statement for an ANOVA.

We found heart rate to differ significantly among individuals with low, normal, and high blood pressures ($F = 33.14$; $df = 2{,}12$; $P < 0.05$).

You may be wondering why we state that these heart rates are different. We just showed with a *t*-test that heart rates of patients with low and normal blood pressures are not statistically different. So why does the ANOVA tell us that there are differences among these patients? Remember that an ANOVA determines if there are any differences among groups. We know that patients with high blood pressure have significantly different heart rates (Figure 1-2). Thus, the ANOVA gives us a *P*-value less than 0.05, indicating that at least one group's mean heart rate is statistically different from the others. Now you can use your 95% CIs to determine which mean is different. A multiple comparisons test could also be used to determine which means are different. Ask your lab instructor to give further instruction on multiple comparisons tests if interested.

Please note that the *df* in an ANOVA are calculated as follows: *df* = number of groups minus 1, number of observations minus the number of groups. Thus $df = 3 - 1, 15 - 3 = 2, 12$.

EXERCISE 1.5 Regression Analysis

Let us imagine that you were interested in investigating the effects of testosterone treatment duration on bone density in males over 65 years of age. You collect the following data set for analysis. Enter these data into the computer, perform a regression analysis, and report your results.

A regression analysis will generate an ANOVA table as you have seen in earlier analyses. A regression analysis will:

a. Determine if there is a cause-and-effect relationship between the dependent and independent variables.
b. Determine the nature of the relationship as defined by a mathematical equation.
c. Evaluate how accurately the mathematical equation defines the relationship.

Duration of Testosterone Treatment (Months)	Bone Mineral Density (% Increase)
0	0
6	0.8
6	1.0
6	0.6
12	1.8
12	2.4
12	1.5
18	2.3
18	2.7
18	1.9
24	2.9
24	3.4
24	2.6
30	3.6
30	3.7
30	3.1
36	4.2
36	5.1
36	3.3

Data taken from Snyder et al. 1999.

To properly report the results of a regression analysis, you should report the *F*-statistic, the two degrees of freedom, the *P*-value from the ANOVA table, and the r^2 value following a statement of your findings.

Note that the *df* in a regression are always 1 and the number of observations minus the number of variables (2 in a simple linear regression).

BOX 1.10 Example result statement for a regression.

We found a significant relationship between testosterone treatment duration and bone density in men over 65 years of age ($F = 137.4$; $df = 1,17$; $P < 0.05$). The duration of testosterone treatment explains nearly 88.3% of the variation in bone density ($r^2 = 0.883$).

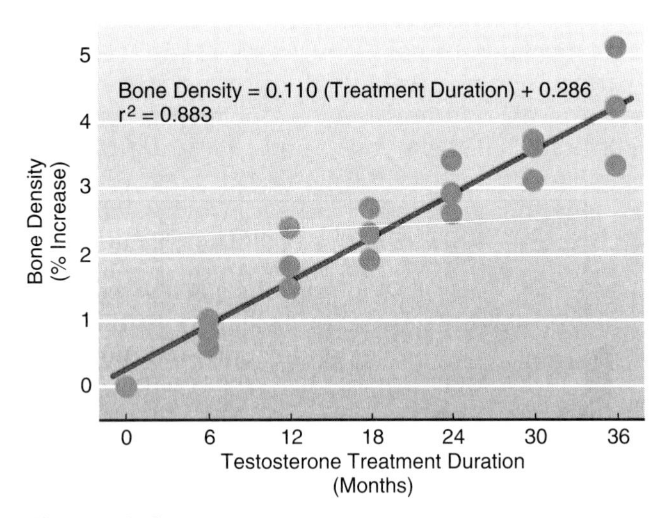

Figure 1-3 Example plot of a regression.

Now to complete the analysis, use your statistics program or a graphing program to plot these data showing the independent variable (treatment duration) on the abscissa (x-axis) versus the dependent variable (bone density) on the ordinate (y-axis). Recall that the dependent variable is the variable that is affected or is dependent upon the independent variable. Plot the best-fit line defined by the regression equation and report the regression equation on the graph. This equation is in the form $y = m(x) + b$ where m is the slope and b is the y-intercept. Your graph should look something like Figure 1-3.

EXERCISE 1.6 Correlation

There are times in biology when an investigator is interested in the possible *association* between two variables but is uncertain if there is a *cause-and-effect* relationship between the variables. In this case, the investigator performs a correlation analysis rather than a regression analysis. A correlation assumes no cause-and-effect relationship between the variables and, rather, attempts to find possible correlates. Think about this in the following way: Do wet roads after a rainfall cause automobile accidents? The answer is no. Wet roads do not cause accidents, or else no one would ever attempt to drive on wet roads. However, wet roads increase the likelihood that some people may get into accidents because they attempt to drive on wet roads as they would on dry roads. Thus, there may be a correlation between wet roads and automobile accidents, but wet roads do not cause automobile accidents.

For our purposes, a correlation analysis is calculated just like a regression. However, your question and hypothesis do not imply a cause-and-effect relationship. A correlation is simply the measure of association between two variables. In this test you also report the same results as a regression except for the r^2 value. A correlation coefficient is reported, which is the square root of the r^2 from the regression. If $r = 0.0$ there is no association. If $r = -1.0$ or $+1.0$, there is a strong negative or positive association, respectively. Therefore, r ranges from -1.0 to $+1.0$. Most importantly, remember never to use the word *relationship* when performing a correlation analysis. We say either that there is an *association* or that there is *no association* between the variables being investigated.

Let us imagine that you are interested in investigating the association between age and testosterone levels in men. You collect data from 20 individuals ranging from 5 to 30 years of age. Is there a significant association between age (independent variable) and testosterone level (dependent variable)?

You may be tempted to analyze these data using a regression analysis. However, you cannot assume any causal effect of age on testosterone levels because you have not directly tested the effects of age on testosterone. There may be other reasons for the observed changes in your response variable. For example, other hormones produced by the brain may be driving the change in testosterone levels. Thus, although increasing age is associated with changing testosterone levels, age may or may not directly cause these changes. Note that in many situations, the direct effect of a variable (e.g., age, body size, brain size) cannot be manipulated. Thus, no cause-and-effect relationship can be implied. In such instances, you may only address the association between two variables, and the most appropriate way to do so is with a correlation analysis.

Now complete a correlation analysis using the following data set and create a graph showing the association between testosterone and age. Your results and graph should look like Figure 1-4. Notice that there is no best-fit line shown in the correlation graph, for this would indicate a cause-and-effect relationship and an ability to calculate y from x.

BOX 1.11 Example result statement for a correlation.

We found a significant association between age and testosterone level in men ($F = 62.29$; df = 1,18; $P < 0.05$).

Testosterone level in men is strongly correlated with age ($r = 0.873$).

(Note that your lab instructor may request that you use a Pearson-Moment correlation. Many statistical software packages now commonly run correlation analyses and may be more appropriately used for such analyses.)

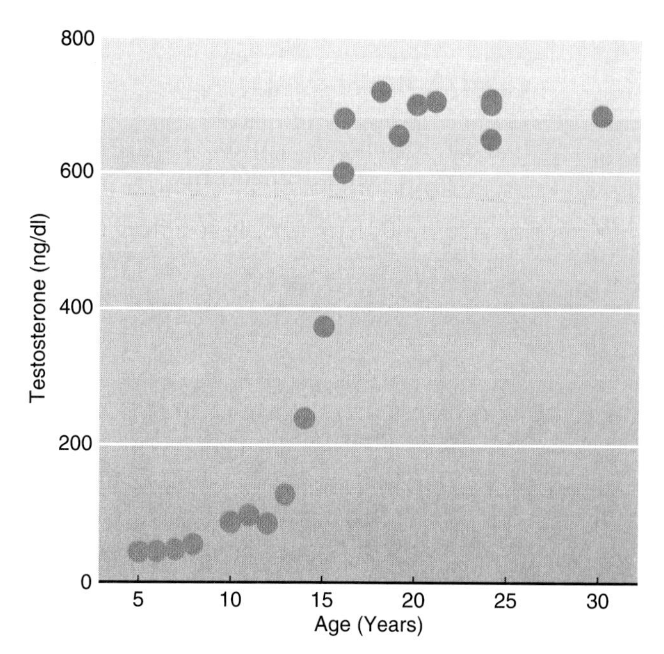

Figure 1-4 Example plot of a correlation.

Age (years)	Testosterone (ng/dl)
10	85
6	42
5	40
8	52
14	240
16	680
18	720
24	708
15	375
12	83
21	705
24	650
7	45
11	95
20	700
13	125
16	600
19	655
24	702
30	685

EXERCISE 1.7 Statistical Concepts and Tests

After completing the previous exercises, you should be able to define and discuss the following topics and terms. You should also be able to decide which tests are appropriate for a particular data set and what descriptive statistics are necessary for their calculation. If you feel you have a good understanding of the following concepts, congratulations—you have successfully completed this exercise.

Offer a short explanation of each of the following terms and state when their use would be appropriate for describing physiological data.

Descriptive statistics -

Normal distribution -

Probability and the 95% CI -

Student's *t*-test -

One-way analysis of variance -

Regression -

Correlation -

RESULTS AND DISCUSSION

LABORATORY REPORT 1

Statistical Problem Sets

Now that you have been introduced to some basic statistical procedures, you are prepared to complete the following problems. The following data sets will require you to think about the scientific method and hypothesis testing. After looking at a particular data set, you will want to do the following for each problem:

a. Use the data set presented to identify what question is being investigated by the researcher. Formally state this question.

b. State the null hypothesis (H_0). Remember that the H_0 is a statement of no difference. We use the H_0 because we can never prove anything in science, we can only disprove. Review the scientific method if you find this point confusing.

c. State which statistical test should be used to analyze the data and state why this test is appropriate.

d. Analyze the data set with the appropriate statistical test. Record all results and determine if you would accept or reject the H_0.

e. Properly state your conclusions with your statistical results as demonstrated in the laboratory exercises.

f. Produce a graph that best illustrates the data.

Be sure to show all work. These problems may be discussed at the end of the period or collected by your lab instructor during your next laboratory period.

Problem 1.1

An experimenter investigated the effects of three newly developed drugs on the human glomerular filtration rate of the kidneys. Glomerular filtration rates (GFR) were determined with the use of inulin and are expressed in ml/min. Which drug has the greatest effect on GFR? Which drug produces the greatest variability in GFR?

Control	Drug 1	Drug 2	Drug 3
125	139	130	110
130	145	124	112
118	155	119	113
122	153	120	101
128	141	128	103

Results:

Problem 1.2

An experimenter measured the blood hematocrit levels of males and females and compiled the following data. Blood hematocrit is the percentage of the whole blood sample consisting of red blood cells.

Males	Females
45	48
50	46
48	39
45	37
48	40
51	36
50	37
46	46
49	38
47	44

Results:

Problem 1.3

An experimenter investigated the amount of evaporative water loss in humans at different environmental temperatures. Evaporative water loss was determined by the amount of perspiration produced (ml/h).

Temperature (°C)	Water Loss (ml/h)
23	4.6
18	3.2
35	58.1
25	5.7
30	42.3
32	40.9
15	2.9
20	3.9

Results:

Problem 1.4

From your results and observations in solving problems 1.1 through 1.3, how do you determine if groups are statistically different or similar? What does a statistical test compare in order to determine whether groups are statistically different or similar? You may want to carefully look at the descriptive statistics for each of your problems and the results that follow. This should help you to answer the question in this problem.

LABORATORY 2

Homeostasis

PURPOSE

This laboratory will introduce you to the principle of homeostasis and demonstrate how regulatory mechanisms of the body help to maintain a state of dynamic constancy.

Learning Objectives

- Define the term *homeostasis.*
- Identify and define the components of feedback loops.
- Discuss the role of negative feedback in maintaining homeostasis.
- Discuss how physiological systems regulate a homeostatic environment dynamically and not statically.
- Discuss the concept of a set point and how physiological mechanisms regulate a process around that set point.

Laboratory Materials

Experiment 2.1: Resting and Active Heart Rates Among Athletes and Nonathletes

1. Stopwatch or watch with a second hand
2. Stethoscope

Introduction and Pre-Lab Exercises

One of the most important concepts in physiology is that the internal environment of an organism must remain relatively constant. This was first observed by the pioneer of modern comparative physiology, Claude Bernard (1813–1878), who stated that the internal environment (*milieu interieur*) of an organism remains constant despite changes in the external environment surrounding that organism.

Physiological processes attempt to maintain this dynamic internal constancy or **homeostasis.** Let us consider body temperature as an example. Normal human body temperature fluctuates or oscillates slightly around 37°C (98.6°F). Thus, we can state that 37°C is the set point or set temperature for humans. Our body monitors internal temperature through temperature sensors. Any deviation from the set temperature will be detected by the sensors, which will then stimulate mechanisms that will return body temperature to the set temperature. Thus, the body maintains homeostasis with a set of sensors and physiological mechanisms that can be activated or inhibited accordingly.

Let us use an abstract example to demonstrate these principles. In many ways, you can compare our homeostatic body temperature control to the thermostat that is controlling the temperature of the room where you are sitting. If the room becomes too warm, the sensor will detect the "too warm" stimulus. The sensor (i.e., the thermostat) will then activate mechanisms that will return the room temperature to its set temperature (the selected thermostat temperature). Thus, the thermostat will turn on the air conditioner to cool the room and return the room temperature to its set temperature.

The maintenance of temperature homeostasis just described is an example of a **feedback mechanism.** All feedback mechanisms use a series of sensors and effectors in maintaining homeostasis around a set point. A **sensor** (also called a receptor) is the component of feedback mechanisms that detects a change in the internal or external environment. Sensors may be single cells or collections of cells located in a gland or organ. For example, single-celled osmoreceptors in the hypothalamus of the brain monitor changes in the concentration of solutes (i.e., the osmolarity) of cerebrospinal fluid. The sensor then relays information to an integrating center. An **integrating center** receives information from many different types of sensors. The net result of all integrated input is then relayed from the integrating center to the effector. A change in the activity of an

effector constitutes the response of the feedback system. An effector may also be a single cell, a gland, or an organ. For example, in response to increased osmolarity of the cerebrospinal fluid, the osmoreceptors in the hypothalamus elicit a neural response that causes you to feel thirsty. Drinking water restores the osmolarity of cerebrospinal fluid to normal homeostatic levels. Figure 2-1 can be used to generalize almost all feedback mechanisms you will encounter.

Notice that the ultimate change induced by the effector (the response) also becomes the next stimulus for the sensor. This is why feedback mechanisms are commonly called **feedback loops.** The term *loop* emphasizes that the feedback mechanism is a continuous cycle that maintains homeostasis.

Let us return to the example of maintaining our body temperature at the set temperature of 37°C. If you become cold, you may perform a set of behaviors to prevent heat loss. These may include pulling your limbs close to your body, crossing and holding your arms, rubbing your hands together, or donning additional clothing. If these behavioral activities fail in maintaining set temperature, your body then activates physiological effectors to both conserve and generate heat. These may include vasoconstriction of peripheral blood vessels and high-frequency muscle contractions that comprise the response known as shivering. Both of these physiological responses attempt to raise body temperature and return the body to its set temperature. We can observe a similar phenomenon when the body begins to overheat.

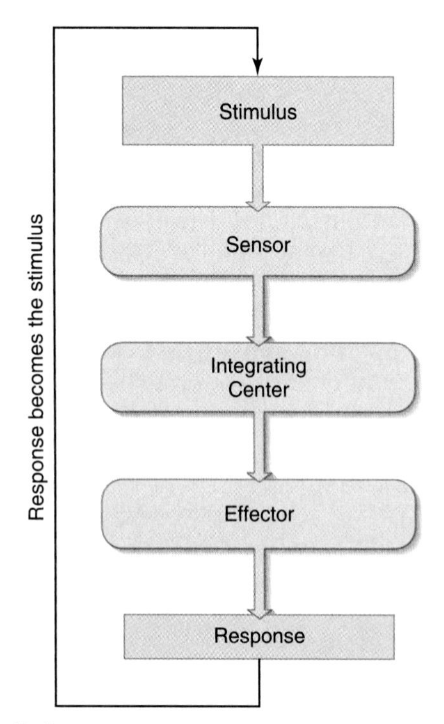

Figure 2-1 Components of a feedback loop.

2-1 List the behaviors and physiological mechanisms that occur in humans to cool the body when it is overheated.

Behaviors:

Physiological mechanisms:

The behaviors and physiological mechanisms you have just listed are important in regulating body temperature. You may not realize it, but you have just described a cause-and-effect sequence known as a **negative feedback** mechanism. It is important to note that negative feedback loops ultimately induce changes that are opposite from the initial stimulus or disturbance in homeostasis. Thus, in our abstract example of room temperature homeostasis, the effector (the air conditioner) initiates changes that oppose the "too warm" stimulus that activated the sensor (the thermostat).

2-2 Using a physiological example (hormone secretion, blood pressure regulation, etc.), draw a schematic that demonstrates a negative feedback mechanism. Use your textbook to help you.

■

2-3 Can set points ever be changed? (Hint: What happens to set temperature during a fever?)

■

2-4 Prostaglandin, a type of hormone, influences the temperature-control center in the hypothalamus. Interestingly, bacterial infections increase prostaglandin synthesis.

a. What effect do you think prostaglandin has on set temperature?

b. Which basic component of the feedback loop (as shown in Figure 2-1) for body temperature control does prostaglandin affect?

c. What effect does acetaminophen (Tylenol®, a fever-reducing or antipyretic drug) have on prostaglandin synthesis?

■

Another type of feedback mechanism is called a **positive feedback loop.** Positive feedback loops contain the same components (sensors, integrating centers, effectors) as negative feedback loops, but the ultimate changes induced by positive feedback loops are quite different. This type of feedback mechanism is called *positive* because the physiological responses are in the same direction as the original stimulus. Thus, the most important difference between negative and positive feedback loops is that negative feedback loops tend to move a physiological measure toward the set point, while positive feedback loops tend to move a physiological measure away from the set point. Positive feedback loops are inherently unstable and always move a system farther away from homeostasis. Examples of true positive feedback in physiology are rare. Most positive feedback loops function within larger negative feedback systems; this provides some control and stability to positive feedback mechanisms. The best examples of positive feedback in human physiology are the steps involved in blood clotting and the enhancement of labor contractions during birth.

2-5 For each of the following feedback mechanisms, indicate whether it is an example of a negative or a positive feedback loop. Explain your reasoning. Although some of the examples are not physiological processes, the basic principles are the same.

a. A snowball rolling down a hill becomes larger and larger, making it roll faster and become larger still.

b. Iron is absorbed from digested foodstuffs through receptor-mediated pinocytosis in the small intestine. When iron concentrations in the blood rise above the set point, excess iron accumulates in the cells of the small intestine. This excess iron itself functions to suppress the further active absorption of iron.

c. Determined to earn an A in Human Physiology, you decide to seek help from a tutor. The tutor is helpful, but very expensive. In order to pay the tutor, you must start working nights. Working reduces your study time, and your grades fall in the course. Thus, you spend more time with the tutor, meaning you have to earn more money. As a result of working more, your grades in the course fall even more. You decide to spend even more time with the tutor.

d. A decrease in the set-point concentration for thyroid hormones results in a greater inhibitory signal from the thyroid gland (effector).

e. During the initial stages of a nerve impulse, sodium channels in the cell membrane are opened. These opened channels allow sodium ions to diffuse into the cell, which in turn creates a slight change in voltage across the cell membrane. In response to this change in voltage, more sodium channels in the cell membrane are opened. This allows more sodium to diffuse into the cell and causes an even greater change in voltage across the membrane. ■

COMPARATIVE NOTE *Laboratory 2*

Body Temperature Homeostasis: Fuzzy Physiology versus Slimy Behavior

Because of the pervasive influence of body temperature on the rate of all physiological processes, body temperature is closely monitored and regulated in most organisms. Body temperature is constantly monitored by internal temperature receptors and compared to a set point in the hypothalamus of the brain. If deviations from the set point occur, the appropriate physiological and/or behavioral systems are activated to maintain homeostasis.

Although all animals can rely on both physiology and behavior to control body temperature, some animals rely more on physiological control and others rely more on behavioral control. Mammals and birds are endotherms (*endo* = from within) and obtain heat through metabolic processes. Endotherms can regulate body temperature physiologically by perspiring and shivering. Behaviorally, we can add or remove clothing, select a "habitat," or even control the temperature of that "habitat" with technology. More interesting than humans, bats and hummingbirds can reduce body temperature and metabolic rate during periods of daily torpor. Many duck species can reduce heat loss while swimming in icy water by constricting blood flow to their legs and feet. This physiological mechanism, called blood shunting, is commonly used to maintain core body temperature.

Ectothermic vertebrates (fishes, amphibians, and reptiles) have fewer options for regulating body temperature. Their metabolic rates are low, and they cannot produce significant metabolic heat to maintain body temperature. Ectotherms must then rely on behaviorally exploiting environmental heat sources. These behaviors may include shuttling among different microhabitats, burrowing, changing posture and orientation toward the sun, and coordinating activity patterns with environmental temperatures.

Although ectotherms may be subject to variations in daily and seasonal body temperatures, the behavioral regulation of body temperature is frequently modulated according to the set-point temperature for the task at hand. A basking lizard may demonstrate a need to increase body temperature and metabolic rate for digestion. Even pythons behaviorally increase their body temperature during egg brooding to insure timely hatching of their eggs (Hutchison et al. 1966). Although a slimy salamander does not generate its own heat, its ability to behaviorally use the environment to regulate body temperature can be as precise as the physiological temperature control of a fuzzy mammal.

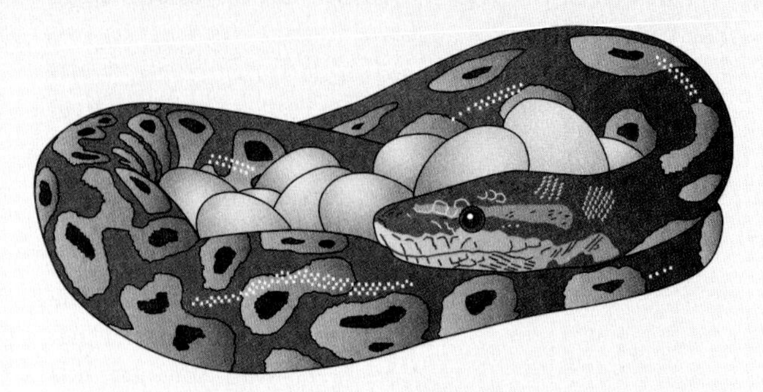

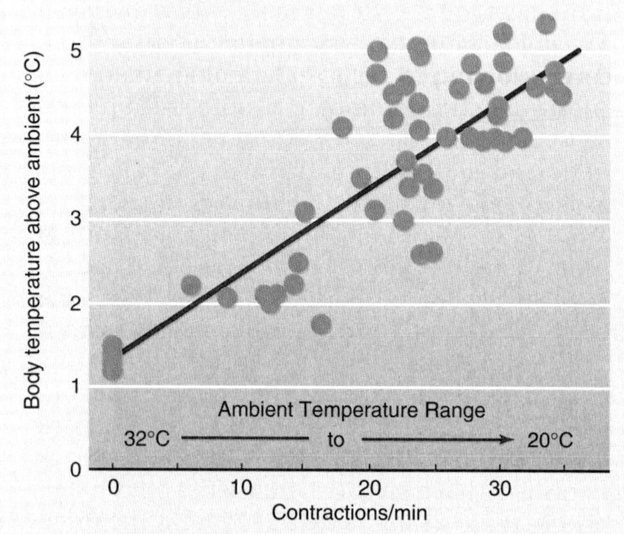

During egg brooding, a female python generates numerous isometric muscle contractions to elevate body temperature and provide warmth to the eggs around which it is coiled.

RESEARCH OF INTEREST

Diaz, J.A. 1997. Ecological correlates of the thermal quality of an ectotherm's habitat: a comparison between two temperate lizard populations. Functional Ecology 11:79–89.

Dzialowski, E.M. and M.P. O'Connor. 2004. Importance of the limbs in the physiological control of heat exchange in *Iguana iguana* and *Sceloporus undulates*. Journal of Thermal Biology 29:299–305.

Huey, R.B. 1974. Behavioral thermoregulation in lizards: importance of associated costs. Science 184:1001–1002.

Hutchison, V.H., H.G. Dowling, and A. Vinegar. 1966. Thermoregulation in the female Indian python *Python molurus bivittatus*. Science 151:694–696.

Sartorius, S.S., J.P.S. do Amaral, R.D. Durtsche, C.M. Deen, and W.I. Lutterschmidt. 2002. Thermoregulation accuracy, precision, and effectiveness in two sand-dwelling lizards under mild environmental conditions. Canadian Journal Zoology 80:1966–1976.

Methods and Materials

EXPERIMENT 2.1 Resting and Active Heart Rates Among Athletes and Nonathletes

In this exercise you will collect data to investigate the recovery time of resting heart rate between two experimental groups (athletes and nonathletes) and determine if there is a significant difference between these groups' recovery times.

- Select five volunteers from the class who perform some type of cardiovascular exercise (running, swimming, aerobics, etc.) for 40 to 50 minutes at least three times per week.
- Select five volunteers from the class who do not engage regularly in cardiovascular activities.
- Have each subject lie down and rest quietly for five minutes. At the end of the five-minute resting period (while the subject is still lying down), record the subject's resting heart rate in beats per minute (BPM) by monitoring his or her pulse. A stethoscope can be used if available.
- Each volunteer should now perform the designated exercise indicated by your instructor (e.g., jumping jacks, stair climbing, running in place). Sustained activity for 5 to 10 minutes works best for showing noticeable differences between groups.
- After exercising, record each volunteer's heart rate immediately following the exercise and then every two minutes thereafter. Continue to record the volunteer's active heart rate every two minutes until it is within 5 beats per minute of the resting heart rate.
- Calculate the time in seconds required for each volunteer's heart rate to return to resting levels.
- Share your data with the class so Results Table 2.1 may be completed on page 19.

EXPERIMENT 2.2 Homeostatic Regulation of Glucose Concentration

Glucose concentrations were measured in a fasting individual every 30 minutes for a total of four hours. The measurements had no adverse effects on the person. The results follow. Use these data to answer the questions in Problem Set 2.2 of your laboratory report.

Time (min)	Glucose (mg/dl)
0	75
30	111
60	92
90	128
120	98
150	107
180	99
210	126
240	68

RESULTS AND DISCUSSION

LABORATORY REPORT 2

EXPERIMENT 2.1

Resting and Active Heart Rates Among Athletes and Nonathletes

Results Table 2.1 Class Data for Experiment 2.1: Resting and Active Heart Rates Among Athletes and Nonathletes.

				Recovery Heart Rate (BPM)					
Athletes	Resting Heart Rate (BPM)	Active Heart Rate (BPM)	Active–Resting Heart Rate (BPM)	2 min	4 min	6 min	8 min	10 min	Recovery Time (Seconds)
1.									
2.									
3.									
4.									
5.									
NonAthletes									
1.									
2.									
3.									
4.									
5.									

Problem Set 2.1: Resting and Active Heart Rates Among Athletes and Nonathletes

Use the data from Results Table 2.1 to answer the following questions.

a. Using the appropriate statistical test, determine if there is a difference in recovery times between athletes and nonathletes.

State the null hypothesis.

Should you accept or reject the null hypothesis? Support your conclusions with the appropriate statistical results.

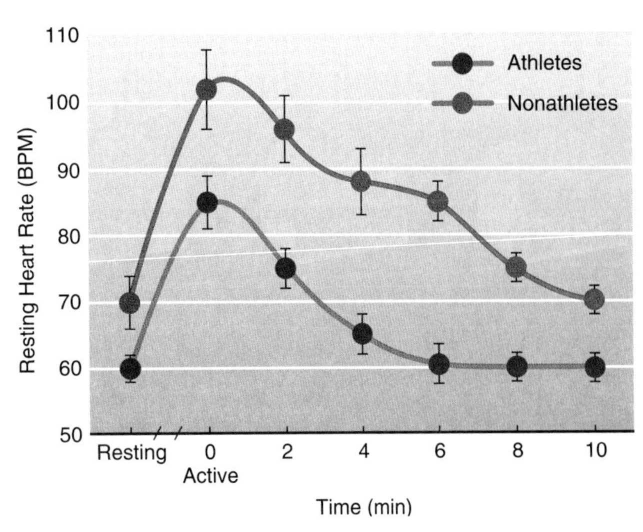

Figure 2-2 Plot of resting, active, and recovery heart rates following exercise.

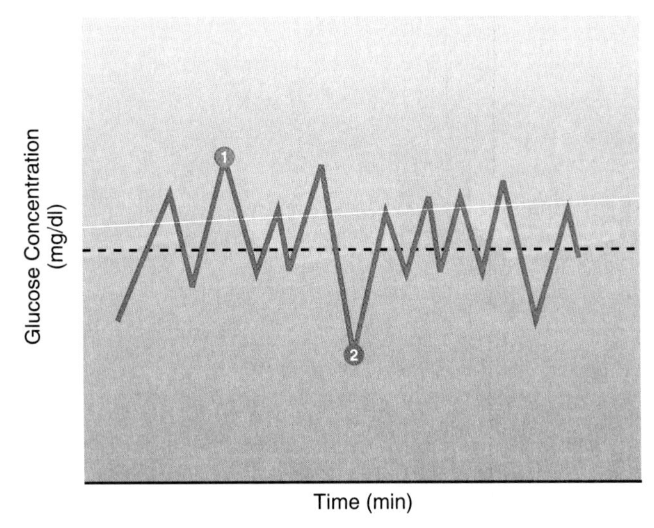

Figure 2-3 Changes of glucose concentration over time in a fasting individual.

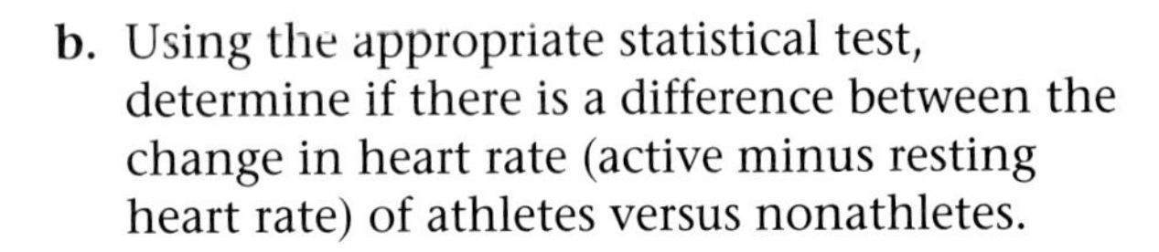

b. Using the appropriate statistical test, determine if there is a difference between the change in heart rate (active minus resting heart rate) of athletes versus nonathletes.

State the null hypothesis.

Should you accept or reject the null hypothesis? Support your conclusions with the appropriate statistical results.

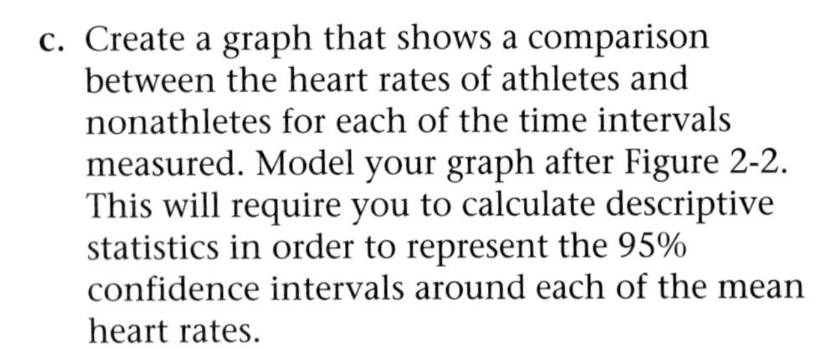

c. Create a graph that shows a comparison between the heart rates of athletes and nonathletes for each of the time intervals measured. Model your graph after Figure 2-2. This will require you to calculate descriptive statistics in order to represent the 95% confidence intervals around each of the mean heart rates.

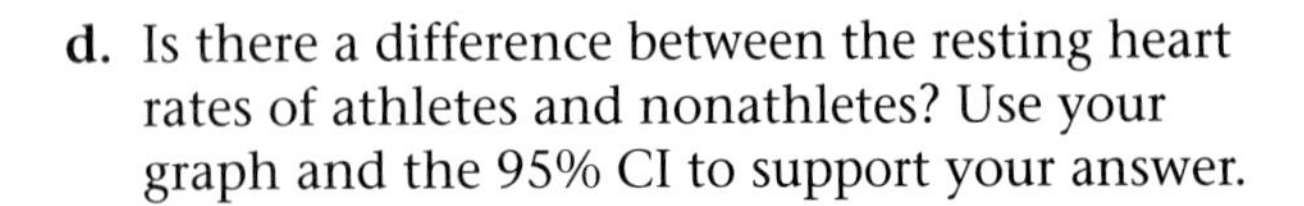

d. Is there a difference between the resting heart rates of athletes and nonathletes? Use your graph and the 95% CI to support your answer.

e. Is there a difference between resting heart rate and active heart rate in nonathletes? Use your graph and the 95% CI to support your answer.

f. Is there a difference between resting heart rate and active heart rate in athletes? Use your graph and the 95% CI to support your answer.

g. Why does heart rate return to resting levels soon after exercise is terminated?

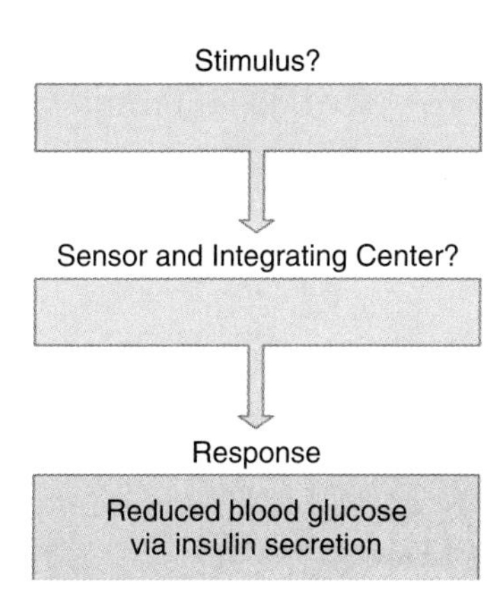

Figure 2-4 Simplified feedback mechanism controlling blood glucose concentrations.

Problem Set 2.2: Homeostatic Regulation of Glucose

Using the data listed under Experiment 2.2, create a simple line graph showing the glucose concentration on the y-axis versus time on the x-axis.

a. Which type of regulatory mechanism do you think is controlling the glucose concentrations?

b. Figure 2-3 shows a similar graph of glucose concentrations versus time. What do you think the dashed line represents in this figure?

c. What type of signal (positive or negative) is the response sending to the sensor at point 1 in Figure 2-3?

d. What type of signal (positive or negative) is the response sending to the sensor at point 2 in Figure 2-3?

e. Within this feedback loop, do the signals from the response change in magnitude (lesser, greater) or direction (positive, negative)?

f. Complete the simplified feedback mechanism that is controlling blood glucose concentrations shown in Figure 2-4. Be sure to draw in the signal to indicate that the response becomes the stimulus. As shown in the figure, sometimes the sensor also performs the functions of the integrating center. Insulin reduces blood glucose concentrations by stimulating the transport of glucose into adipocytes and muscle cells as well as inhibiting glucose release from the liver.

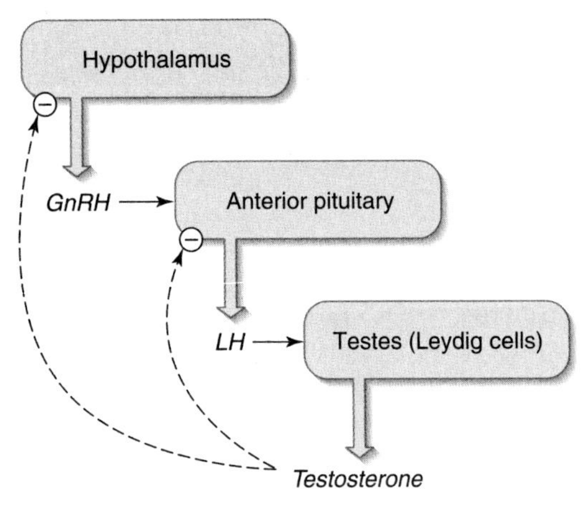

Figure 2-5 Diagram of the feedback loop for the hormone testosterone. The hypothalamus, located in the brain, produces gonadtropin releasing hormone (GnRH). GnRH in turn stimulates the production of luteinizing hormone (LH) from the anterior pituitary. LH stimulates the production of testosterone from the Leydig cells of the testes.

Problem Set 2.3: Homeostatic Regulation of Testosterone

a. A well-recognized side effect of anabolic steroid use in males is a decrease in testes size. Before you answer the following questions, do you think this is a fact or an exaggeration?

b. Is this a negative or positive feedback loop? Use Figure 2-5 to explain your reasoning.

c. What structures from Figure 2-5 represent the sensors?

d. What structure from Figure 2-5 represents the effector?

e. Anabolic steroids are really just synthetic testosterone. What do you think taking anabolic steroids will do to the amount of testosterone naturally produced by the testes? Use Figure 2-5 to help you with your answer.

f. Many attributes of the human body follow a "use it or lose it" principle. Do you think there is a scientific basis for the well-recognized side effect of anabolic steroid use in males? Explain your reasoning.

LABORATORY 3

Diffusion, Osmosis, and Tonicity

PURPOSE

This laboratory will introduce you to the principles of diffusion, osmosis, and tonicity and will discuss the importance of these physical phenomena to basic physiological processes.

Learning Objectives

- Understand the terms *solute, solvent,* and *solution.*
- Define the terms *diffusion* and *osmosis.*
- Understand the difference between mechanisms of diffusion and osmosis.
- Define and understand *osmotic pressure* and *tonicity.*
- Define the terms that indicate relative concentration of solute: *hypotonic, isotonic,* and *hypertonic.*
- Understand how the concentration of solute as well as the size of solute influences the processes of diffusion and osmosis.
- Be able to calculate and distinguish between the molarity and osmolarity of solutions.

Laboratory Materials

Experiment 3.1: Diffusion Experiment

1. 5 ml pipettes
2. Agar
3. Disposable pipettes and pipette bulbs
4. Solutions: 0.10 M potassium permanganate (MW 158 g/mol), 0.10 and 0.01 M methylene blue (MW 374 g/mol), 0.10 M Congo red (MW 697 g/mol)
5. Erlenmeyer flask

Experiment 3.2: Tonicity and Osmosis Experiment

1. Osmosis chamber with semipermeable membrane (Appendix 4).
2. Solutions: 4 M glucose (MW 180 g/mol), 0.001 M methylene blue (MW 374 g/mol).
3. 250 ml beaker
4. Disposable glass Pasteur pipettes and pipette bulbs
5. Spectrophotometer and cuvettes

Introduction and Pre-Lab Exercises

Nearly every part of the human body experiences the movement of molecules from one area to another. As we learned about homeostasis in Laboratory 2, the balance of molecules and their movement are extremely important for normal body function. For example, if a person receives intravenous fluids because of dehydration, the solution being infused into the bloodstream must have the proper concentration of electrolytes to ensure that the blood cells will not rupture (lyse) or shrivel (crenate). You may not have considered this until now, but the wrong concentration of intravenous fluids can greatly harm or kill a patient.

A **solution** is a uniform mixture of solute and solvent. The **solute** is the dissolved and suspended molecules and the **solvent** is the liquid in which those molecules are dissolved. In our bodies, as in all animals, water is the most abundant or universal solvent. Water is so universal that it can make up as much as 95% of an animal's body. Water comprises nearly 60% of human body mass. Figure 3-1 shows the fluid compartments within the human body.

Within a solution, the solute is in a constant dynamic state in which the molecules are always moving unless they are at absolute zero (−273 °C or 0 K). The molecules within a solution move from an area where they are close together to an area with fewer molecules until these molecules are evenly distributed throughout the solution. The movement of molecules is **diffusion. Net diffusion** can be formally defined as the net movement of molecules from an area of high concentration to an area of low

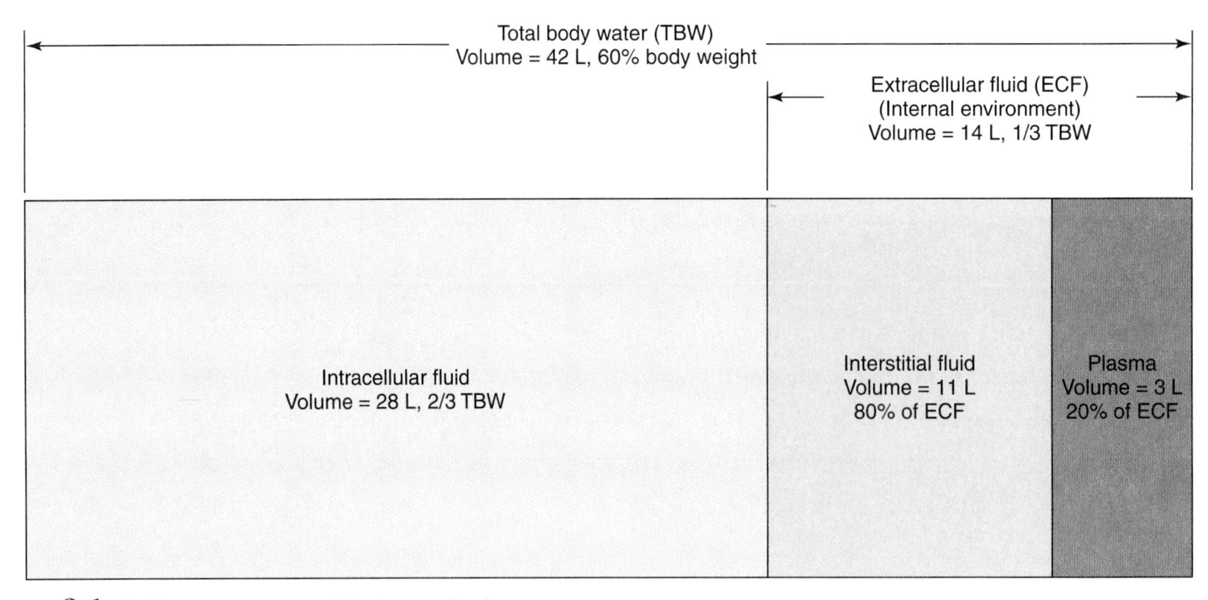

Figure 3-1 Fluid compartments of the human body.

concentration of molecules or solute. In many cases, diffusion occurs across a semipermeable (or selectively permeable) membrane. The rate at which molecules can diffuse is dependent upon: (1) the concentration gradient across the selectively permeable membrane, (2) the permeability and thickness of the selectively permeable membrane (usually cell membranes), (3) the surface area of the membrane available for diffusion, (4) the size of the molecules that are diffusing, and (5) temperature.

3-1 If diffusion rate is affected by the surface area of the selectively permeable membrane, what is the advantage of red blood cell shape?

Similarly, what is the functional significance of mitochondrial cristae?

■

Most cell membranes are selectively permeable. This means that unless a molecule can diffuse directly across the cell membrane, it will be prohibited from diffusing into or out of a cell. Because of the amphipathic (having both polar and nonpolar regions) nature of the cell membrane, only small, nonpolar molecules can diffuse directly across the cell membrane. All other molecules must be moved across the cell membrane with the assistance of a protein transporter. Glucose, for example, is an extremely important nutrient for all of our cells, but it is too large to cross the cell membrane on its own. Instead, glucose is transported across the cell membrane by a protein transporter. This process of molecular transport is called **facilitated diffusion.** Note that such protein transporters merely facilitate diffusion by providing a pathway for large and/or polar molecules to cross the cell's selectively permeable membrane. Thus, facilitated diffusion does not require energy because it still involves molecules moving from an area of high solute concentration to an area of low solute concentration.

3-2 Another mechanism of molecular transport used by cells is called **active transport.** However, active transport always requires energy to move substances to the other side of the cell membrane. Does active transport move molecules from high to low *or* from low to high solute concentration? Explain your answer.

■

While diffusion refers only to the movement of solute, **osmosis** refers to the diffusion of the solvent. If a selectively permeable membrane separates a concentration gradient of molecules and the molecules cannot pass through the membrane, water (the solvent), which is a relatively small molecule (H_2O = MW 18 g/mol), may pass through the membrane to dilute the concentration of solute on the other side. You may think of this in the following way: Suppose you were in a small and crowded room that makes everyone extremely uncomfortable. This high concentration of people may make you want to leave (diffuse from) the room. However, the door is locked and no one can exit. What may make you feel more comfortable would be for the room to somehow increase in size so you do not feel so crowded. This is exactly what osmosis does; it makes "the room" larger. If you have 10 marbles in a water balloon that has a 10 cm diameter and fill the balloon with more water, making it expand to a 20 cm diameter, the marbles will be less crowded, or concentrated. Thus we may formally define **osmosis** as the net movement of water (solvent) from an area of low solute concentration to an area of high solute concentration across a selectively permeable membrane.

Osmotic pressure describes the tendency of a solution to elicit osmosis. The amount of osmotic pressure a solution has depends greatly on the number of solute particles contained in that solution. If we revisit the balloon example just described, a balloon having a 10 cm diameter and containing 30 marbles will have a greater osmotic pressure than a balloon of the same diameter containing 10 marbles. The greater the number of particles a solution contains, the larger the osmotic pressure (or the greater will be the tendency for water to move into that solution). **Osmotic pressure** is the force or pressure that must be exerted on the solution to prevent the movement of water from an area of low solute concentration to an area of high solute concentration.

When determining how much osmotic pressure a solution has, it is important to know how many molecules that solution contains. For example, a 1 molar (M) solution of glucose contains 1 mole of glucose molecules in 1 liter of solution. A mole of anything (glucose, NaCl, H_2O) contains 6.02×10^{23} molecules. A 1 M glucose solution has a larger osmotic pressure than a 0.5 M glucose solution because it contains more glucose molecules. However, a 1 M glucose solution has the same osmotic pressure as a 0.5 M NaCl solution because they have the same number of particles in solution. Remember that electrolytes such as NaCl dissociate in solution, so every molecule of NaCl generates two particles: a Na^+ and a Cl^-. This is the concept of **osmolarity,** which quantifies the number of particles in a solution that will generate osmotic pressure. Thus, a solution's osmolarity, or the amount of osmotic pressure it has, is determined by the number of particles in solution.

BOX 3.1 Example calculation of molarity and osmolarity.

a. 20 grams of glucose in 100 ml of solution

Atomic mass of 1 mole of glucose ($C_6H_{12}O_6$):

$6\ C \times 12\ \text{grams} = 72\ \text{grams}$

$12\ H \times 1\ \text{gram} = 12\ \text{grams}$

$6\ O \times 16\ \text{grams} = 96\ \text{grams}$

180 grams in 1 mole of glucose

Molarity:

$$\frac{20\ \text{grams glucose}}{100\ \text{ml solution}} \times \frac{1\ \text{mole glucose}}{180\ \text{grams}} \times \frac{1000\ \text{ml}}{1\ \text{L}}$$

$= 1.1$ moles glucose/liter of solution $= 1.1$ M glucose

Osmolarity

= the sum of all the moles of solute in a liter of solution

$= 1.1$ moles glucose/liter $= 1.1$ osmoles glucose/liter

$= 1.1$ Osm

b. 2 Moles of NaOH in 1.0 l of solution

Molarity:

$$\frac{2\ \text{moles NaOH}}{1.0\ \text{liter of solution}} = 2\ \text{M NaOH}$$

Osmolarity:

2 moles Na^+ + 2 moles OH^-/liter of solution

$= 4$ moles solute/liter

$= 4$ osmoles solute/liter

$= 4$ Osm

3-3 Fill in the following table by calculating the molarity and osmolarity of the solutions. Be sure to show your calculations.

Contents of Solution	Molarity	Osmolarity
3 moles of sucrose in 1 liter of solution		
90 grams of glucose in 1 liter of solution		
1 mole of NaCl in 1 liter of solution		
117 grams of NaCl in 1 liter of solution		
15 grams of NaOH in 500 milliliters of solution		

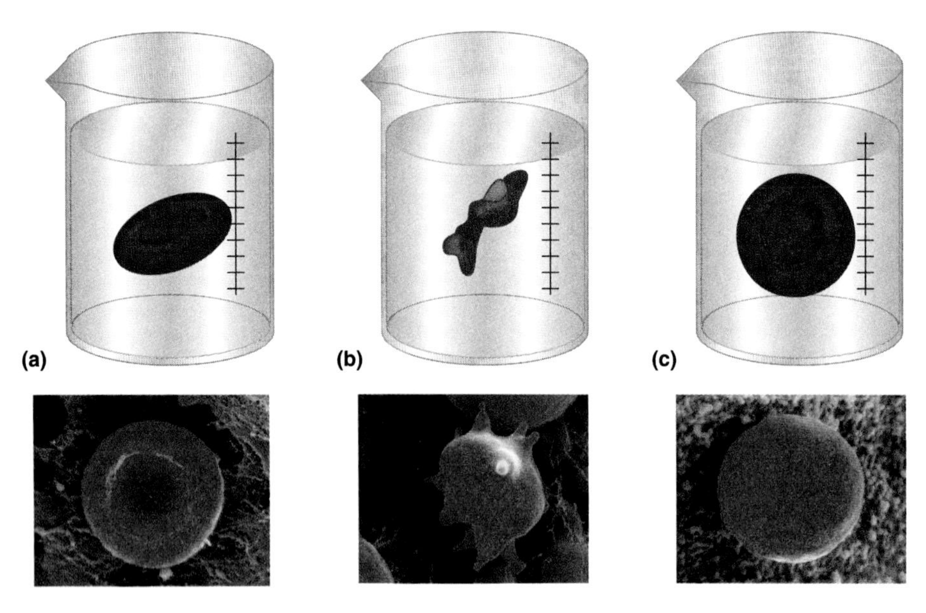

Figure 3-2 Effect of solution tonicity on human red blood cells.

 **3-4** Which of the solutions from Question 3-3 would create the greatest osmotic pressure?

■

Solutions having different solute concentrations have different tonicities relative to one another. **Tonicity** describes *relative* differences in osmotic pressure among different solutions. For example, a swimming pool is **hypotonic** to our body's cells. This means that the swimming pool has a lower concentration of solutes (and therefore lower osmotic pressure) than our cells. In contrast, a solution that is **hypertonic** to our cells (e.g., seawater) has a higher concentration of solutes and therefore a higher osmotic pressure. The difference in osmotic pressure between two compartments will cause water to move from an area of low solute concentration to an area of high solute concentration. In this way, our body's cells may experience swelling or shrinking if they are surrounded by a solution that is not **isotonic** (having the same solute concentration) to them. Note that tonicity describes relative differences in concentrations among solutions. A solution can only be described as being hypotonic, isotonic, or hypertonic with reference to another solution or cell.

3-5 What type of solution (hypotonic, isotonic, or hypertonic) is each of the red blood cells in Figure 3-2 submerged in?

■

3-6 Describe the movement of water with respect to osmotic pressure in each of the conditions in Figure 3-2.

■

3-7 One of your patients needs an infusion of fluids intravenously. What solution tonicity do you think would best facilitate water uptake by your patient's cells?

■

Methods and Materials

EXPERIMENT 3.1 Diffusion Experiment

In this exercise you will test how the concentration of a solution affects the rate of diffusion. You will also examine how the size of a molecule affects the rate of diffusion.

- Obtain four 5 ml pipettes filled with hardened agar and place them in an Erlenmeyer flask to hold them upright. Label one of the four pipettes containing hardened agar for each of the following four solutions:

    0.10 M Potassium Permanganate (MW 158 g/mol)

    0.10 M Methylene Blue (MW 374 g/mol)

    0.01 M Methylene Blue (MW 374 g/mol)

    0.10 M Congo Red (MW 697 g/mol)

- Using a Pasteur pipette, add 1 ml of dye to the top of the agar layer of the appropriately labeled 5 ml pipette. Repeat for each of the other three dye solutions.
- Every 10 minutes, measure the cumulative distance, in milliliters, each dye has diffused down the agar. This is measured most accurately in milliliters because although the dye is diffusing linearly, it is diffusing across a volumetric area.
- Record these data in Results Table 3.1a of your Laboratory Report.

EXPERIMENT 3.2 Tonicity and Osmosis Experiment

In this exercise you will examine the movement of water molecules across a selectively permeable membrane during osmosis.

- Obtain an osmosis chamber fitted with a selectively permeable membrane (Figure 3-3).
- Pour 100 ml of 4 M glucose solution into a beaker. Add 3 ml of 0.001 M methylene blue to this glucose solution. The methylene blue in this solution will serve as a color indicator of osmosis.

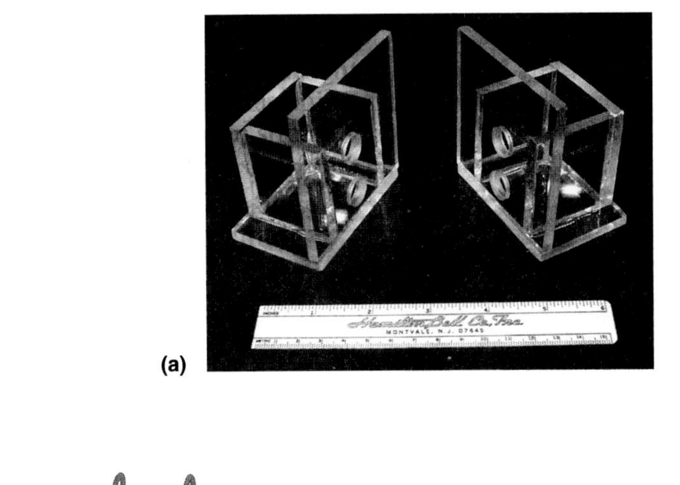
(a)

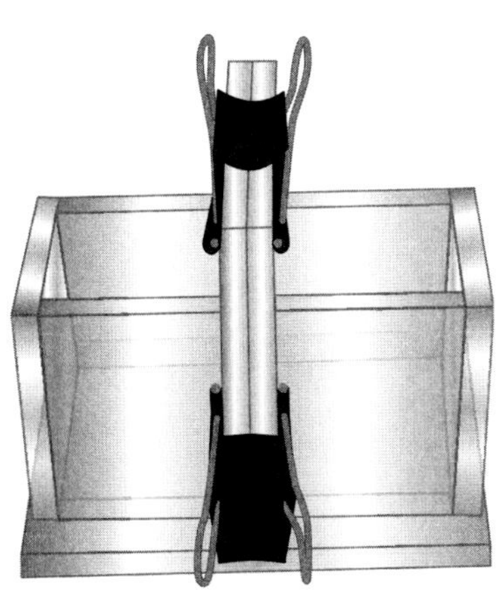
(b)

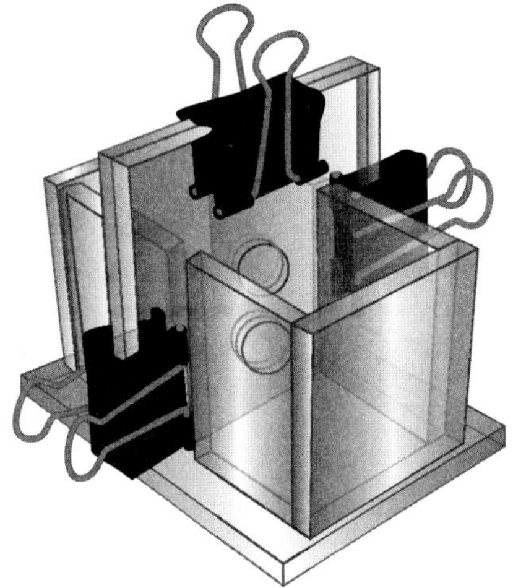

Figure 3-3 (a) Picture of the osmosis chamber. (b) Illustration of osmosis chamber fitted with a selectively permeable membrane between the two halves of the chamber.

- Pour approximately 80 ml of the glucose + methylene blue solution into the left side (Side A) of the osmosis chamber.
- Pour approximately 80 ml of distilled water into the right side (Side B) of the osmosis chamber.

  Note: Once both the left and right sides of the osmosis chamber are filled, make sure that the selectively permeable membrane is completely covered by fluid.
- Every 10 minutes, remove approximately 1-2 ml of the glucose + methylene blue solution from Side A and pipette this sample into a cuvette. Also remove approximately 1-2 ml of the distilled water from Side B and pipette this sample into a cuvette.
- Measure the absorbance of each sample from Side A and B in a spectrophotometer using a wavelength of 670 nanometers (nm). This is the wavelength of light that will be emitted by the spectrophotometer. The spectrophotometer will then measure what percentage of this light is absorbed by the solution in the cuvette. This is your absorbance reading. Note that because this reading is a percentage of the light absorbed, absorbance has no units.
- Before measuring the absorbance of your solutions, zero (or blank) the spectrophotometer using the distilled water blank that can be found near the spectrophotometer.
  - — Place the water blank into the spectrophotometer and zero the machine so that it reads 0% absorbance.
  - — Now place one of your cuvettes into the spectrophotometer and record its absorbance.
- When you have finished measuring the absorbance of the two samples, pour them back into their respective sides of the osmosis chamber. Rinse your cuvettes with distilled water after each sample to ensure accurate readings.
- Record these absorbance data in Results Table 3.2 of your Laboratory Report.

RESULTS AND DISCUSSION

LABORATORY REPORT 3

EXPERIMENT 3.1
Diffusion Experiment

Results Table 3.1a. Individual Group Data for Experiment 3.1: Diffusion Experiment.

Dye Solution	Cumulative distance moved at each 10-minute interval (in milliliters)												Total Distance Moved (ml)
	10 min	20 min	30 min	40 min	50 min	60 min	70 min	80 min	90 min	100 min	110 min	120 min	
0.10 M Potassium Permanganate													
0.10 M Methylene Blue													
0.01 M Methylene Blue													
0.10 M Congo Red													

Results Table 3.1b. Class Data for Experiment 3.1: Diffusion Experiment.

Dye Solution	Group data of total distance moved (in milliliters)								Mean Total Distance Moved (ml) ± Standard Error
	Group 1	Group 2	Group 3	Group 4	Group 5	Group 6	Group 7	Group 8	
0.10 M Potassium Permanganate									
0.10 M Methylene Blue									
0.01 M Methylene Blue									
0.10 M Congo Red									

Problem Set 3.1: Diffusion Experiment

Using the data from Results Table 3.1a, create a simple line graph showing the cumulative distance moved by each dye versus time. This graph will thus show the diffusion rate (volumetric distance/time) for each dye. Note that your line graph will have only one observation to represent each dye per time interval, and thus a 95% CI cannot be calculated or shown.

Using the data from Results Table 3.1b, perform the appropriate statistical test to examine possible differences in total diffusion distances among the four different dye solutions and answer the following questions. Note that your mean distance moved is calculated from several observations collected by the different laboratory groups. Thus, you should calculate a 95% CI for each of the four dyes to evaluate which dyes have significantly different mean diffusion rates.

a. State the null hypothesis.

b. Calculate a 95% CI for the mean total diffusion distance for each of the four dyes.

c. Should you accept or reject the null hypothesis? Support your conclusions with the appropriate statistical results.

d. How does the concentration of a solution affect volumetric diffusion distance? Explain.

e. How does the size (or mass) of a molecule affect volumetric diffusion distance? Explain.

f. Suppose you performed this experiment using agar-filled pipettes with twice the diameter of those that were used in this experiment. How would this larger diameter affect the total diffusion distance?

EXPERIMENT 3.2
Tonicity and Osmosis Experiment

Results Table 3.2. Individual Group Data for Experiment 3.2: Tonicity and Osmosis Experiment.

Time (minutes)	Absorbance	
	Chamber Side A 4 M glucose + 0.001 M methylene blue	Chamber Side B Distilled Water
10		
20		
30		
40		
50		
60		
70		
80		
90		
100		
110		
120		

Problem Set 3.2: Tonicity and Osmosis Experiment

Using the data from Results Table 3.2, create a graph showing the percent absorbance of each side of the osmosis chamber versus time and answer the following questions:

a. At time zero, or at the start of the experiment, is Side B (distilled water) hypotonic, isotonic, or hypertonic to Side A?

b. At the end of the experiment, is Side B (distilled water) hypotonic, isotonic, or hypertonic to Side A?

c. If Side A and B could reach equilibrium, would Side A be hypotonic, isotonic, or hypertonic to Side B?

d. In which direction are water molecules moving during the experiment? Explain why.

e. In which direction are glucose and methylene blue molecules moving during the experiment? Explain why.

f. How would increasing the concentration of the glucose and methylene blue solution affect osmotic pressure and the rate of osmosis? Explain your answer.

g. Does the size of a molecule affect osmotic pressure? If not, what does affect osmotic pressure?

h. Suppose you repeat this experiment, but this time the selectively permeable membrane has transport channels for glucose and methylene blue. How would this new membrane change your results?

COMPARATIVE NOTE *Laboratory 3*

Switching Osmoregulatory Mechanisms: Who's Fresh and Who's Salty?

Fishes that live in freshwater continuously lose ions to their environment and gain water through osmosis, primarily through their highly permeable gills. The high permeability of the gills is critical for gas exchange, so reducing gill permeability to decrease water influx would consequently impair gas exchange. To compensate for this inflow of water, most freshwater fishes produce large quantities (as much as one-third of body mass per day) of dilute urine. Lost solutes are replaced in the blood via active transport by ion pumps located in the gill epithelium. Conversely, the major osmoregulatory problem facing saltwater fishes is the gain of ions and the loss of water. Saltwater fishes therefore drink seawater, excrete highly concentrated urine, and use their gills to actively transport ions from the blood to the marine environment.

Interestingly, not all fishes spend their entire life in one type of osmotic habitat. The common eel, for example, is a catadromous (the Greek word *kata* = down; *dramein* = to run) teleost fish that grows to adult size in freshwater but migrates to the sea to reproduce. Such a migration from hypoosmotic freshwater to hyperosmotic seawater results in dramatic changes in osmoregulatory demands. In hyperosmotic seawater, substantial amounts of water loss to the environment will lead to severe dehydration and death. To compensate for this water loss, the migrating eel drinks seawater. However, drinking seawater now presents the problem of eliminating the excess salt that is ingested and absorbed along with the water from the intestinal tract. Again, the ion pumps located in the gill epithelium will play the major role in maintaining osmotic balance via active transport of ions. However, in order to maintain osmotic balance in seawater, the direction of ion transport in the gills must reverse! We do not yet understand how this change in osmoregulatory mechanisms occurs in the catadromous eel as well as in anadromous (the Greek word *ana* = up) fish such as shad and salmon that ascend from the sea to spawn in freshwater. Because migration into the sea or freshwater, respectively, occurs just prior to reproduction, the switch in osmoregulatory mechanisms is likely regulated by endocrine and neuroendocrine mechanisms. For example, it is likely that transcription and/or translation of a second type of ion pump is activated by the endocrine mechanisms that regulate reproduction.

RESEARCH OF INTEREST

Hansen, H. J. M. and M. Grosell. 2004. Are membrane lipids involved in osmoregulation? Studies *in vivo* on the European eel, *Anguilla anguilla,* after reduced ambient salinity. Environmental Biology of Fishes 70:57–65.

Wilson, J. M., J. C. Antunes, P. D. Bouca, and J. Coimbra. 2004. Osmoregulatory plasticity of the glass eel of *Anguilla anguilla:* freshwater entry and changes in brachial ion-transport protein expression. Canadian Journal of Fisheries and Aquatic Sciences 61:432–442.

Tam, W. L., W. P. Wong, A. M. Loong, K. C. Hiong, S. F. Chew, J. S. Ballantyne, and Y. K. Ip. 2003. The osmotic response of the Asian freshwater stingray (*Himantura signifer*) to increased salinity: a comparison with marine (*Taeniura lymma*) and Amazonian freshwater (*Potamotrygon motoro*) stingrays. Journal of Experimental Biology 206:2931–2940.

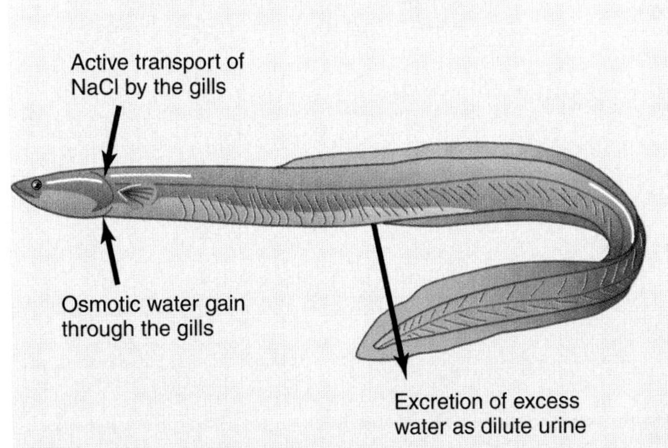

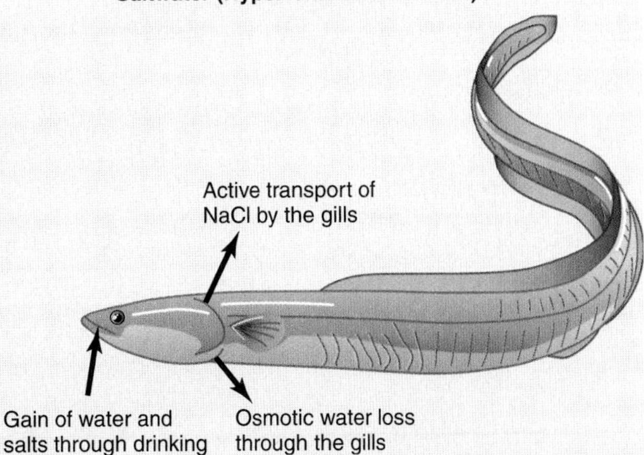

Osmoregulatory mechanisms utilized by the catadromous eel in response to freshwater and saltwater.

LABORATORY 4

Enzyme Activity

PURPOSE

This laboratory will introduce you to the biochemical and structural significance of enzymes and how their activity can be influenced by pH and temperature.

Learning Objectives

- Describe the differences between the lock-and-key and induced-fit models for enzyme–substrate interaction.
- Recognize that most enzymes are proteins, and identify the differences between primary, secondary, tertiary, and quaternary protein structure.
- Understand the calculation of pH and how pH indicates H^+ concentration.
- Understand temperature as a measure of kinetic energy or movement of molecules.
- Understand that enzymes serve as biological catalysts that lower the activation energy of a chemical reaction.
- Describe how pH and temperature can influence enzyme structure.
- Understand how changes in protein structure affect enzyme activity.

Laboratory Materials

Experiment 4.1: Effects of pH on Enzyme Activity

1. Para-nitrophenyl phosphate (the substrate), 10 mg/ml distilled water
2. Alkaline Phosphatase (the enzyme), 0.1 mg/ml distilled water
3. pH buffers: 7, 8, 9, 10, 11
4. Spectrophotometer and cuvettes

Experiment 4.2: Effects of Temperature on Enzyme Activity

1. Para-nitrophenyl phosphate (the substrate), 10 mg/ml distilled water
2. Alkaline Phosphatase (the enzyme), 0.1 mg/ml distilled water
3. pH buffer
4. Ice bath (0°C)
5. Water baths (23, 37, and 45°C)
6. Spectrophotometer and cuvettes

Introduction and Pre-Lab Exercises

Enzymes are specialized molecules that function as biological catalysts. Most enzymes are proteins, although some forms of RNA can also function as enzymes. Simplistically, we can think of anabolic enzymes as proteins that combine molecules by bringing them closer together so that they may react with one another. Similarly, catabolic enzymes break molecules apart by "stressing" or changing the stability of the molecule.

The structure of an enzyme is unique. This allows each enzyme to bind to a specific substrate just as a key is specific for a particular lock. This enzyme property is called **specificity,** and it determines the enzyme's particular function in the body. This **lock-and-key model** explains the association between an enzyme and its substrate, which is determined by the three-dimensional (3-D) shape of the enzyme. Just as a key has a specific shape, so does an enzyme. However, sometimes a substrate induces a change in the shape of the enzyme's active site when it binds to the enzyme. This conformational change in turn improves the fit or binding of the enzyme to its substrate. This model of enzyme–substrate interaction is referred to as the **induced-fit model** (Figure 4-1).

Recall that proteins (and therefore most enzymes) have a structural hierarchy. The 3-D shape of an enzyme is referred to as its **tertiary structure** and is determined by both its primary structure and its secondary structure, as explained in Figure 4-2. It is important to understand that it is

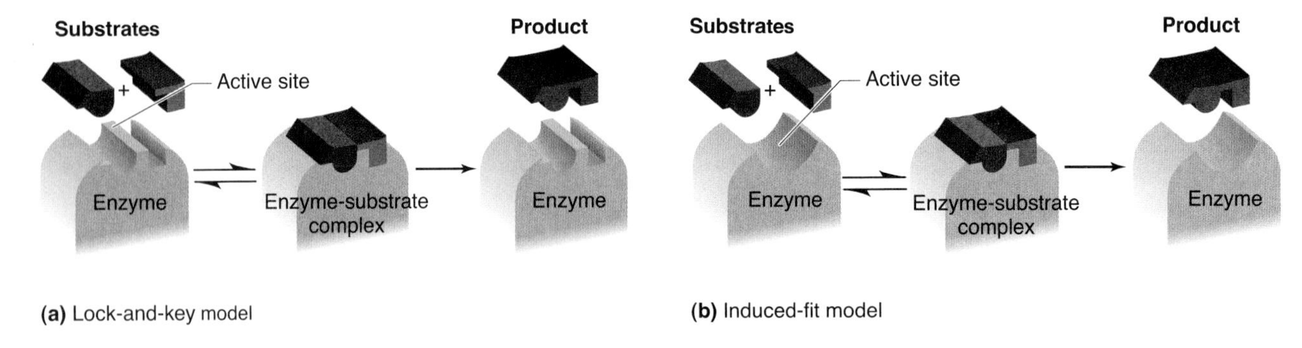

Figure 4-1 An enzyme catalyzing the reaction of two substrates in the (a) lock-and-key and (b) induced-fit models for enzyme–substrate interactions.

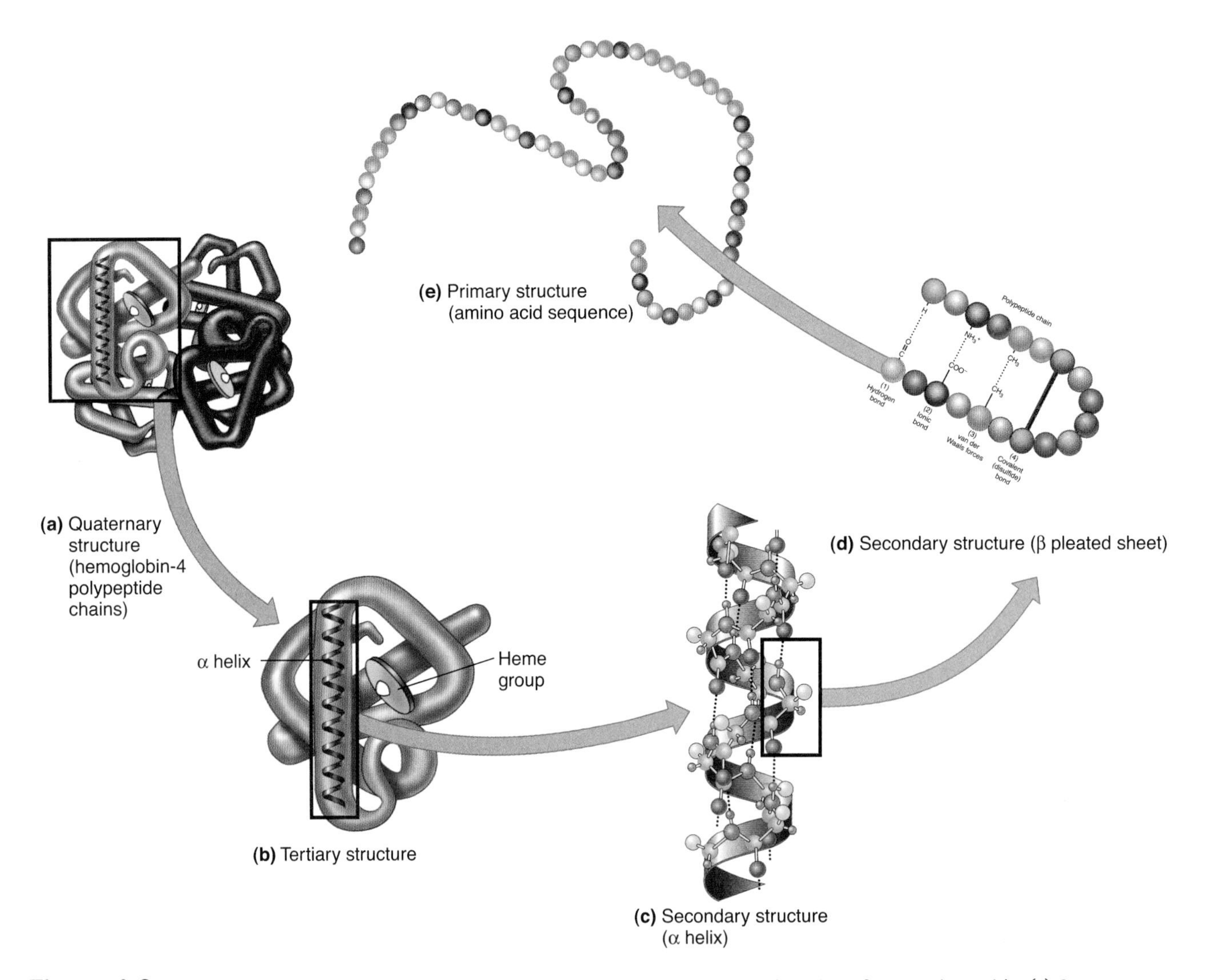

Figure 4-2 The structural hierarchy of proteins from quaternary structure to a simple string of 200 amino acids. (a) Quaternary structure—the bonding of two or more polypeptide chains; (b) tertiary structure—3-D structure of one polypeptide chain; (c and d) secondary structure—the conformation of the polypeptide chain created by hydrogen bonding between amino acids; and (e) Primary structure—the sequence of amino acids in the polypeptide chain.

this specific 3-D shape that allows an enzyme to function as a biological catalyst.

4-1 Suppose the 3-D shape of an enzyme was changed. What would happen to its ability to function? (Hint: If you bent a key in half, would you still be able to open a lock with it?)

■

If you said that the enzyme would not be able to easily attach to its substrate, you are correct. As an enzyme's 3-D shape changes, the enzyme begins to lose its **affinity** or attraction for its substrate, and the enzyme's activity decreases. If we change the enzyme's 3-D shape enough, the enzyme may completely lose its affinity for substrate and stop functioning altogether.

Both pH and temperature are physiological parameters that can greatly affect the 3-D shape of enzymes. pH is the measure of H^+ concentration and is calculated using the following equation:

$$pH = \log \cdot \frac{1}{[H^+]}$$

The 3-D structure of an enzyme is determined, in part, by its secondary structure, which involves how a single polypeptide chain interacts with itself, primarily through hydrogen bonding between nearby amino acids. Common examples of secondary structures include an α-helix and a β-pleated sheet. Because hydrogen bonding among amino acids is affected by changes in the concentration of H^+, changes in pH greatly influence both the secondary and tertiary structures of proteins.

4-2 The pH of blood and body fluids is very tightly controlled by buffer systems and negative feedback loops. In fact, pH is so highly controlled that human blood pH normally fluctuates between only 7.35 and 7.45. Indeed, a patient whose blood pH falls to 7.2 is experiencing a serious medical condition known as acute acidosis. Why do you think it is so important to keep pH at a relatively constant value?

■

4-3 A patient has a blood $[H^+]$ of 5.25×10^{-8}. What is the blood pH of this patient? Is this patient experiencing metabolic acidosis or alkalosis?

■

4-4 The patient's blood pH increases by 0.10 after treatment. What is the new $[H^+]$ of the treated patient? Has this increase in blood pH changed the patient's condition?

■

Similar to these effects of pH, changes in temperature also influence the 3-D shape of enzymes. Temperature is a measure of the kinetic energy of molecules. As temperature decreases or increases, the kinetic energy of molecules in a substance also decreases or increases. An increase in the kinetic energy of an enzyme will cause an increase in the movement of the enzyme's molecules. If the increase in temperature (and thus kinetic energy) is great enough, then the associations between and the alignment of the enzyme's molecules will be affected. If the molecules move so quickly that they are rarely in the proper alignment, then the enzyme will not function properly. Thus, temperature affects the shape of a protein by changing the kinetic energy of the molecules in that protein. At very high temperatures, the hydrogen bonds between nearby amino acids within a polypeptide chain are broken. Thus, the enzyme's secondary structure is destroyed, and the denatured enzyme becomes nonfunctional.

4-5 As the kinetic energy of molecules increases with increasing temperature, reaction rates also increase. Explain what happens to chemical and enzymatic reactions as temperature decreases. Why are tissue and DNA samples commonly stored in −70°C freezers?

■

Enzymes are catalysts for chemical reactions. In order for chemical reactions to occur, reactants must have sufficient energy (called the **activation energy**) to allow chemical bonds to be formed and broken. Enzymes catalyze reactions by lowering the activation energy for a particular reaction, thus increasing the number of molecules at the energy level required for activation (Figure 4-3). Enzymes allow reactions to occur with less energy input or endergonic energy, as shown by the shift in the energy needed for a reaction from the right to the left (Figure 4-3). If we wanted to break down a polysaccharide such as starch into simple sugars, we could heat the starch and allow the increased kinetic energy of the molecules to break the chemical bonds in the starch molecules. However, this requires considerable heat (> 200°C). We break down starches in our mouths and intestines through the use of enzymes rather than extreme heat. Our bodies function at an optimal temperature of 37°C and would begin to experience heat shock if we approached 40.6°C (105°F). It is obvious that we cannot increase our body temperature to a level that allows for the breakdown of starch. Thus, the lowering of activation energy by enzymes allows the starch molecule to be broken down or catabolized at low temperatures such as body temperature

Figure 4-4 shows the results of an experiment investigating the effects of high temperature and the presence of an enzyme on the breakdown of starch. One way to determine if starch has been catabolized into glucose is to use Benedict's reagent. When you add Benedict's reagent to a solution containing glucose, it reacts with the glucose and forms a yellow solution with a red precipitate.

4-6 Explain why the test for the presence of sugar is positive in experimental condition (b) (Figure 4-4).

Explain why the test for the presence of sugar is positive in experimental condition (c) (Figure 4-4).

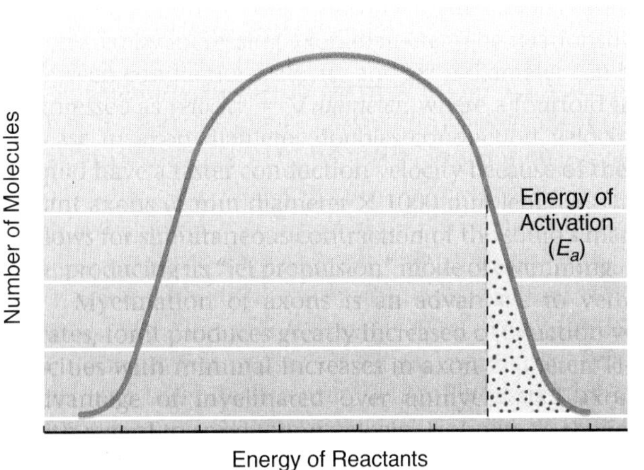

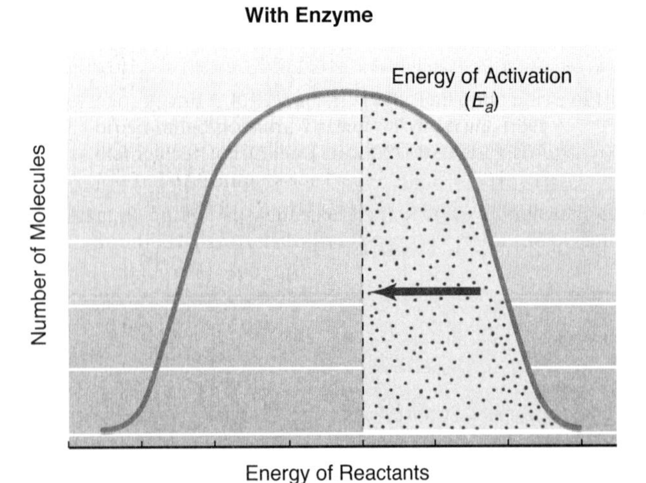

Figure 4-3 Shift in energy of activation due to the presence of an enzyme.

Enzymes are found in all organisms and are extremely important in physiological function and homeostasis. Because enzymes exhibit substrate specificity, most enzymes are named for the specific substrate they act on and their particular function. For example, the enzyme lactate dehydrogenase removes an H^+ from lactate, which is then oxidized in the mitochondria to form a water molecule (thus "dehydrating" lactate, its substrate). Alcohol dehydrogenase also catalyzes the removal of H^+ from alcohols.

The experiments you will be performing in lab today examine the effects of pH and temperature on the enzyme phosphatase, an important enzyme found in all cells. This enzyme specifically cleaves a phosphate group from its substrate. The substrate you will be using in lab today is para-nitrophenyl phosphate, which is colorless in solution. Although this substrate does not occur naturally in cells, its property as a color indicator makes it a useful substrate for today's laboratory experiments. When phosphatase cleaves the phosphate group from para-nitrophenyl phosphate, the resulting product is para-nitrophenol. Because this product is yellow in solution, you will be able to visually witness the action of the phosphatase enzyme on its substrate.

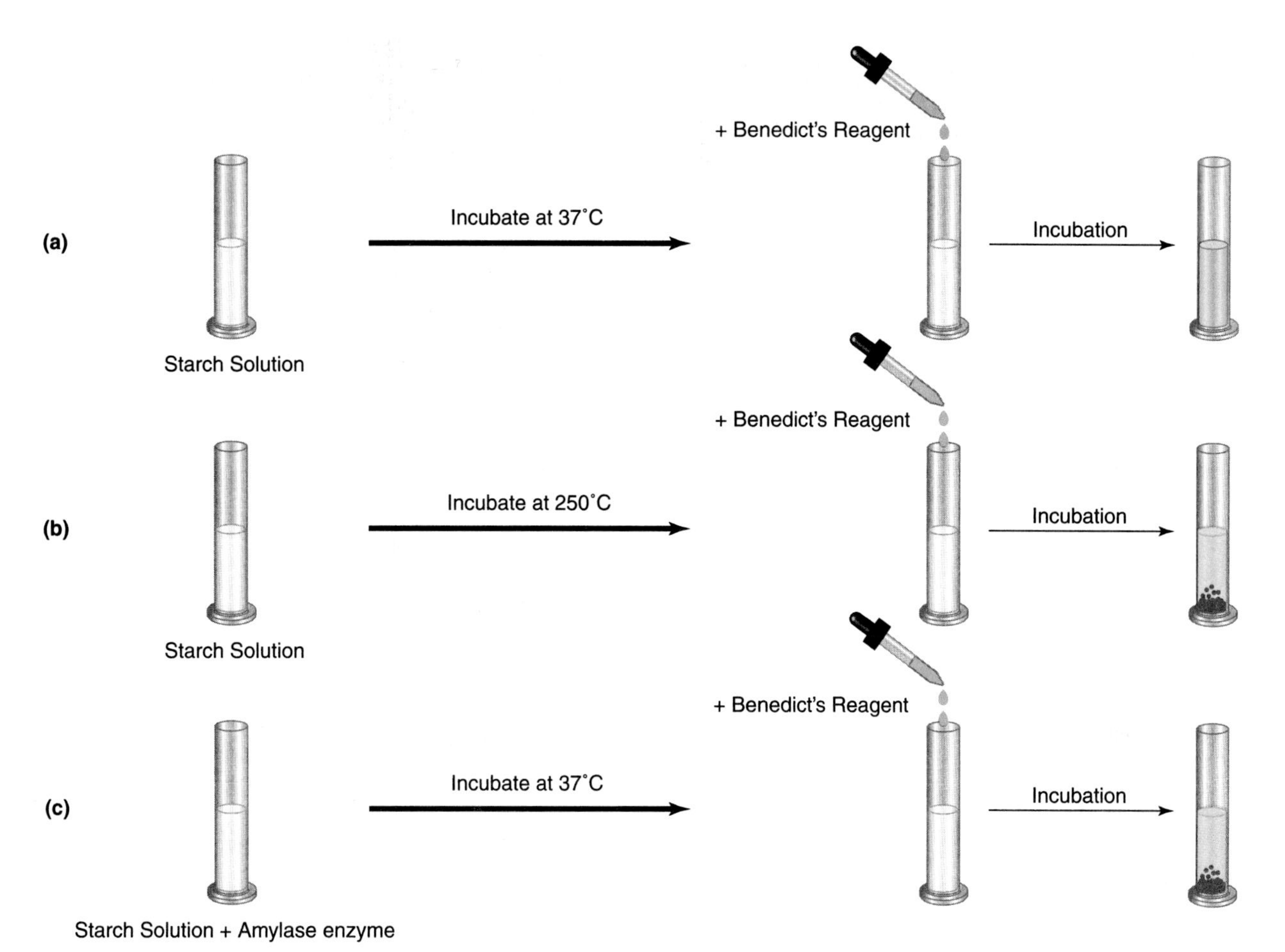

Figure 4-4 Chemical reactions for the breakdown of starch to glucose under three different experimental conditions: (a) at body temperature without enzyme; (b) at high temperature without enzyme; (c) at body temperature with enzyme.

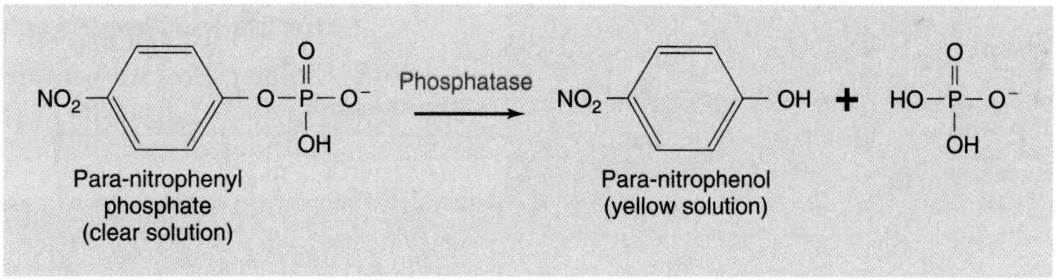

$$NO_2-C_6H_4-O-P(=O)(OH)-O^- \xrightarrow{\text{Phosphatase}} NO_2-C_6H_4-OH + HO-P(=O)(OH)-O^-$$

Para-nitrophenyl phosphate (clear solution)

Para-nitrophenol (yellow solution)

Figure 4-5 Chemical reaction of para-nitrophenyl phosphate with the enzyme phosphatase.

This change in color will also allow you to more accurately measure enzyme activity by measuring the change in light absorbance of the reaction solutions in a spectrophotometer. In general, the darker a solution's color, the more light a solution will absorb and the higher the absorbance of the solution. The equation for today's enzyme reaction is shown in Figure 4-5.

COMPARATIVE NOTE Laboratory 4

Some Like It Hot: Genetic Techniques from Hot Springs

All living organisms contain enzymes that function optimally under a certain range of temperatures. An organism's activity and even survival depend upon optimal enzyme activity matching the range of body temperatures typically experienced by the organism. Large deviations from these temperatures would reduce enzyme activity, with dramatic consequences for the organism. However, some organisms, and their enzymes, can function at temperatures that would kill most other organisms.

Dr. Kary Mullins revolutionized the field of molecular biology and was awarded the Nobel Prize in Chemistry in 1993 for developing an amplifying technique known as the **polymerase chain reaction (PCR).** This technique is used to amplify small quantities of DNA by exploiting the enzymes of a hot-springs bacterium that function optimally at high temperatures (75–80°C). The use of PCR in amplification involves raising the temperature of a double-stranded DNA template until the DNA is denatured. At the denaturation point, the two strands separate, and the single-stranded DNA sequences can then be used as templates for making copies of short DNA sequences. The temperature is decreased slightly to allow specific primers (chosen and added to the mixture by the researcher) to anneal to the single-stranded DNA sequences. Extracted polymerase enzyme from the bacterium *Thermus aquaticus* is also included in the mixture. This enzyme, called Taq polymerase (the name is derived from the bacterium's scientific name), directs the DNA copying process using excess nucleotides that are added to the sample. Because the bacteria's polymerase enzyme functions optimally at high temperatures, the polymerase readily copies the single-stranded DNA where the primers have annealed. In contrast, using a polymerase from a mammal would require decreasing the temperature of the mixture significantly. At these lower temperatures, single-stranded DNA readily forms complimentary base pairs with other single-stranded DNA. Thus, the sequences of interest are unavailable for copying at lower temperatures. Repeated cycles of raising the temperature to denature the newly copied, double-stranded DNA, followed by a slight decrease in temperature to allow re-annealing of primers and DNA synthesis, can significantly increase the quantity of a particular sequence as much as a billionfold. Such quantities of DNA were previously unobtainable, so many genetic analyses could not be readily performed. The polymerase chain reaction is now widely used in many fields of research for a variety of research goals, including synthesis of DNA from specific messenger RNA sequences, creating cDNA probes, analyzing DNA sequences, and screening for mutations of specific genes.

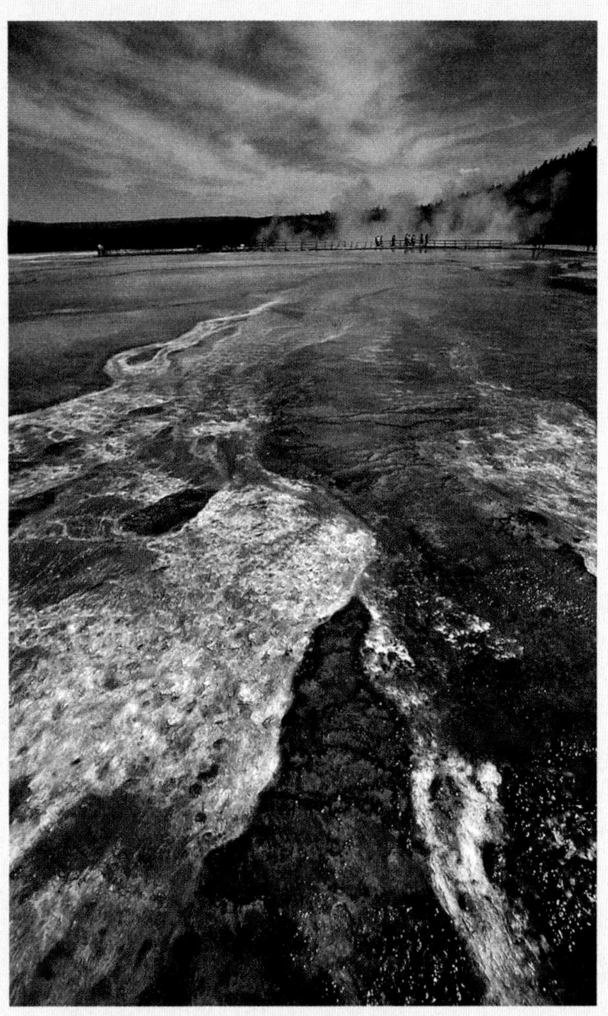

A hot spring in Yellowstone National Park.

RESEARCH OF INTEREST

Pearson, A., Z. Huang, A.E. Ingalls, C.S. Romanek, J. Wiegel, K.H. Freeman, R.H. Smittenberg, and C.L. Zhang. 2004. *Nonmarine crenarchaeol* in Nevada hot springs. Applied and Environmental Microbiology 70:5229–5237.

Simbahan, J., R. Drijber, and P. Blum. 2004. *Alicyclobacillus vulcanalis* sp. nov., a thermophilic, acidophilic bacterium isolated from Coso Hot Springs, California, USA. International Journal of Systematic and Evolutionary Microbiology 54:1703–1707.

Breitbart, M., S. Leeds, T. Schoenfeld, and F. Rohwer. 2004. Phage community dynamics in hot springs. Applied and Environmental Microbiology 70:1633–1640.

Methods and Materials

EXPERIMENT 4.1 The Effects of pH on Enzyme Activity

In this experiment you will examine how different pH buffers affect the rate of enzyme reactions.

- Obtain five cuvettes and label them pH 7, 8, 9, 10, and 11.
- Add 3 ml of the appropriate pH buffer to your labeled cuvettes.
- Add 1 ml of substrate (para-nitrophenyl phosphate) to each of your five cuvettes.
- Add 100 μl of enzyme (alkaline phosphatase) to the first cuvette containing pH 7 buffer.
- Cover the cuvette with parafilm and invert gently to mix.
- Incubate the reaction at room temperature for approximately 3 minutes. You must measure the absorbance of each reaction precisely 3 minutes following the addition of enzyme.
- Measure the absorbance of this solution in the spectrophotometer at a wavelength of 405 nanometers (nm). This is the wavelength of light that will be emitted by the spectrophotometer. The spectrophotometer will then measure what percentage of this light is absorbed by the solution in the cuvette. This is your absorbance reading. Note that because absorbance is a percentage of light absorbed, absorbance has no units.
  - For each of your tubes containing a particular pH buffer, you will need to zero (or blank) the spectrophotometer. This zeroing procedure will correct for any additional absorbance caused by the buffer solutions. You will find a set of blanking solutions (buffer + substrate) by the spectrophotometer.
  - Place the pH 7 blank solution into the spectrophotometer. Zero the spectrophotometer so that it reads 0% absorbance.
  - Now place your experimental enzyme reaction incubated in pH 7 buffer into the spectrophotometer. Record the absorbance.
- Starting with the addition of enzyme, repeat these steps until you have recorded the absorbance of all of your pH-dependent enzyme reactions in Results Table 4.1a of the Laboratory Report.

EXPERIMENT 4.2 The Effects of Temperature on Enzyme Activity.

In this experiment you will examine how different incubation temperatures affect the rate of enzyme reactions.

- Obtain 4 cuvettes and label them 0, 23, 37, and 45° C.
- Determine the optimal pH for the phosphatase enzyme from Experiment 4.1. Add 3 ml of this optimal pH buffer to all cuvettes.
- Add 1 ml of substrate (para-nitrophenyl phosphate) to all cuvettes.
- Incubate the solutions at the appropriate temperatures for 20 minutes.
- Add 100 μl of enzyme (alkaline phosphatase) to the first cuvette incubated at 0°C.
- Cover the cuvette with parafilm and invert gently to mix.
- Return the reaction to the water bath and incubate at 0°C for approximately 3 minutes. You must measure the absorbance of each reaction precisely 3 minutes following addition of enzyme.
- Measure the absorbance of this reaction tube in the spectrophotometer at a wavelength of 405 nm. Because all tubes contain the same pH buffer, you will only need to zero the spectrophotometer one time. You will find the blanking solution (buffer + substrate) for the temperature experiment by the spectrophotometer.
  - Place the blanking solution into the spectrophotometer. Zero the spectrophotometer so that it reads 0% absorbance. This is the only time you must zero the spectrophotometer in this experiment.
  - Now place your reaction incubated at 0°C into the spectrophotometer. Record the absorbance.
- Starting with the addition of enzyme, repeat these steps until you have recorded the absorbance of all of your temperature-dependent enzyme reactions in Results Table 4.2 of the Laboratory Report.

RESULTS AND DISCUSSION

LABORATORY REPORT 4

EXPERIMENT 4.1:
Effects of pH on Enzyme Activity

Results Table 4.1a Individual Group Data for Experiment 4.1: Effects of pH on Enzyme Activity.

pH Buffer	Absorbance
7	
8	
9	
10	
11	

Results Table 4.1b Class Data for Experiment 4.1: Effects of pH on Enzyme Activity.

pH Buffer	Absorbance								Mean Abs.	SE
	Group 1	Group 2	Group 3	Group 4	Group 5	Group 6	Group 7	Group 8		
7										
8										
9										
10										
11										

Problem Set 4.1: Effects of pH on Enzyme Activity

Using the data from Results Table 4.1b, create a bar graph showing the mean absorbance for the reactions incubated in each of the five different pH buffers. Include the 95% CI for each of the five pH buffers in your graph. Using the data from Results Table 4.1b, perform the appropriate statistical test to examine possible differences in absorbance among the five different pH buffers and answer the following questions:

a. State the null hypothesis.

b. Should you accept or reject the null hypothesis? Support your conclusions with the appropriate statistical results.

c. How does pH affect the activity of the phosphatase enzyme? What is the optimal pH for this enzyme?

d. Explain the mechanism underlying the effects of pH on enzyme activity.

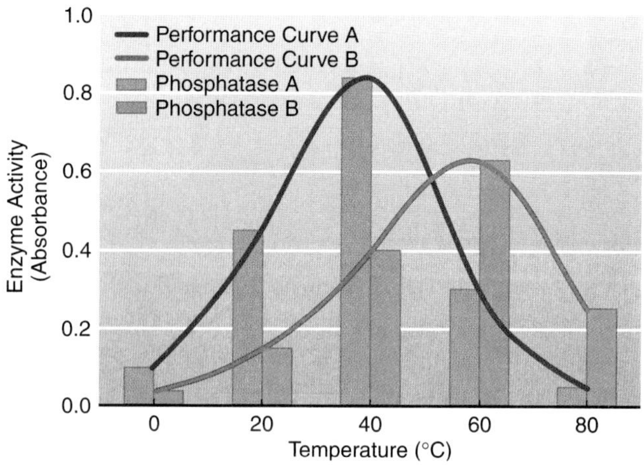

Figure 4-6 Example graph showing the effect of temperature on enzyme activity for two different phosphatase enzymes.

a. How does temperature affect the activity of alkaline phosphatase? Explain your answer.

b. What is the mechanism underlying the effects of temperature on enzyme activity? [*Hint:* Which enzyme structures (primary, secondary, tertiary, or quaternary) do you think temperature affects?]

c. Does your graph look similar to Figure 4-6? Can you determine the optimal temperature of alkaline phosphatase? What is the optimal temperature of the newly-isolated phosphatase?

d. Which of these enzymes do you think could best function in a human? Explain your answer.

e. How might the properties of this newly-isolated phosphatase benefit an organism's ecology?

f. What would happen to alkaline phosphatase if it were incubated at 100°C?

g. How would doubling the concentration of the enzyme affect the rate of these reactions?

EXPERIMENT 4.2

Effects of Temperature on Enzyme Activity

Results Table 4.2 Individual Group Data for Experiment 4.2: Effects of Temperature on Enzyme Activity.

	Absorbance	
Temperature (°C)	Alkaline Phosphatase	A Newly-Isolated Phosphatase
0		0.321
23		0.382
37		0.464
45		0.892
80	No Data	0.537

Problem Set 4.2: Effects of Temperature on Enzyme Activity

Using the data from Results Table 4.2, create a bar graph showing the absorbance of the reactions at each of the incubation temperatures for both alkaline phosphatase and the newly-isolated phosphatase shown in Table 4.2. Note that your bar graph will have only one observation to represent the absorbance at each incubation temperature, and thus 95% CI cannot be calculated or shown.

LABORATORY 5

Action Potentials

PURPOSE

This laboratory will introduce you to the principles of membrane potentials in both individual cells and whole nerves. The major principles discussed here are resting membrane potential, the electrical events that occur during an action potential, refractory periods, and compound action potentials. This laboratory simulation uses crawfish muscle cells to experimentally manipulate the ion concentrations in the extracellular fluid and a frog sciatic nerve to investigate the effects of electrical stimuli.

Learning Objectives

- Define *resting membrane potential* and observe how the membrane potential is influenced by changes in Na^+ and K^+ concentrations in the extracellular fluid.
- Understand the electrical events that occur during an action potential.
- Discuss how an action potential in a single axon differs from a compound action potential in a nerve.
- Investigate how stimulus strength influences the amplitude of a compound action potential generated in a nerve.
- Define *relative* and *absolute refractory* periods for an action potential.
- Understand the chemical properties of the voltage-gated sodium ion channels that result in refractory periods.
- Observe how the amplitude of a second compound action potential is relative to the period between two stimuli.

Laboratory Materials

Experiment 5.1: Influence of Extracellular Ion Concentration on Resting Membrane Potential

Ph.I.L.S. Membrane potential simulation

Experiment 5.2: Compound Action Potentials

Ph.I.L.S. Nerve function simulation

Experiment 5.3: Refractory Periods

Ph.I.L.S. Nerve function simulation

Introduction and Pre-Lab Exercises

The nervous system is highly specialized for communication through electrical impulses known as action potentials. An **action potential** is caused by the movement of ions, particularly Na^+ and K^+. Before discussing the particular ionic events that occur during an action potential, however, we must first discuss why Na^+ and K^+ diffuse.

Every cell in your body maintains ion concentration gradients across the cell membrane. These ion gradients are established by active solute pumps, which use ATP to move ions against their concentration gradient. One of the most important ion pumps found in all cells of the body is the Na^+/K^+ pump. Because the Na^+/K^+ pump moves both ions against their concentration gradients (from a low concentration to a higher concentration), the ion pump must consume energy in the form of ATP to perform the work of moving these ions. The Na^+/K^+ pump moves 3 Na^+ out of the cell while moving 2 K^+ into the cell for every 1 ATP molecule.

This movement of Na^+ and K^+ by the ion pump establishes and maintains both a **concentration gradient** and an **electrical gradient** across the plasma membrane of a cell. The concentration gradients are established because of the relative difference in the number of Na^+ and K^+ present in the intra- and extracellular fluids. Likewise, an electrical

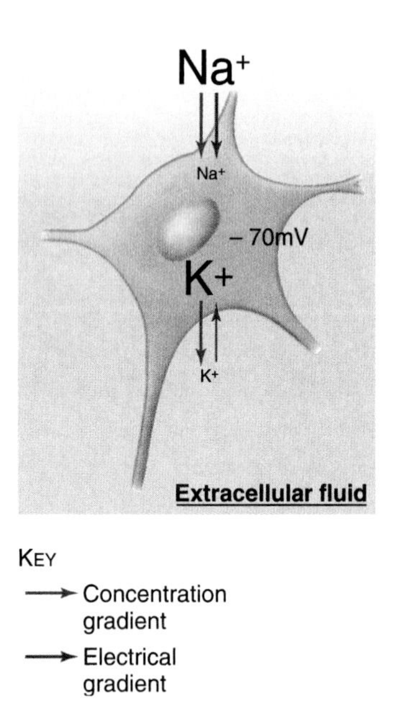

Figure 5-1 Forces influencing sodium and potassium ions at the resting membrane potential.

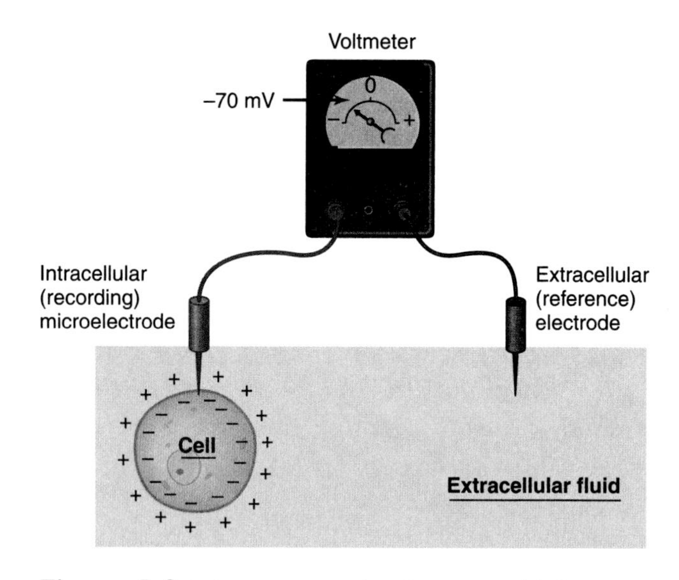

Figure 5-2 Voltmeter measuring the potential difference across the cell membrane.

gradient exists because the Na^+/K^+ pump contributes to the separation of electrical charges (ions) across the cell membrane. Potential energy is contained in both the concentration and electrical gradients, and both types of gradients produce forces that influence the movements of Na^+ and K^+ while a cell is at rest (Figure 5-1).

For many cells, if you inserted a tiny voltmeter into the cell through the cell membrane, the voltmeter would read −70 millivolts (mV) (Figure 5-2). This is the **resting membrane potential** of a cell, or the voltage charge across the cell membrane while the cell is at rest. Different cell types have different resting membrane potentials, and the magnitude of the resting membrane potential can vary from −5 to −100 mV. A typical nerve cell has a resting membrane potential of approximately −70 mV, meaning that the intracellular fluid has a charge of −70 mV relative to the extracellular fluid.

The resting membrane potential is created, in part, by the differences in the ion concentrations on the outside (O) versus the inside (I) of the cell membrane. Because the intracellular fluid contains an excess of negative charges (e.g., anions and negatively charged proteins) relative to the extracellular fluid, the resting membrane potential is negative. This means that the intracellular fluid is negatively charged relative to the extracellular fluid. Other factors that contribute to the maintenance of resting membrane potential include the relative permeability, or leakiness, of the plasma membrane to Na^+ and K^+ ions as well as the concentration gradients of other ions such as Cl^-. Although many factors contribute to the unique characteristics of resting membrane potential and action potential generation, this lab will focus primarily on the roles of Na^+ and K^+ in neurons.

Figure 5-2 demonstrates how the resting membrane potential of a cell can be measured experimentally using a voltmeter. However, we can also determine theoretically the resting membrane potential of this cell using the Goldman-Hodgkin-Katz equation. This equation uses the known extracellular and intracellular ion concentrations and the cell's relative permeability to each ion to calculate the potential difference across the cell membrane.

Goldman-Hodgkin-Katz Equation

$$V_{membrane} = \frac{RT}{Fz} \cdot \log\left(\frac{P_K[K^+]_O + P_{Na}[Na^+]_O + P_{Cl}[Cl^-]_I}{P_K[K^+]_I + P_{Na}[Na^+]_I + P_{Cl}[Cl^-]_O}\right)$$

5-1 The relative permeability and extracellular and intracellular ion concentrations for a human nerve cell are listed in the following table. Rather than experimentally measuring the resting membrane voltage (V_m) of a neuron, use the Goldman-Hodgkin-Katz equation to mathematically prove that the resting membrane potential is indeed −70 mV. For our general purposes, note that the relationship between the universal gas constant (R), absolute temperature (T), Faraday's constant (F), and ion valence (z) is equal to 60.

Ion	Relative Permeability	Extracellular (Outside) Concentration (mM)	Intracellular (Inside) Concentration (mM)
K^+	1.00	5	150
Na^+	0.035	150	15
Cl^-	0.001	110	7

$$V_{membrane} = 60 \cdot \log\left(\frac{1.00 \cdot [K^+]_O + 0.035 \cdot [Na^+]_O + 0.001 \cdot [Cl^-]_I}{1.00 \cdot [K^+]_I + 0.035 \cdot [Na^+]_I + 0.001 \cdot [Cl^-]_O}\right)$$

$$V_{membrane} = \_\_\_\_\_\_\_\_\_\_$$

■

Now that you have demonstrated that the cell's resting membrane potential can be calculated, let us consider an even easier calculation using the **Nernst equation.** The Nernst equation is used to calculate the equilibrium potential of a cell due to a particular ion when the cell membrane is completely permeable to that ion. **Equilibrium potential** is simply the electrical potential necessary to balance an ionic concentration gradient across a membrane so that the net flux of that ion is zero.

Nernst Equation

$$E_{Na^+} = \frac{RT}{Fz} \cdot \log\left(\frac{[Na^+]_O}{[Na^+]_I}\right)$$

$$E_{Na^+} = 60 \cdot \log\left(\frac{[Na^+]_O}{[Na^+]_I}\right)$$

$$E_{Na^+} = \_\_\_\_\_\_\_\_\_\_$$

5-2 Why don't we use ion permeability to calculate the equilibrium potential of sodium?

■

5-3 Which ion's equilibrium potential contributes most to the resting membrane potential? (*Hint:* Use the Nernst equation for K^+ and Cl^- and demonstrate which ion has an equilibrium potential closer to -70 mV.)

■

5-4 Based upon your calculations and the information provided, is the resting cell most permeable to Na^+, K^+, or Cl^-? Explain your answer.

■

The resting membrane potential of a cell will remain relatively stable unless the electrical current of that cell changes. In cells, electrical currents are created by the movement of ions into or out of a cell. Because ions are polar molecules, they cannot readily diffuse through the nonpolar region of the plasma membrane. Thus, ions require a protein channel to facilitate their diffusion across the plasma membrane. Note that these protein channels permit only passive diffusion of ions. (These protein channels are quite different from the ion pumps that transport ions against their concentration gradient via active transport.) Most ion channels are "gated," and a specific chemical, electrical, or mechanical stimulus is required to open the channels and allow ions to diffuse across the plasma membrane.

The cell membranes of neurons contain many specialized ion channels specific for Na^+ and K^+. Because the opening of these channels is regulated by voltage changes in the cell, these channels are called **voltage-gated Na^+ and K^+ channels.** When a neuron receives an electrical stimulus, it induces voltage-gated Na^+ and K^+ channels in the cell membrane to open. Voltage-gated Na^+ and K^+ channels have different speeds of reaction to a stimulus: Na^+ channels are the first to open in response to a stimulus. In fact, most Na^+ channels are already closed by the time K^+ channels open. In contrast, K^+ channels respond to changes in membrane voltage more slowly (Figure 5-3).

While the voltage-gated Na^+ and K^+ channels are open, Na^+ and K^+ are free to diffuse through their respective channels. Because of the concentration and electrical gradients that exist across the cell

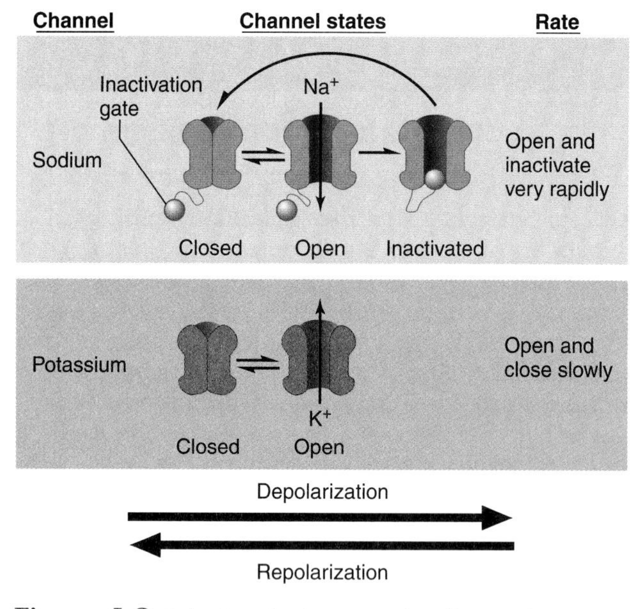

Figure 5-3 Behavior of voltage-gated sodium and potassium ion channels in the cell membrane of a neuron. In addition to having a faster rate of response to voltage changes across the membrane, sodium channels have an extra intracellular structure known as an inactivation gate (sometimes visualized as a "ball-and-chain"). The inactivation gate limits the influx of sodium ions into the cell by blocking the channel shortly after depolarization opens it.

membrane, both Na^+ and K^+ diffuse down their concentration gradients while the voltage-gated channels are open. This movement of ions in response to a stimulus creates the ion current (i.e., the movement of charged particles) characteristic of action potentials. The voltage-gated protein ion channels are not stable in their open position, however, and so they automatically close (resume a stable conformation) soon after the stimulus occurs.

Both the ion concentration gradients and the electrical gradient across the cell membrane are required for the generation of an action potential. In the case of Na^+, both the electrical and concentration gradients move Na^+ into the cell. The portion of the action potential that corresponds to this inward motion of Na^+ is called **depolarization.** During depolarization, the cell's membrane potential changes from about −70 mV to about +30 mV as positively charged Na^+ enter the cell. The membrane potential at the end of depolarization is usually close to the equilibrium potential for Na^+ (as previously calculated).

At the peak of depolarization (+30 mV), when voltage-gated K^+ channels are opening, the electrical gradient and the concentration gradient move K^+ out of the cell. Thus, there is a net efflux of K^+ during the action potential. The portion of the action potential that corresponds to this outward motion of K^+ is called **repolarization.** During repolarization, K^+ efflux moves the cell membrane potential back toward its original resting membrane potential of −70 mV. Because voltage-gated K^+ channels are also slow to close, K^+ efflux usually "overshoots" the resting membrane potential, and the membrane potential of the cell is therefore usually lower (more negative) than −70 mV at the end of repolarization. A decrease in membrane potential below resting membrane potential is called **hyperpolarization.** Hyperpolarization is a common phenomenon that can occur in neurons even when they are not generating action potentials (e.g., during an inhibitory postsynaptic potential [IPSP]). Thus, the "overshoot" of the resting membrane potential by K^+ efflux that occurs at the end of an action potential is just one type of hyperpolarization and is called **after-hyperpolarization** (Figure 5-4).

5-5 What property of voltage-gated K^+ channels causes afterhyperpolarization?

■

5-6 What would happen to a cell's resting membrane potential if potassium ion channels were suddenly opened (but Na^+ channels remained closed)? Explain your answer.

■

5-7 Which of the following processes require the use of ATP? Fill in the following table.

Movement of Ions	Does Movement Occur Against or With the Concentration Gradient?	Is Energy Required?
K^+ movement into the cell		
Na^+ movement into the cell		
K^+ movement out of the cell		
Na^+ movement out of the cell		

■

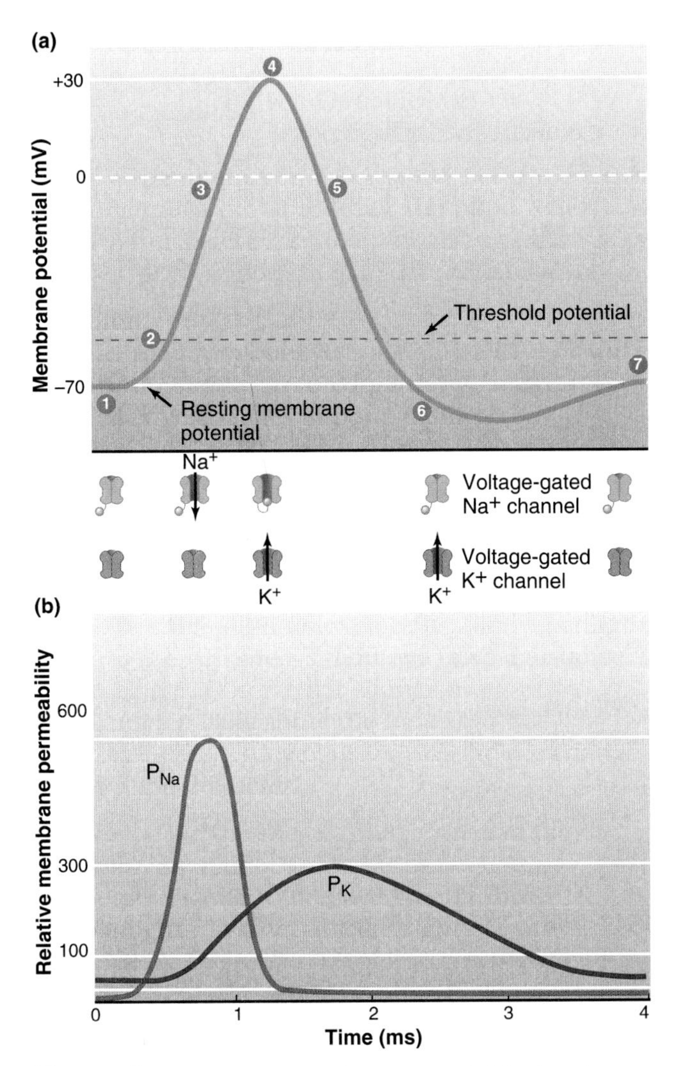

Figure 5-4 Changes in membrane potential (mV) and membrane permeability (P) during an action potential. Voltage-gated Na^+ (blue) and K^+ (red) channels are also shown.

5-8 You add a drug to a neuron that poisons the Na^+/K^+ pumps. One hour later you try to stimulate the neuron to generate an action potential, but the neuron is unresponsive. Why? Explain these results.

An action potential is an all-or-none phenomenon. If the initial stimulus opens enough voltage-gated Na^+ channels to depolarize a neuron to its **threshold** voltage (typically around −55 mV) then an action potential will be generated. It is very important to understand that the amplitude of an action potential does not change in response to changing stimulus strength and duration. The amplitude of depolarization is determined by the chemical properties of the voltage-gated ion channels and the concentration and electrical gradients of the cell (recall that a cell is depolarized to approximately the equilibrium potential of sodium). Thus, if the cell is depolarized to threshold voltage, an action potential will be generated, and the cell will depolarize to approximately +30 mV. Any stimulus that does not sufficiently depolarize the cell is considered a **subthreshold** stimulus and will not generate an action potential. Any stimulus greater than the threshold voltage is considered a **suprathreshold** stimulus.

During the initial stages of action potential generation, a stimulus creates a voltage change across the cell membrane. This stimulus induces voltage-gated Na^+ channels to open, which then allow an increase in sodium current across the membrane. The additional voltage change caused by the movement of Na^+ into the cell then induces more voltage-gated Na^+ channels to open. This is one of the few examples of positive feedback that occurs under normal physiological conditions.

It is important to understand that during an action potential a population of Na and K ion channels is responding to the initial stimulus. As shown in Figure 5-3, voltage-gated Na^+ channels become inactivated almost immediately after opening. During this inactivated state, Na^+ channels cannot be reactivated (opened) until they return to their original, stable (closed) conformation. At any given time during depolarization, some Na^+ channels are closed and are unresponsive to a stimulus (inactivated), some are open and allowing Na^+ movement, and some may be closed but ready to respond to a voltage stimulus. It is the proportion of Na^+ channels that are able to respond to a second stimulus that determines how soon after the first action potential a second action potential may be generated. This property is known as the refractory period, and it allows for the generation of distinct nerve impulses. The **absolute refractory period** is the duration of time following an action potential during which a second action potential cannot be generated, no matter how strong the stimulus. During the **relative refractory period** a second action potential may be generated, but it requires a stimulus that is stronger than the one that generated the first action potential.

5-9 Why do you think it is important to be able to generate separate, distinct action potentials, rather than just having one action potential lead directly into the next?

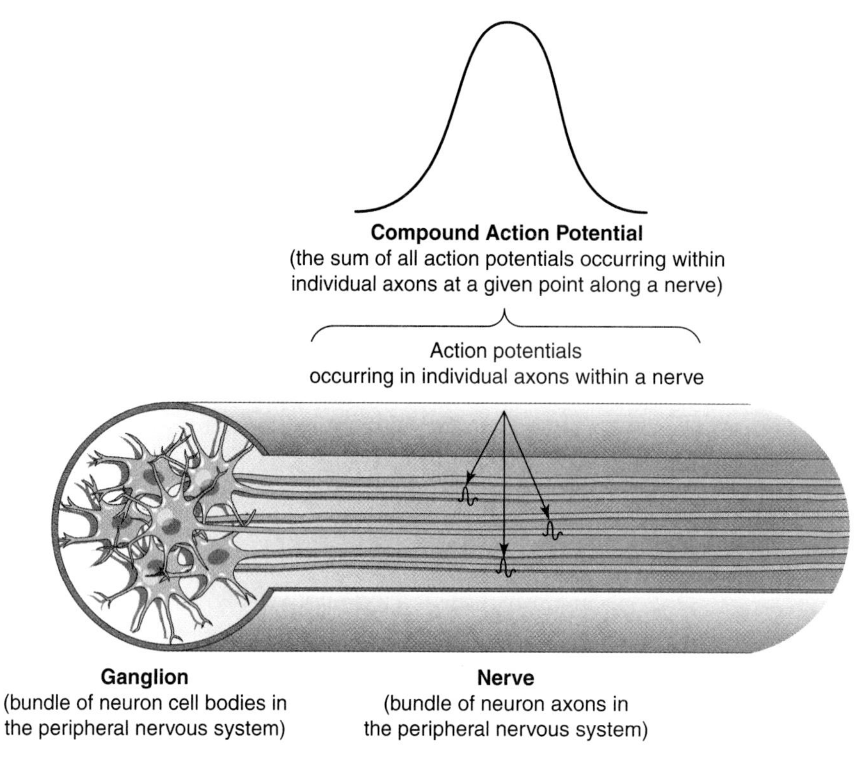

Figure 5-5 Cross section of a nerve showing individual axons generating action potentials.

In the peripheral nervous system, axons of neurons traverse the body in bundles called nerves. Each axon in a nerve is capable of generating action potentials. In many laboratory experiments, the properties of action potentials are investigated along a single axon. It is possible, however, to examine collectively the properties of all the action potentials generated by the axons in a nerve. Such action potentials are called **compound action potentials** because they are the sum of all the action potentials being generated in individual axons in the nerve (Figure 5-5). Compound action potentials measure the voltage change that occurs across the nerve itself, rather than across the membrane of a single axon. Thus, the amplitude of a compound action potential varies with the number of axons within the nerve that are generating action potentials. As more axons within the nerve are recruited into the compound action potential, the amplitude of the compound action potential increases. In Experiment 5.2 of today's laboratory (Compound Action Potentials), you will investigate the influence of increasing stimulus strength on the amplitude of the resulting compound action potential. In Experiment 5.3 (Refractory Periods), you will use compound action potentials to determine the relative and absolute refractory periods of action potentials.

Methods and Materials

EXPERIMENT 5.1 Influence of Extracellular Ion Concentration on Resting Membrane Potential

In this exercise you will conduct experiments to investigate the influence of extracellular K^+ and Na^+ concentrations on the resting membrane potential of muscle cells in crayfish. Because muscle cells also generate action potentials, this larger, more resilient tissue can be used as an experimental model for studying membrane potentials.

Extracellular K^+ Concentration

- Launch the Physiology Interactive Lab Simulations (Ph.I.L.S.). In the section entitled ***Resting Potentials,*** select Experiment ***8. Resting Potential and External [K⁺].***
- Read the *Objectives* of this simuation.
- Click the **Continue** button in this window to review the introductory material. Be sure to click on and view all informational links presented within the *Introduction* window of this simulation. These links present helpful information and animations to assist you in understanding the laboratory exercise.
- Once you have completed your review of the introductory material, the **Continue** button will open the next tab entitled *Pre-Lab Quiz*. You may also click the *Pre-Lab Quiz* tab directly to enter this section of the simulation. Answer all questions in the Pre-Lab Quiz to test your comprehension and understanding of the introductory material. *Note*: The printed Lab Report will indicate your correct and incorrect responses on the *Pre-Lab Quiz* and report the number of correct responses out of the total number of questions possible. The simulation will only allow one attempt at the quiz, so answer each question carefully.
- Once you have completed the *Pre-Lab Quiz*, the **Continue** button within this window will open the next tab entitled *Wet Lab*. You may also click the *Wet Lab* tab directly to enter this section of the simulation. Read the material presented and be sure to click on and view all informational links presented within the *Wet Lab* window of this simulation. These links provide valuable information and videos to introduce the apparatus and procedures necessary to conduct this experiment. *Note*: The printed Lab Report will indicate the number of video clips you viewed in the *Wet Lab* section.
- Once you have completed the *Objectives & Introduction, Pre-Lab Quiz*, and *Wet Lab* sections of the simulation, you are now ready to conduct the experiment. Click the ***Continue*** button to open the next tab entitled *Laboratory Exercise*, or click on the *Laboratory Exercise* tab directly to enter this section of the simulation. Follow the instructions presented at the bottom of the simulation window and click all informational links to view procedural details. Note that all of the experimental procedures are organized into 11 steps. The simulation will indicate which step you are currently conducting. You may also click on **View All** to view the entire experimental procedure.

Setting Up the Apparatus

- Click on the **Power Switch** of the Data Acquisition Unit.
- Click on the **Power Switch** of the Electrometer.
- Plug the blue output cable attached to the Electrometer into Recording Input-1 of the Data Acquisition Unit.
- Click the **Start Button** in the Control Panel to display the red horizontal line on the virtual monitor. This red trace is a real-time recording of voltage.
- Zero the Electrometer by clicking the **Zero Button** on the right side of the instrument. The digital display on the Electrometer will now read approximately 0.
- Move the microelectrode tip into a muscle fiber by clicking the **Down** control on the Micromanipulator. Notice that the recorded voltage increases slightly as the electrode tip hits the surface of the muscle cell. You will know you have penetrated a muscle fiber when you observe a dramatic decrease in your voltage reading. Notice that your voltage reading on the Electrometer's digital display is approximately –75 mV.
- Record this value in your journal by clicking the **Journal** icon in the lower right of your instruction window. Notice that at the bottom of the journal page membrane potential (Em) is recorded (Em = –75 mV) with a note stating that this value is for normal saline. Close the Journal.
- **Note:** During this experiment, you will be making several recordings of resting potential from the muscle preparation. You will be alerted to make recordings from fresh, untested muscle fibers by moving the microelectrode left, right, anterior, and posterior using the Micromanipulator controls.

Changing the External Ion Concentration

- Move the electrode tip out of the muscle fiber by clicking the **Up** control on the Micromanipulator.
- Now move the electrode to a different position on the muscle preparation by pressing the **Anterior/Posterior** and **Right/Left** controls on the Micromanipulator. This will allow you to collect data from a fresh muscle fiber.
- Again zero the Electrometer.
- Now perfuse the muscle preparation with the 5 mEq/LK^+ solution by clicking on the valve for the vial labeled 5. Empty the entire contents of the vial into the muscle preparation. *Note*: For monovalent ions such as potassium and sodium with a net charge of +1, a solution containing 5 milliequivalents per liter (mEq/L) is equal to a 5 millimolar (mM) solution. Although the solutions are labeled as mEq/L in this part of the simulation, they will be recorded in your Journal as mM solutions.
- Move the electrode tip into a new muscle fiber by clicking the **Down** control on the Micromanipulator. You will know when you have penetrated a muscle fiber because there will be a dramatic decrease in your voltage reading. Notice your voltage reading on the Electrometer's digital display and the red recording line in the Virtual Monitor.

- Record this value in your journal by clicking the **Journal** icon. Notice your recording has been written into the Journal for observation #1 of 5 mM. Using the same procedure, collect four more data points for this 5 mM K^+ concentration.
- Once you have completed your data collection for the 5 mM concentration of K^+, follow the same procedures to collect data for K^+ concentrations of 10, 20, 50, and 100 mM.

Data and Graphs

- Once you have collected all five data points for each of the five K^+ concentrations, click the **CALC** button to calculate the average resting membrane potential for each of the K^+ concentrations.
- A graph page automatically opens. Click on the **PLOT** button to view your data. The plot reports the regression equation and the r^2 value from a regression analysis.
- Copy the data from your completed Journal in the computer simulation to Results Table 5.1a of the results section in your Laboratory Report.
- Click on the *Post-Lab Quiz* and *Lab Report* tab and answer the questions.
- Once you have completed the *Post-Lab Quiz*, a *Conclusion* will be presented for this experiment. Read the conclusion carefully and click the **Finish Lab** button.
- The *Print Your Lab Report* window will allow you to enter your name and course details. Click the **Print Lab** button at the bottom of the window. Once you have successfully printed your laboratory report, you have completed the laboratory simulation.

Extracellular Na^+ Concentration

- Launch the Physiology Interactive Lab Simulations (Ph.I.L.S.). In the section entitled ***Resting Potentials,*** select Experiment ***9. Resting Potentials and External [Na⁺].***
- Read the *Objectives* of this simulation.
- Follow the procedures outlined previously to obtain similar data for the influence of extracellular Na^+ concentration on resting membrane potential. Make sure your data and results from your completed Journal in the computer simulation are recorded in Results Table 5.1b of the results section of your Laboratory Report.

EXPERIMENT 5.2 Compound Action Potentials

In this exercise you will examine the relationship between stimulus strength and the amplitude of a compound action potential in a frog sciatic nerve.

- Launch the Physiology Interactive Lab Simulations (Ph.I.L.S.). In the section entitled ***Action Potentials,*** select Experiment ***10. The Compound Action Potential.***

  Repeat the same steps found in Experiment 5.1 until the section for setting up the Apparatus.

Setting Up the Apparatus

- Click on the **Power Switch** of the Data Acquisition Unit.
- Attach the red, blue, and green recording cables to their corresponding colored posts on the nerve chamber. These cables are connected to the recording inputs of the Data Acquisition Unit and will record the compound action potential generated in the excised nerve.
- Attach the orange and black stimulator cables to their corresponding colored posts on the nerve chamber. These cables are connected to the stimulator outputs of the Data Acquisition Unit and will deliver the voltage stimulus to the excised nerve.
- Drain the saline from the nerve chamber by clicking on the valve located by the top left of the chamber.
- Click the **Shock** button in the Control Panel. Notice that the stimulator voltage is preset to 1.0 V. Observe the Compound Action Potential (CAP) that has been generated on the virtual computer monitor.
- Record the maximum amplitude of this CAP by clicking on the peak of the voltage trace. Notice that a black inverted triangle has marked where you have clicked. You may adjust this pointer to the highest peak of the voltage trace.
- Once you have marked the peak of the voltage trace, click on the baseline of the voltage trace to determine the peak's amplitude in millivolts (mV). This measured amplitude now appears in the Data box labeled AMP.
- Click on the **Journal** icon to record this data point in your Journal data sheet. Close the Journal.
- You may erase prior voltage traces by clicking on the **Erase** button in the Control Panel.
- Repeat these procedures to collect data for all voltages ranging from 0.0 to 1.6 V. Increase the stimulus voltage in 0.1 volt increments. *Note*: If the stimulus voltage produces no CAP, simply click on the red voltage trace two times to record an amplitude of 0 mV in the Data box.
- Once your Journal's data sheet is completed, make sure you record these data in Results Table 5.2 of the results section of your Laboratory Report.
- Click on the *Post-Lab Quiz* and *Lab Report* tab. Answer the Post-Lab questions.
- Once you have completed the *Post-Lab Quiz*, a *Conclusion* will be presented for this experiment. Read the conclusion carefully and click the **Finish Lab** button.
- The *Print Your Laboratory Report* window will allow you to enter your name and course details. Click the **Print Lab** button at the bottom of the window. Once you have successfully printed your laboratory report, you have completed the laboratory simulation.

EXPERIMENT 5.3 Refractory Periods

In this exercise you will use compound action potentials to determine the absolute and relative refractory periods. Because compound action potentials are being used to conduct this exercise, you will use a change in the amplitude of the compound action potential to indicate when voltage-gated Na^+ channels are responsive to a stimulus.

- Launch the Physiology Interactive Lab Simulations (Ph.I.L.S.). In the section entitled ***Action Potentials,*** select Experiment ***12. Refractory Periods.***

  Repeat the steps found in Experiment 5.1 until the section for setting up the Apparatus.

Setting Up the Apparatus

- Click on the **Power Switch** of the Data Acquisition Unit.
- Attach the red, blue, and green recording cables to their corresponding colored posts on the nerve chamber. These cables are connected to the recording inputs of the Data Acquisition Unit and will record the compound action potential generated in the excised nerve.
- Attach the orange and black stimulator cables to their corresponding colored posts on the nerve chamber. These cables are connected to the stimulator outputs of the Data Acquisition Unit and will deliver the voltage stimulus to the excised nerve.
- Drain the saline from the nerve chamber by clicking on the valve located by the top left of the chamber.
- Click the **Shock** button in the control panel. Notice that the simulator voltage is preset to 0.7 V. Observe the Compound Action Potential (CAP) that has been generated on the virtual computer monitor.
- Change the stimulus voltage to 0.8 V and click the **Shock** button. Notice that the peak of the voltage trace has increased. Continue to increase the stimulus voltage by 0.1 volt and shock the sciatic nerve until you no longer observe an increase in the peak of the CAP.
- Measure the amplitude of this maximum CAP by clicking on the peak of the voltage trace. Notice that a black inverted triangle has marked where you have clicked. You may adjust this pointer to the highest peak of the voltage trace.
- Once you have marked the peak of the voltage trace, click on the baseline of the voltage trace to determine the peak's amplitude in millivolts (mV).
- Click the **Journal** icon to record this data point. Notice that the stimulus voltage and CAP amplitude are recorded at the top of the Journal page. What is the lowest stimulus voltage that produces the maximum compound action potential amplitude? Fill in the blanks of the following sentence using the voltages reported at the top of your Journal page. **A stimulus voltage of______V produces a maximum compound action potential amplitude of______mV.**
- Close your Journal window.
- Click on the **# SHOCKS** button in the Control Panel to increase the number of shocks delivered to the nerve to 2. Do not alter the stimulus voltage. At this point in the experiment, you will only change the interval or time period between the first and second stimuli. Notice that the interval is currently 7 ms.
- Click the **Shock** button. Measure the amplitude of the second CAP by clicking on and marking the peak of the CAP followed by its baseline. Record this amplitude by clicking on the **Journal** icon. Notice that the amplitude of the second CAP has been recorded in the Journal's data sheet. Close the Journal.
- You may erase prior voltage traces by clicking on the **Erase** button in the control panel.
- Change the interval between the two stimuli to 7.5 ms.
- Click the **Shock** button. Again measure the amplitude of the second CAP and record this amplitude in the Journal.
- Repeat these procedures until you have collected the amplitudes of the second CAP for all stimulus intervals ranging from 1.0 to 9.0 ms (in 0.5 ms increments).
- Once your Journal's data sheet is completed, make sure you record these data in Results Table 5.3a of the results section of your Laboratory Report. Use these data to complete Results Table 5.3b.
- Click on the *Post-Lab Quiz* and *Lab Report* tab. Answer the Post-Lab questions.
- Once you have completed the *Post-Lab Quiz*, a *Conclusion* will be presented for this experiment. Read the conclusion carefully and click the **Finish Lab** button.
- The *Print Your Laboratory Report* window will allow you to enter your name and course details. Click the **Print Lab** button at the bottom of the window. Once you have successfully printed your laboratory report, you have completed the laboratory simulation.

COMPARATIVE NOTE Laboratory 5

Axon Myelination: Faster but Not Fatter

The speed at which an action potential travels along an axon (conduction velocity) is obviously important. The faster the conduction velocity, the faster the reaction time to a stimulus. Generally, invertebrates have much slower conduction velocities than vertebrates. For example, a fish can have a conduction velocity as great as 36 m/s, 18 times that of a cockroach (2 m/s). This is because the axons of most invertebrates are ensheathed by loosely associated glial cells that do not form true myelin. Remember that myelin increases the conduction velocity of axons due to the *nodes of Ranvier* and saltatory conduction.

If invertebrates do not have functionally-equivalent myelin, what governs their nerves' conduction velocity? The conduction velocity of unmyelinated axons can be increased by increasing axon diameter. The relationship between conduction velocity and axon diameter can be expressed as $velocity = \sqrt{diameter}$, where a fourfold increase in axon diameter doubles conduction velocity. Squid have a faster conduction velocity because of their giant axons (1 mm diameter × 1000 mm length). This allows for simultaneous contraction of the squid's mantle, producing its "jet propulsion" mode of swimming.

Myelination of axons is an advantage to vertebrates, for it produces greatly increased conduction velocities with minimal increases in axon diameter. The advantage of myelinated over unmyelinated axons with regard to conduction velocity is shown in the following figure. Why is it efficient to have unmyelinated axons at smaller axon diameters? At what axon diameter is it beneficial to have myelinated axons?

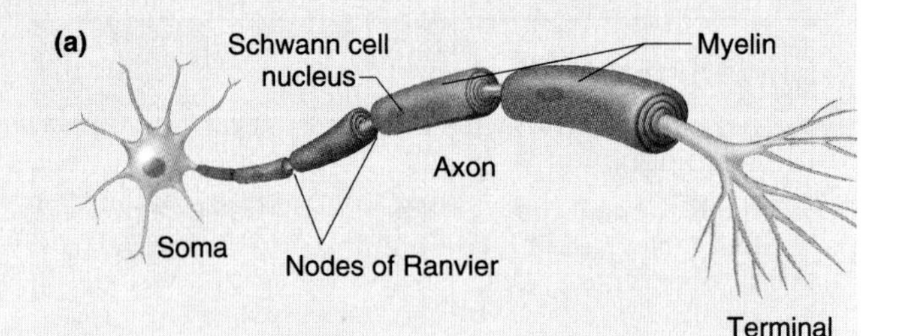

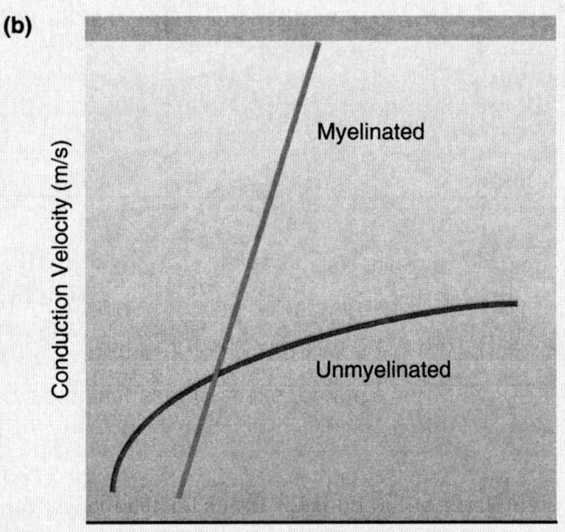

(a) A neuron in the peripheral nervous system showing a myelinated axon. (b) Relationship between axon diameter and conduction velocity in myelinated and unmyelinated axons.

RESEARCH OF INTEREST

Freeman, B. and C.R.R. Watson. 1978. The optic nerve of the brush-tailed possum, *Trichosurus vulpecula*, fiber diameter spectrum and conduction latency groups. Journal of Comparative Neurology 179:739–752.

Necker, R. and C. C. Meinecke. 1984. Conduction velocities and fiber diameters in a cutaneous nerve of the pigeon *Columba livia*. Journal of Comparative Physiology A: Sensory Neural and Behavioral Physiology 154:817–824.

Sato, T. M. Ohkusa, T. Miyamoto, and Y. Okada. 1989. Centrifugal decrement in diameter of myelinated afferent fibers innervating frog taste organ. Comparative Biochemistry and Physiology A 92:435–440.

RESULTS AND DISCUSSION

LABORATORY REPORT 5

EXPERIMENT 5.1

Influence of Extracellular Ion Concentration on Resting Membrane Potential

Results Table 5.1a Individual Group Data for Experiment 5.1: Influence of Extracellular K^+ Concentration on Resting Membrane Potential

	Resting Membrane Potential (mV)						
K^+ Concentration Added (mM)	Cell 1	Cell 2	Cell 3	Cell 4	Cell 5	Mean Resting Membrane Potential (mV)	Standard Error

Results Table 5.1b Individual Group Data for Experiment 5.1: Influence of Extracellular Na^+ Concentration on Resting Membrane Potential

	Resting Membrane Potential (mV)						
Na^+ Concentration Added (mM)	Cell 1	Cell 2	Cell 3	Cell 4	Cell 5	Mean Resting Membrane Potential (mV)	Standard Error

Problem Set 5.1a: Influence of Extracellular K^+ Concentration on Resting Membrane Potential

Using the data from Results Table 5.1a, create a graph showing the resting membrane potential in response to each of the five different K^+ concentrations. Be sure to include the 95% CI in your graph. Using these data, perform the appropriate statistical test to examine the relationship between extracellular K^+ concentration and resting membrane potential.

a. State the null hypothesis.

b. Should you accept or reject the null hypothesis? Support your conclusions with the appropriate statistical results.

c. How does the extracellular K^+ concentration affect resting membrane potential? Why did you observe these results? Explain the mechanism thoroughly, using diagrams of a resting cell if necessary.

d. A large quantity (>100 mM) of K^+ are added to the extracellular fluid of a neuron. The neuron is then stimulated to generate an action potential. During this action potential, how will the addition of K^+ to the extracellular fluid influence the repolarization and afterhyperpolarization phases of the action potential? Explain.

Following this action potential, how will the neuron's resting membrane potential compare to its resting membrane potential prior to the addition of K^+?

How will the addition of K^+ to the extracellular fluid affect the neuron's overall function? Why? (*Hint:* Injections of potassium chloride solution are lethal.)

Problem Set 5.1b: Influence of Extracellular Na^+ Concentration on Resting Membrane Potential

Using the data from Results Table 5.1b, create a graph showing the resting membrane potential in response to each of the five different Na^+ concentrations. Be sure to include the 95% CI in your graph. Using these data, perform the appropriate statistical test to examine the relationship between extracellular Na^+ concentration and resting membrane potential.

a. State the null hypothesis.

b. Should you accept or reject the null hypothesis? Support your conclusions with the appropriate statistical results.

c. How does the extracellular Na^+ concentration affect resting membrane potential? Why did you observe these results? Explain the mechanism thoroughly, using diagrams of a resting cell if necessary.

EXPERIMENT 5.2
Compound Action Potentials

Results Table 5.2 Individual Group Data for Experiment 5.2: Compound Action Potentials

Stimulus Amplitude (V)	Amplitude of Compound Action Potential (mV)

Problem Set 5.2: Compound Action Potentials

Using the data from Results Table 5.2, create a graph showing the relationship between stimulus strength and the amplitude of the compound action potential.

a. Define the term *threshold voltage.* At what stimulus strength did you observe the threshold voltage?

b. Why does the amplitude of the compound action potential increase as stimulus strength increases? Explain your answer thoroughly.

c. Why doesn't the amplitude of the compound action potential continue to increase with very large stimulus strengths?

d. If you performed this experiment with a single axon, would you observe the same relationship between action potential amplitude and stimulus strength? Why or why not?

EXPERIMENT 5.3
Refractory Periods

Results Table 5.3a. Individual Group Data for Experiment 5.3: Refractory Periods

Interval (ms)	Amplitude of Second Compound Action Potential (mV)

Problem Set 5.3: Refractory Periods

Using the data form Results Table 5.3a, calculate the beginning, end, and duration of the relative and absolute refractory periods. Record these data in Results Table 5.3b.

a. What causes the absolute refractory period? Explain.

b. What causes the relative refractory period? Explain.

c. The change in the amplitude of the compound action potentials observed in this experiment is not due to recruitment of axons. Why?

d. If you performed this experiment with a single axon, would you observe the same relationship between stimulus frequency (or the interval between stimuli) and refractory periods? Why or why not?

Results Table 5.3b. Individual Group Data for Experiment 5.3: Refractory Periods

Refractory Period	Beginning (ms)	End (ms)	Duration (ms)
Relative			
Absolute			

LABORATORY 6

Reflexes

PURPOSE

This laboratory will introduce the principles of reflexes and reflex arcs. You will also observe the importance of stretch reflexes in maintaining balance and examine the differences among reflexes, learned responses, and unlearned responses.

Learning Objectives

- Identify the neural pathway and components of a simple reflex arc.
- Understand the difference between monosynaptic and polysynaptic reflexes.
- Understand the difference between ipsilateral and contralateral reflexes.
- Learn the proper techniques for examining the monosynaptic patellar tendon stretch reflex and understand its clinical significance.
- Understand how stretch reflexes adjust muscle length and aid in maintaining balance.
- Investigate the difference in reaction times between learned and unlearned responses.

Laboratory Materials

Experiment 6.1: An Overview and Investigation of Reflexes

1. Reflex mallets
2. Penlight
3. Two handheld vibrating massagers

Experiment 6.2: Learned and Unlearned Responses

Wooden 46-cm (18-inch) rulers

Introduction and Pre-Lab Exercises

A **reflex** is an involuntary and stereotypic response to a stimulus that serves in the maintenance of homeostasis. The stimulus for a reflex may be either internal (such as a change in carbon dioxide concentration in your blood) or external (such as the hot surface of a stove touching your finger). Reflexes are innate responses and therefore they never have to be learned. Reflexes are accomplished through a neural pathway called a **reflex arc**. The five components of a reflex arc include: (1) the sensory receptors; (2) the sensory or **afferent** neurons, which enter the spinal cord through the dorsal root; (3) synapses in the spinal cord; (4) motor or **efferent** neurons, which exit the spinal cord through the ventral root; and (5) the effector. Typical effectors are muscles or glands that respond to the original stimulus.

6-1 Fill in the components of a reflex arc on Figure 6-1. ■

Reflexes occur in both the somatic and autonomic nervous systems (Figure 6-2). The effectors of somatic reflexes are skeletal muscles, which are usually under voluntary control. The effect of the somatic motor neuron impulse on its skeletal muscle effector is always excitatory (causing a muscle contraction). Somatic motor neuron fibers are thick (9 to 13 μm), myelinated axons that conduct action potentials very quickly. Thus, somatic reflexes such as the patellar tendon reflex initiate a very rapid response in the skeletal muscle effector.

In contrast, autonomic nervous system reflexes assist in regulating the activity of almost all physiological systems, including the endocrine, cardiovascular, respiratory, urinary, and digestive systems. Typical effectors of autonomic reflexes are those

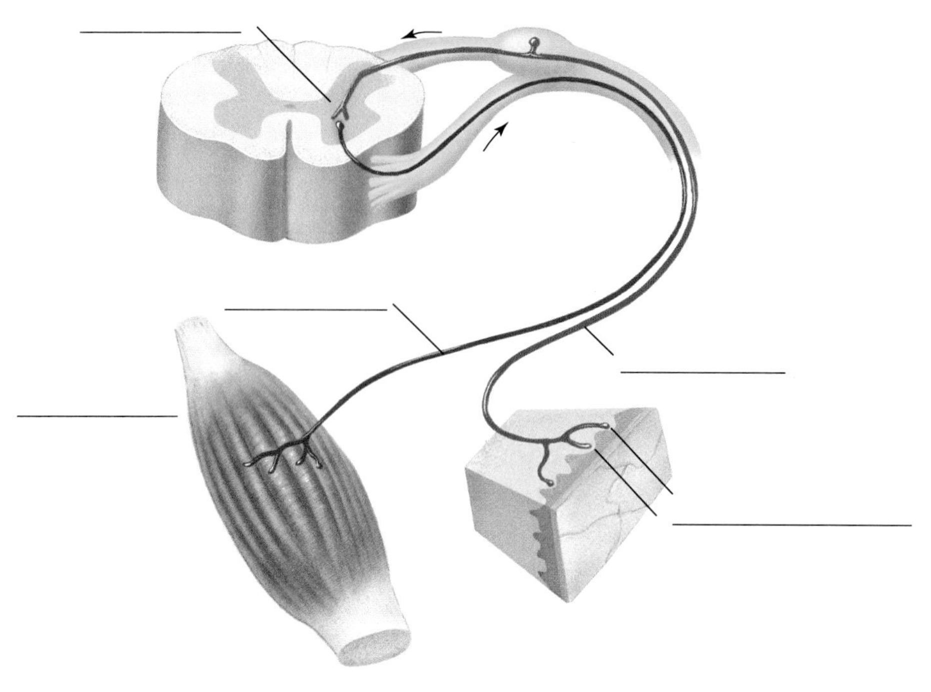

Figure 6-1 Label the somatic reflex arc.

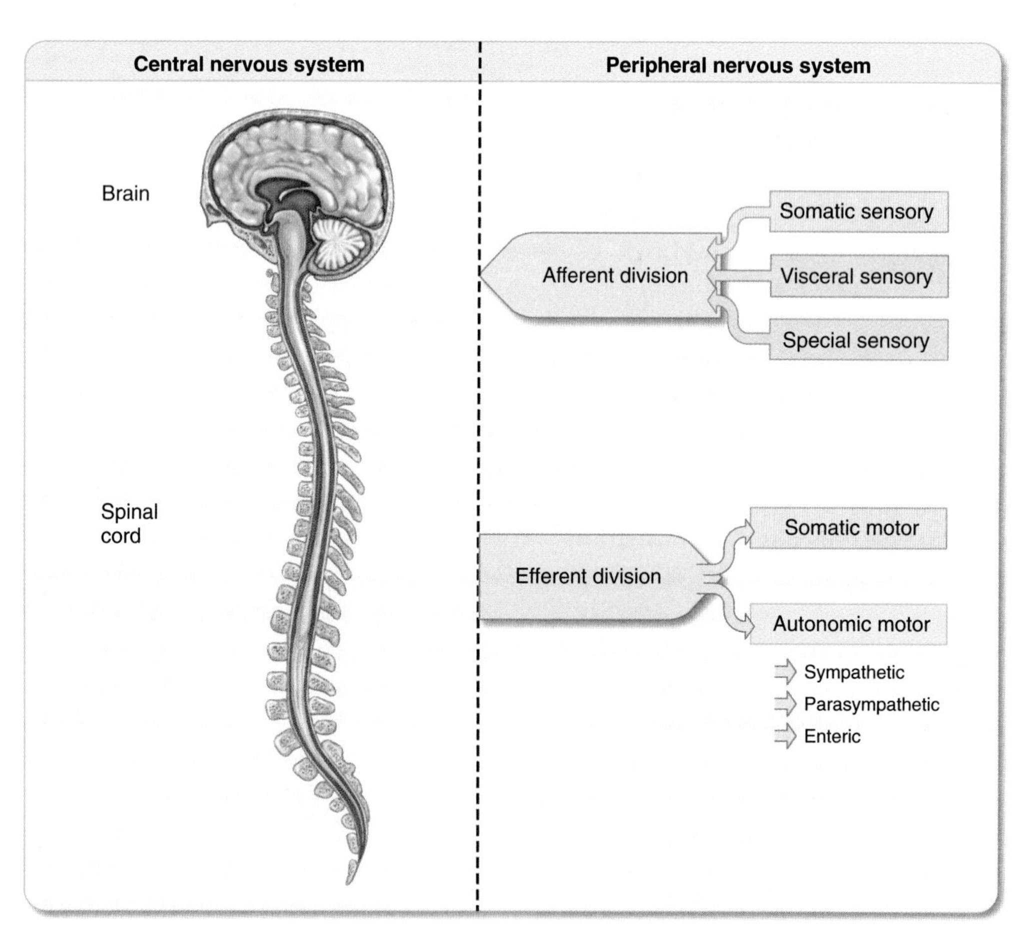

Figure 6-2 Overview of the structural and functional organization of the nervous system. The enteric nervous system is a division of the autonomic nervous system specific to the gastrointestinal tract.

Reflex	A Somatic or Autonomic Reflex?
You quickly pull your hand away after touching a hot pan.	
Your pupils dilate when you are frightened.	
The doctor swabs the back of your throat to obtain a throat culture and you gag.	
When you scratch your dog's belly he begins to reflexively kick his leg.	
Your stomach begins to secrete acid as you consume your lunch.	

structures that are not under voluntary control: smooth muscle, cardiac muscle, and many glands. The effect of the autonomic motor neuron impulse on such effectors can be either stimulatory or inhibitory, depending on the neurotransmitter released by the efferent motor neuron and the receptor type present on the effector. Note that autonomic reflexes occur in both the sympathetic and parasympathetic divisions of the autonomic nervous system (Figure 6-2).

6-2 Are the reflexes in the above table somatic or autonomic nervous system reflexes?

■

Reflexes occur very quickly in response to a stimulus because processing by higher brain centers is not required for the reflexive action. If a stimulus occurs on one side of the body, a reflex action may occur either on the same side as the stimulus (an **ipsilateral reflex**), on the opposite side (a **contralateral reflex**), or on both sides of the body simultaneously. These properties of reflex arcs are generally determined by the connections between sensory and motor neurons in the integration center of the spinal cord.

6-3 What important diagnostic information can be acquired by testing a patient's reflexes?

■

All muscles exhibit reflex contraction in response to stretching of the muscle. This reflex action helps to maintain posture and balance by monitoring the degree of muscle stretch. When a tendon is struck with a rubber reflex hammer, for example, the entire muscle is stretched, which activates sensory receptors located within the muscle spindle (Figure 6-3). The axons of these sensory neurons travel through peripheral spinal nerves and enter the spinal cord through the dorsal root. These sensory neurons terminate in the gray matter and synapse either on interneurons or directly on motor neurons. The axons of motor neurons exit the spinal cord via the ventral root and continue through spinal nerves to reach the effector. When activated, the skeletal muscle effectors of stretch reflexes cause contraction of the stretched muscle.

The patellar tendon stretch reflex is an example of an ipsilateral **monosynaptic reflex,** in which the sensory neuron synapses directly onto the motor neuron (Figure 6-4). In this example, there are no interneurons present in the interface between afferent and efferent neurons and therefore there is only one synapse in the spinal cord. Note that the designation as a monosynaptic reflex arc is determined by the absence of an interneuron between the afferent neuron relaying the signal from the stimulated muscle and the efferent neuron that relays the response to the *same* muscle. Thus, the activation of the monosynaptic patellar tendon reflex results in contraction of the quadriceps femoris muscles in the anterior thigh, producing extension of the leg at the knee. Stretch reflexes have the only known monosynaptic reflex arcs.

As in many stretch reflexes, however, the antagonistic muscle group must be simultaneously inhibited from contracting, a process called **reciprocal inhibition.** In the patellar tendon reflex, the afferent fibers from the muscle spindle divide into branches after entering the spinal cord. In path B of Figure 6-4, the afferent fibers terminate on inhibitory interneurons. When activated, these interneurons inhibit the motor neurons that innervate

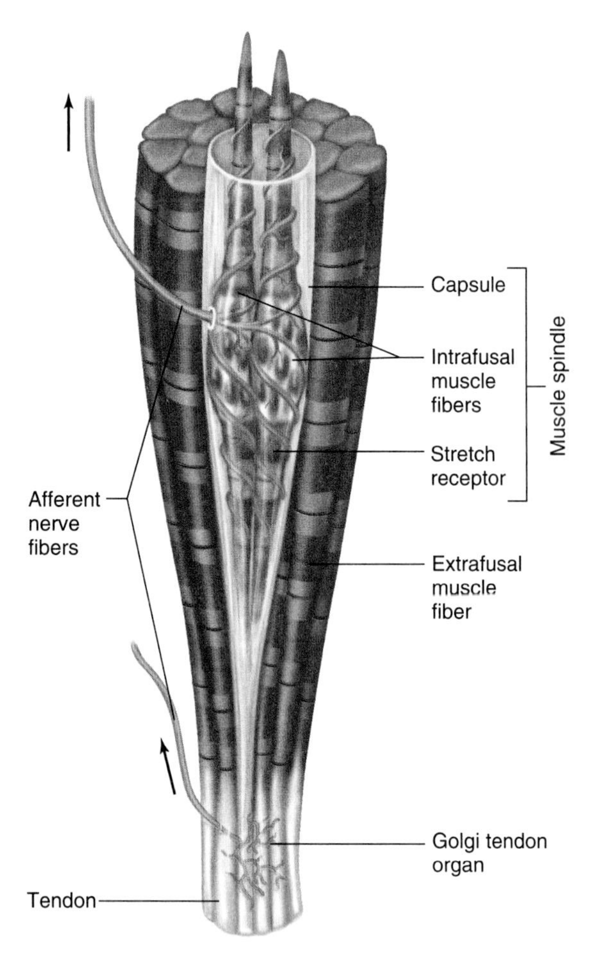

Figure 6-3 A muscle spindle containing stretch receptors. The muscle spindle is exaggerated in size compared to the extrafusal muscle fibers.

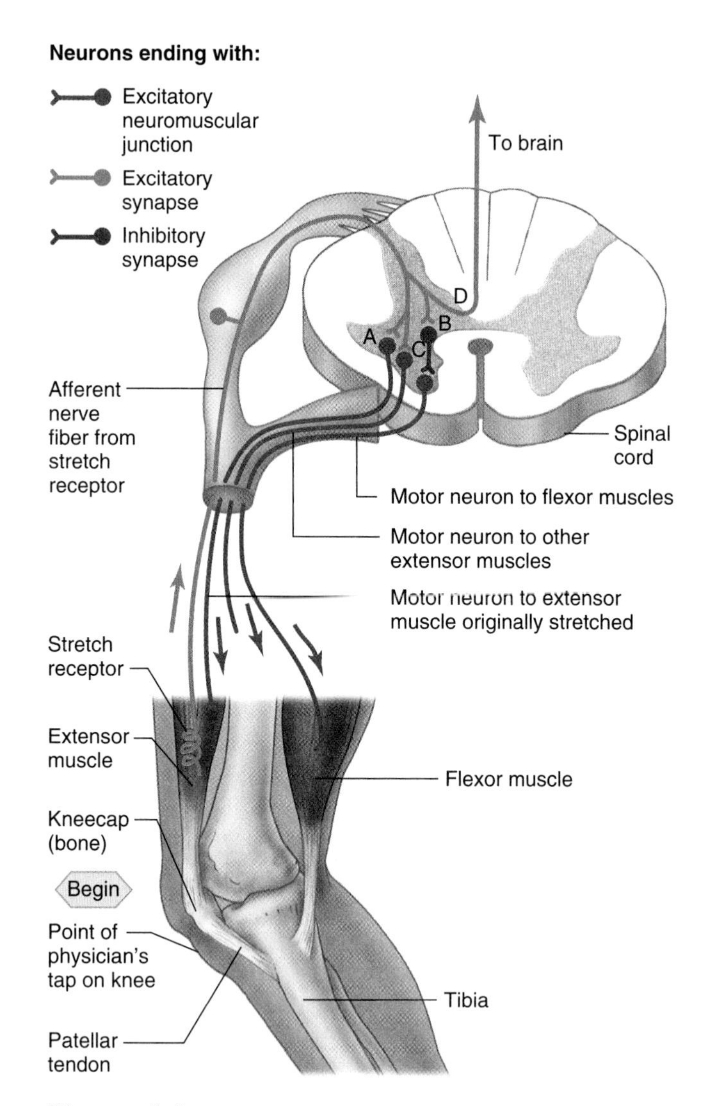

Figure 6-4 The patellar tendon reflex arc.

the hamstring muscles of the posterior thigh, preventing flexion of the lower leg during the reflex. This component of a stretch reflex (as are all other reflexes besides stretch reflexes) is polysynaptic. **Polysynaptic reflex arcs** include interneurons in the integration center between the sensory and motor neurons. There are many different combinations of polysynaptic integration. A common example is the crossed extensor reflex (Figure 6-5), which requires flexion of muscle groups on one side of the body (the ipsilateral side) and extension of muscle groups on the other side of the body (the contralateral side). An example would be pulling your foot away from a tack as you shift your weight to the other side of the body.

In contrast to a truly innate reflex, you can learn to perform a particular task reflexively. For example, when you see a glass falling to the floor, you may reflexively try to catch the glass. This is not an innate reflex action, nor does it travel the pathway of the five reflex arc components we have discussed. Rather, this could be considered a learned response, because you have learned that a glass dropping to the floor will break. Unlike reflexes, learned responses require higher-level processing in the brain. Many different stimuli may initiate a learned response, but some stimuli, such as visual stimuli, generate a response more quickly than other types of stimuli. Any response that you have not previously learned is called an unlearned response. You will examine differences in response times between learned responses and unlearned responses.

6-4 The patellar tendon reflex arc was used as a noninvasive method for determining action potential velocity. It was determined that the action potential required approximately 0.02 seconds to travel the distance of the reflex arc. This reflex arc distance was approximately 1.6 meters from the sensory receptors of the muscle spindle to

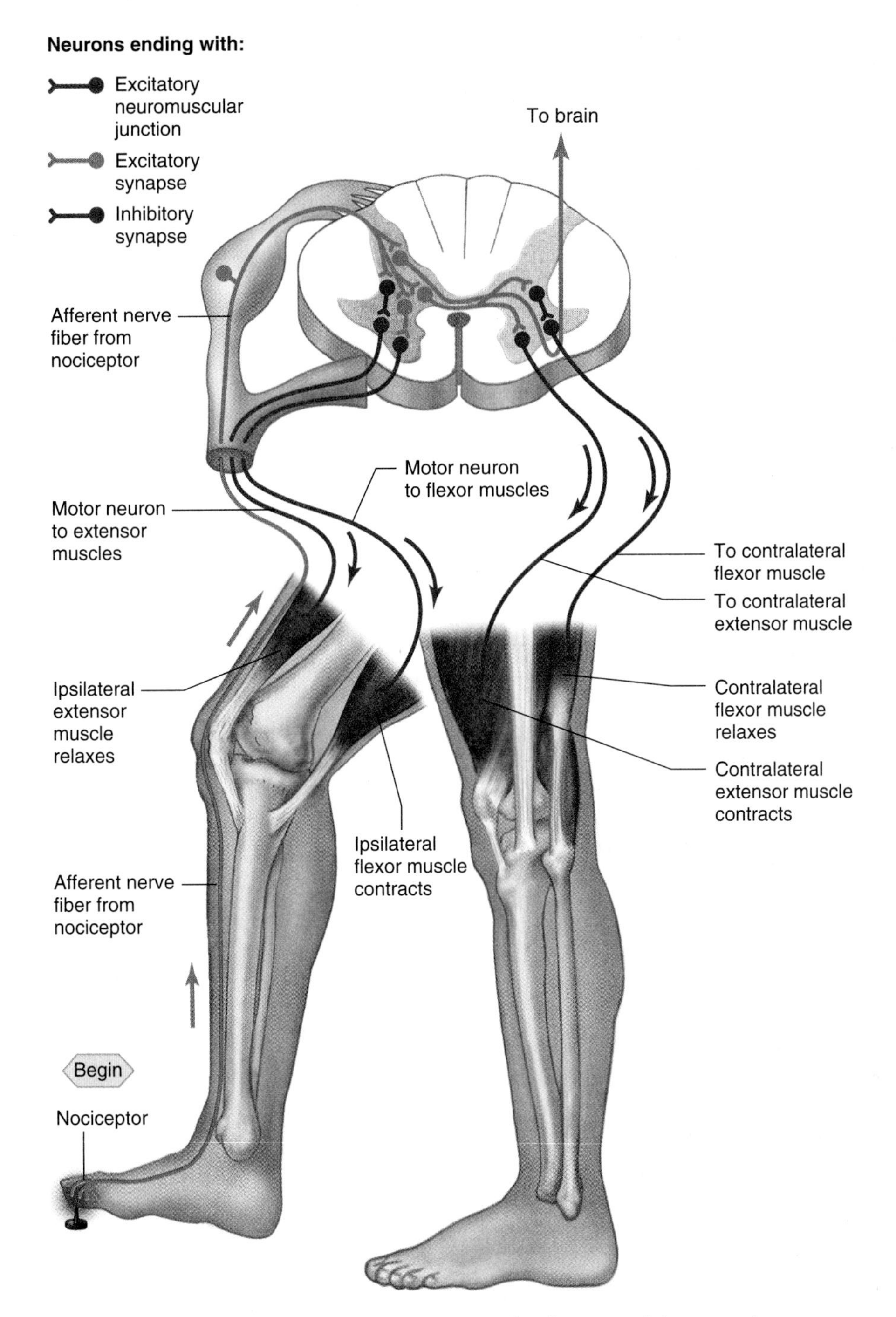

Figure 6-5 A crossed extensor reflex arc. Arrows indicate direction of action potential propagation.

the integration center in the spinal cord and back to the muscle effector. Thus, action potential velocity, as calculated using this reflex arc, was 80 meters per second. However, action potential velocity along a single myelinated axon is known to be approximately 100 to 150 meters per second.

What is the primary reason for this observed difference in action potential velocity? Explain your answer thoroughly. Hints: The differences are not a result of experimental error! Draw pictures of the different methods used for measuring action potential velocity to help answer this question.

COMPARATIVE NOTE *Laboratory 6*

Diving Reflexes: Breathing Deep

Although all mammals have lungs and breathe air, some mammals are adapted to an aquatic life. Some mammalian orders, such as the seals and whales, are entirely aquatic. These mammals must cope with the problems associated with prolonged, deep dives. Of those aquatic mammals that have been studied, the longest dives (120 minutes to a depth of over 900 meters) have been recorded in elephant seals (*Mirounga angustirostris*). The major physiological problem associated with prolonged dives is, of course, the limited availability of oxygen. Some adaptations in diving mammals that circumvent this problem include increased blood volume (and therefore increased oxygen-carrying capacity) and much higher muscle myoglobin (the oxygen-carrying pigment in muscle tissue) content as compared to the muscles of terrestrial mammals. In addition, heart rate decreases dramatically at the beginning of submersion. This decrease in heart rate occurs almost instantly and is regulated by a nerve reflex, called the diving reflex. The diving reflex is an example of an autonomic reflex and is regulated by the vagus nerve. The heart rate of grey seals (*Halichoerus grypus*), for example, decreases from about 120 beats per minute to as little as 4 beats per minute during a dive. This dramatic reduction in heart rate, combined with dramatically reduced blood flow to all organs except the central nervous system, helps to redistribute the limited oxygen supply during prolonged dives. Although this reflex is very pronounced in aquatic mammals, all mammals (including humans) demonstrate this autonomic reflex.

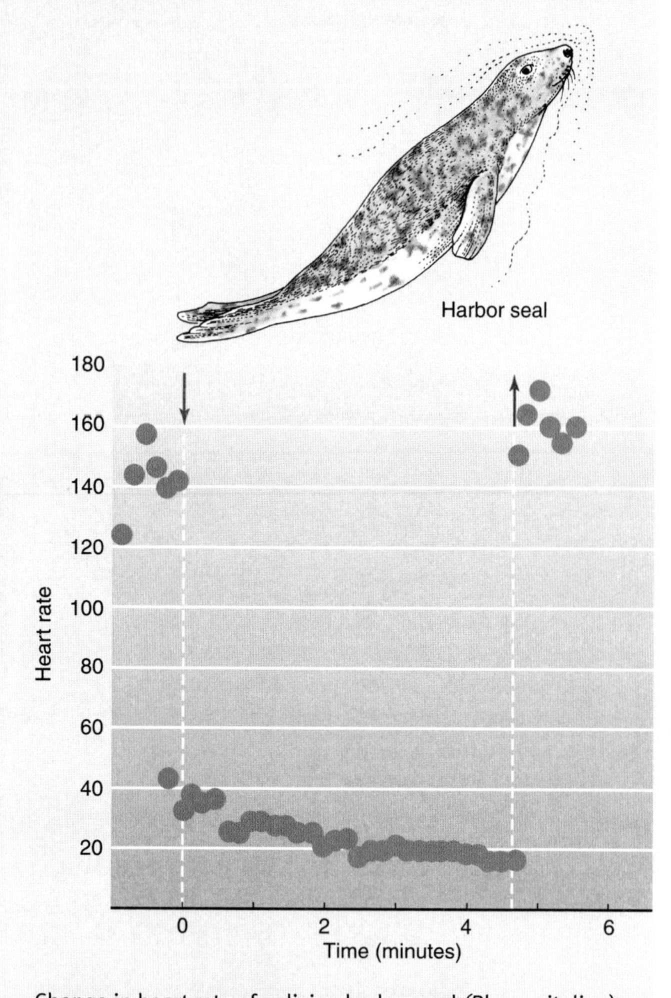

Change in heart rate of a diving harbor seal *(Phoca vitulina)*. Dive duration is indicated by the region between the broken vertical lines. (Data taken from Elsner 1965).

RESEARCH OF INTEREST

Elsner, R. 1965. Heart rate response in forced versus trained experimental dives in pinnipeds. Hvalrådets Skrifter 48:24–29.

Elsner, R., D. Wartzok, N.B. Sonafrank, and B.P. Kelly. 1989. Behavioral and physiological reactions of arctic seals during under-ice pilotage. Canadian Journal of Zoology 67:2506–2513.

Naito, Y. 2004. Recent advances in studies of the diving behavior of marine birds and mammals with micro-data loggers. Journal of the Yamashina Institute for Ornithology 35:88–104.

Thompson, D. and M.A. Fedak. 1993. Cardiac responses of grey seals during diving at sea. Journal of Experimental Biology 174:139–164.

Williams, T.M., D. Noren, P. Berry, J.A. Estes, C. Allison, and J. Kirtland. 1999. The diving physiology of bottlenose dolphins (*Tursiops truncatus*): III. Thermoregulation at depth. Journal of Experimental Biology 202:2763–2769.

Methods and Materials

EXPERIMENT 6.1 An Overview and Investigation of Reflexes

In this experiment you will test the performance of two different types of reflexes: the pupil reflex and muscle stretch reflexes. You will also examine how muscle stretch reflexes help to maintain balance.

Pupil Reflex

- The pupil of your eye is actually a hole in the iris muscle (the colored part of your eyes). Reflexive changes in pupil diameter in response to changes in light are actually accomplished by the contraction (causing constriction) or relaxation (causing dilation) of the circular iris muscle that surrounds the pupil.
- To test the pupil reflex, briefly shine a penlight into the volunteer's right eye from a distance of 20 cm. Observe the change in pupil diameter. Now shine the light into the volunteer's left eye. Again, observe the change in pupil diameter.
- Now have the subject hold a sheet of darkly colored paper between his or her eyes, perpendicular to the face. Briefly shine the penlight into the subject's right eye only. Observe the responses of both the right eye (the eye receiving the stimulus) and the left eye (the eye not receiving the stimulus).

Muscle Stretch Reflexes

- To test the patellar tendon reflex, obtain a reflex mallet. Have the volunteer sit so that his or her feet do not touch the ground. Strike the patellar tendon with the reflex mallet and observe the extension of the lower leg in response to the stretch of the patellar tendon.
- Repeat these steps, but now have the subject close his or her eyes so that the subject cannot see the mallet strike.
- To understand how muscle stretch reflexes aid in maintaining balance, you will simulate the stretch of the gastrocnemius muscle (one of the large muscles in the calf) by stimulating the Achilles tendon.
- Have a volunteer stand with his or her feet shoulder-width apart; this experiment works best if the subject removes his or her shoes to expose both Achilles tendons. Have two lab partners stand around the volunteer in case the subject begins to fall. Using two handheld massagers, stimulate the Achilles tendons of the volunteer by holding the massagers against the posterior ankles for approximately 15 seconds (Figure 6-6). Observe the subject's response to the stimuli while the subject's eyes are closed.

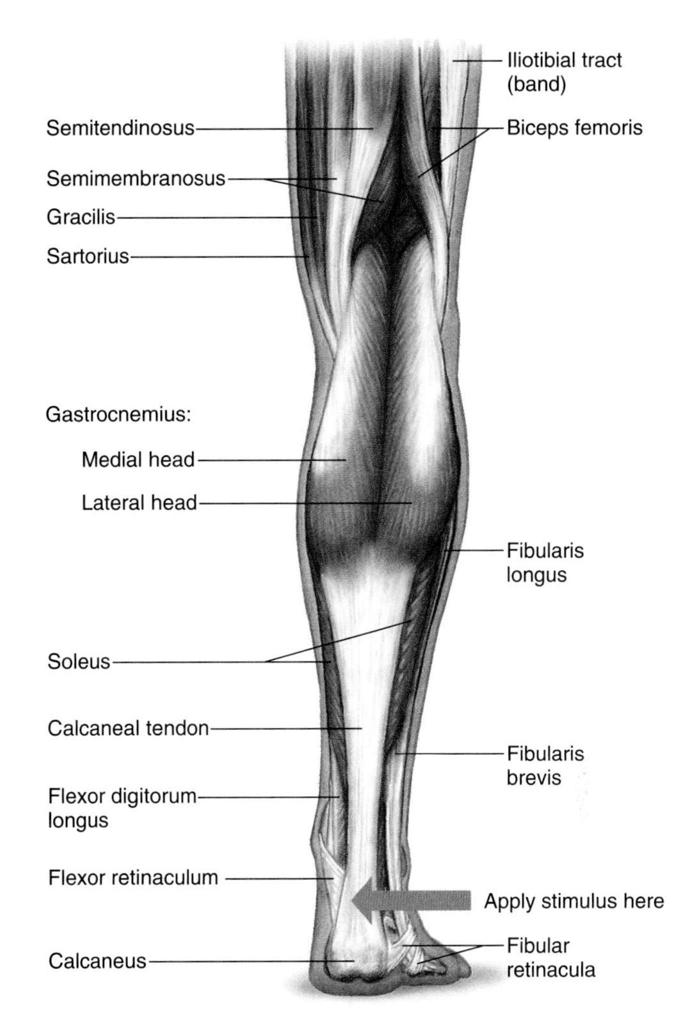

Figure 6-6 Anatomy for the attachment of the gastrocnemius muscle via the Achilles (calcaneal) tendon.

- Repeat this experiment to test the gastrocnemius muscle stretch reflex, but now have the subject keep his or her eyes open during the procedure.

EXPERIMENT 6.2 Learned and Unlearned Responses

In this experiment you will examine the differences in response times between learned and unlearned responses. Although learned responses are generally quick responses that have been learned over time, they are not innate reflex responses. Thus, learned responses require information processing by higher brain centers, and completing the reaction requires

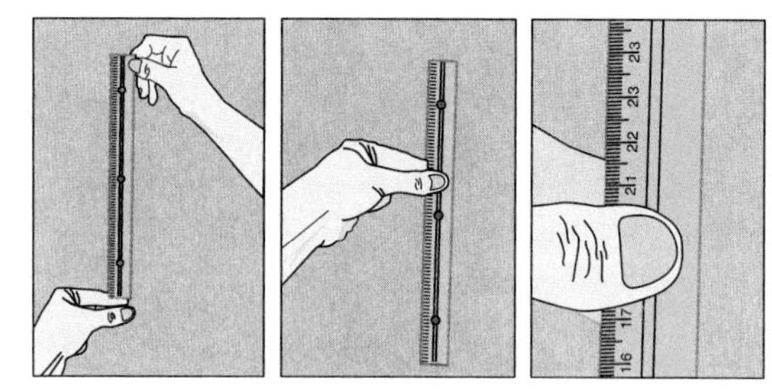

Figure 6-7 Procedure for measuring response distance.

much more time than a reflex. Many different stimuli may initiate a learned response, but some stimuli, such as visual stimuli, generate a response more quickly than other types of stimuli. In this experiment, you will compare the response times of learned responses using two different types of stimulus cues: a visual cue (i.e., seeing the falling object) and an auditory cue (i.e., being told when the object is falling). To test the response time of an unlearned response, you will modify a previously learned response (catching a falling object) to incorporate a level of unlearned information processing. The stimulus cue for this unlearned response will also be an auditory cue, but it will be a word that has not been learned to mean "the object is falling now!" It is important to note that if this experiment is repeated enough times, then the unlearned response will become a learned response.

Learned Response Times Using a Visual Cue

- Suspend a ruler just above your subject's outstretched index finger and thumb. Be sure the ruler's 1 cm mark is at the bottom (i.e., closest to the subject's fingers) as shown in Figure 6-7.
- While allowing the subject to visually observe the ruler, release the ruler and allow the subject to catch the ruler as it falls between his or her thumb and index finger. Record, to the nearest 0.5 cm, where the subject catches the ruler. Note that a slower reaction time will result in a greater response distance (cm). Repeat these steps three times and record these data in Results Table 6.2a.

Learned Response Times Using an Auditory Cue

- The same procedures used for testing the reaction times of learned responses using visual cues will also be used to test the reaction times of learned responses using auditory cues. However, the subject will not be permitted to see when the ruler has been released. The subject must now keep his or her eyes closed.
- For this part of the experiment, the auditory cue should be a word that has already been learned to imply "catch the ruler." Such common cue words include "now" and "start." As the cue word is spoken, release the ruler and allow the subject to catch the ruler as it falls between his or her thumb and index finger. Record, to the nearest 0.5 cm, where the subject catches the ruler. If the ruler drops to the ground, record a distance of 50 cm as the response distance in Results Table 6.2a. Repeat three times.

Unlearned Response Times Using an Auditory Cue

- The procedures used for testing the reaction times of learned responses will now be used to test the reaction times of unlearned responses. However, the learned response of catching a falling object in response to visual cues and common auditory cues such as "now" will be modified so that catching the falling ruler is an unlearned response. Again, the subject will not be permitted to see when the ruler has been released.
- As a group, select any random word that, when spoken aloud, will be the subject's cue that the ruler has been released. The more unrelated the word to words associated with "go," "now," "grab," and "catch," the more difficult it will be for the subject to learn that cue (because it will require more information processing by the brain). This experiment also works best if the cue word (e.g., dog) is spoken by the student releasing the ruler amidst normal conversation with your group members. If the room is quiet, any spoken cue word will elicit a startle response. The point of this exercise is to require information processing in order for your subject to catch the ruler.
- Whenever the cue word is spoken, release the ruler and allow the subject to catch the ruler as it falls between his or her thumb and index finger. Record, to the nearest 0.5 cm, where the subject catches the ruler. If the ruler drops to the ground, record a distance of 50 cm as the response distance in Results Table 6.2a. Repeat three times.

RESULTS AND DISCUSSION

LABORATORY REPORT 6

Problem Set 6.1: An Overview of Reflexes

a. What was the response of the right and left pupils when a light was shone in each of the eyes?

b. What was the response of the left pupil when a light was shone only in the right eye? Did this response differ from that of the stimulated right eye?

c. Based on your observations in a and b, is the pupil reflex an ipsilateral reflex, a contralateral reflex, or both?

d. If a patient's left pupil is dilated more than his right pupil, what significant information would this tell you?

e. During the patellar tendon reflex, did having the subject close his or her eyes affect the response to the mallet strike? Why?

f. Although reflexes do not require the activation of higher brain centers, they can be voluntarily inhibited. Why? Explain your answer.

g. How does visual information processing affect the response to muscle stretch reflexes? Explain.

EXPERIMENT 6.2
Learned and Unlearned Responses

Results Table 6.2a Individual Group Data for Experiment 6.2: Learned and Unlearned Responses.

Student	Learned Response Distance (cm) Using a Learned Visual Cue				Learned Response Distance (cm) Using a Learned Auditory Cue			
	Trial 1	Trial 2	Trial 3	Mean	Trial 1	Trial 2	Trial 3	Mean
1								
2								
3								
4								
5								
6								

Student	Unlearned Response Distance (cm) Using an Unlearned Auditory Cue			
	Trial 1	Trial 2	Trial 3	Mean
1				
2				
3				
4				
5				
6				

Results Table 6.2b Class Data for Experiment 6.2: Learned and Unlearned Responses.

Student (Name)	Mean Response Distance (cm)		
	Learned Visual Cue	Learned Auditory Cue	Unlearned Auditory Cue
1			
2			
3			
4			
5			
6			
7			
8			
9			
10			
11			
12			
13			
14			
15			
16			
17			
18			
19			
20			
21			
22			
23			
24			
Mean of Class Data			
SE of Class Data			

Problem Set 6.2: Learned and Unlearned Responses

Using the data from Results Table 6.2b, create a graph showing the mean response distance (with 95% CI) versus treatment (learned visual cue, learned auditory cue, and unlearned auditory cue). This experiment is designed so that each subject serves as his or her own control (i.e., the same subject was used in all 3 treatments). Thus, a repeated measures ANOVA should be used to determine if there are differences among the response distances of learned and unlearned responses and to answer the following questions.

a. State the null hypothesis.

b. Should you accept or reject the null hypothesis? Support your conclusions with the appropriate statistical results.

c. How does the level of information processing required for a response affect the response time? Explain your answer.

d. Do you see a difference in the response time when using a learned visual cue versus a learned auditory cue? If yes, why might this be the case? (*Hint:* Because both of the cues used in these tests have previously been learned, your answer should not include a discussion of learned versus unlearned responses.)

e. Would you expect to see a difference between the response times of these learned and unlearned responses if you performed 25 trials for each test? Why? Explain your answer.

LABORATORY 7

Sensory Physiology

PURPOSE

This laboratory will introduce you to the principles of sensory receptor density and receptive fields. This lab will also introduce the basic properties of two special senses, hearing and vision.

Learning Objectives

- Examine the two-point threshold test in different regions of the body and understand how the two-point threshold test distance relates to the density of sensory receptors in those regions.
- Learn how to clinically examine subjects for conduction deafness by performing the Rinne's and Weber's tests.
- Understand the physiology and functional anatomy of the eye.
- Understand the phenomena of afterimages and the blind spot.
- Calculate the diameter of the optic disc based upon measures of the blind spot.
- Learn how to clinically evaluate visual acuity.

Laboratory Materials

Experiment 7.1: Two-Point Threshold Test

Calipers

Experiment 7.2: Clinical Investigation of Hearing

1. Tuning forks
2. Reflex mallets

Experiment 7.3: Clinical Investigation of Vision

1. Blind spot cards
2. Afterimage cards
3. Rulers
4. Snellen eye chart

Introduction and Pre-Lab Exercises

External sensory information is **transduced** (or changed into another form of energy) by a variety of sensory receptors throughout the body. The cutaneous sensations of touch, pressure, cold, and nociception, for example, are mediated by the dendritic endings of different sensory receptors. Each of these sensory receptors is a specialized cell that is maximally sensitive to one particular **modality** or type of sensory stimulus. When stimulated, these dendritic nerve endings induce the propagation of action potentials along afferent nerve fibers toward the central nervous system. Once in the central nervous system, all incoming sensory information is integrated so that we may maintain appropriate awareness of our position and surroundings.

7-1 Complete the following table by explaining the major functions of these sensory receptor types.

Receptor Type	Modality and Function
Photoreceptors	
Hair cells of the cochlea	
Merkel's corpuscles	
Meissner's corpuscles	
Pacinian corpuscles	
Ruffini corpuscles	
Free nerve endings	

■

7-2 Although most sensory receptors are maximally sensitive to a particular modality, some sensory receptors are also sensitive to other types of stimuli if these stimuli are strong enough. For example, close your eyes and apply pressure on your eyelids for several seconds with your fingers. Do you see a light spot? Why do you see light if your eyes are closed and your photoreceptors are not being photostimulated?

■

The distribution of sensory receptors throughout the body determines which areas of the body will be sensitive to a particular stimulus. For example, the eye contains a very dense collection of specialized sensory receptors that respond to light. Furthermore, the density of sensory receptors in a particular area will determine how sensitive that area is to a stimulus. In this laboratory, you will investigate differences in the density and receptive fields of somatic sensory receptors among different areas of the body. You will also examine some clinical properties of hearing and vision, two of the three special senses in humans.

The Auditory System

The sense of hearing is based upon the physics of sound and the physiology of the ear itself. **Sound waves** are transmitted through a medium because they cause a disturbance in the molecules of the medium. Anything that can cause movement of molecules is capable of producing sound (such as a book being dropped on a table). Likewise, any substance that contains molecules, including air, liquids, and solids such as bone, may serve as a medium for sound wave conduction.

In this lab you will be using a tuning fork to test the function of the auditory system. When a tuning fork is struck, it vibrates causing molecules in the air around it to also vibrate. These vibrations form a sound wave, which consists of alternating zones of **compression** (where molecules are close together and pressure is increased) and **rarefaction** (where molecules are further apart and pressure is decreased) (Figure 7-1).

7-3 What characteristic of sound waves is responsible for the loudness of a sound?

■

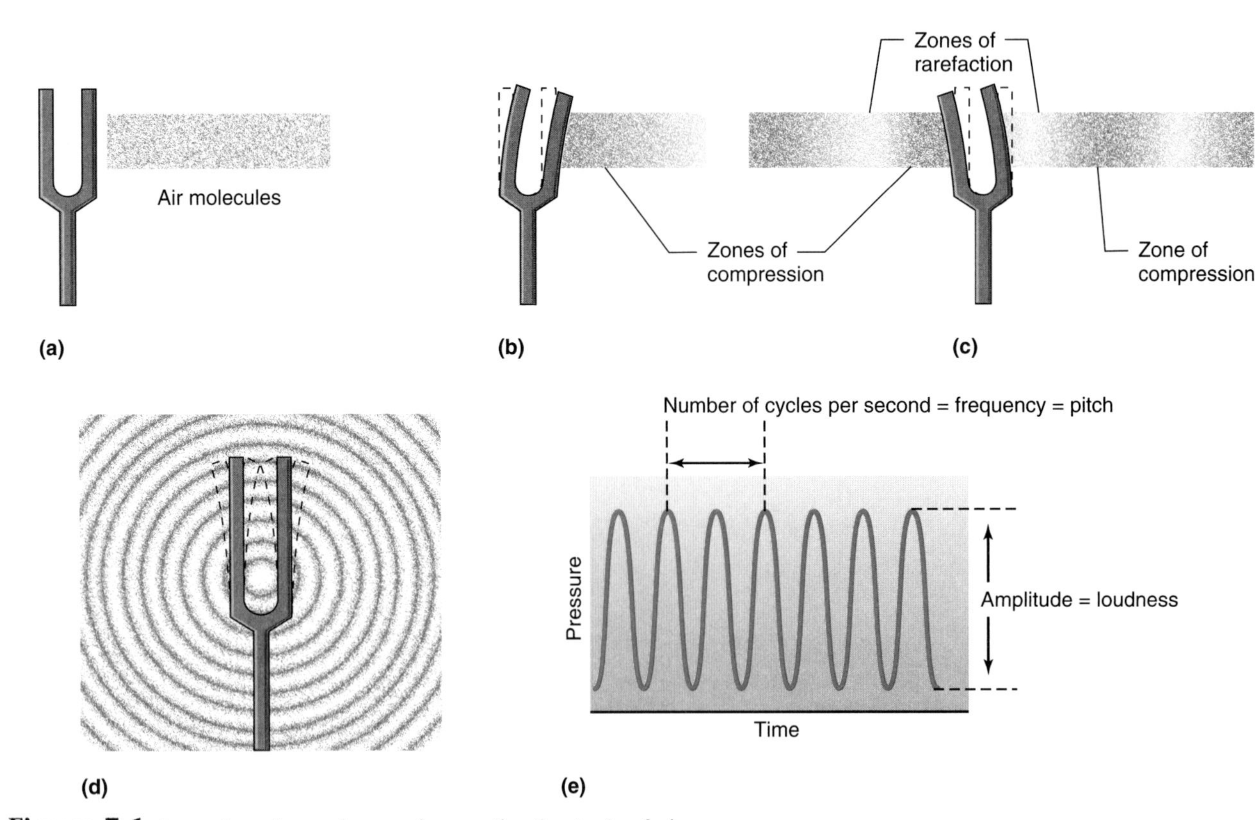

Figure 7-1 Formation of sound waves from a vibrating tuning fork.

7-4 What characteristic of sound waves is responsible for the pitch of a sound?

Sound waves are transmitted through the **external auditory canal** to the middle ear via the **tympanic membrane** (eardrum). The tympanic membrane in turn transmits sound waves to the three small bones of the middle ear (the **malleus, incus,** and **stapes**). These bones amplify the sound pressure before it is transmitted to the fluid medium of the inner ear, or **cochlea,** a fluid-filled, spiral structure in the temporal bone of the skull (Figure 7-2).

The cochlea is the site of the sensory receptors that are specialized for the modality of hearing. These sensory receptors, called **hair cells,** are mechanoreceptors that have hairlike **stereocilia** protruding from one end of the cell. The vibrations caused by the transmission of sound waves in the cochlea cause the stereocilia to bend, which in turn opens cation channels in the cell's membrane and depolarizes the cell. Thus, stereocilia transduce the mechanical stimulus of sound pressure waves in the cochlea into nerve impulses that are ultimately transmitted to the central nervous system.

7-5 In contrast to most other cells, the extracellular fluid surrounding the hair cells of the cochlea contains a higher concentration of K^+ than the intracellular fluid. In response to the bending of the stereocilia, K^+ channels in the cell membrane open. Using your knowledge from the action potential laboratory (Laboratory 5), explain why the opening of K^+ channels depolarizes the hair cell.

The Eye

The outermost layer of the eye is the **sclera,** which can be seen toward the anterior portion of the eyeball as the white of your eye. In the most anterior portion of the eye, the sclera is modified into the transparent **cornea,** which performs much of the

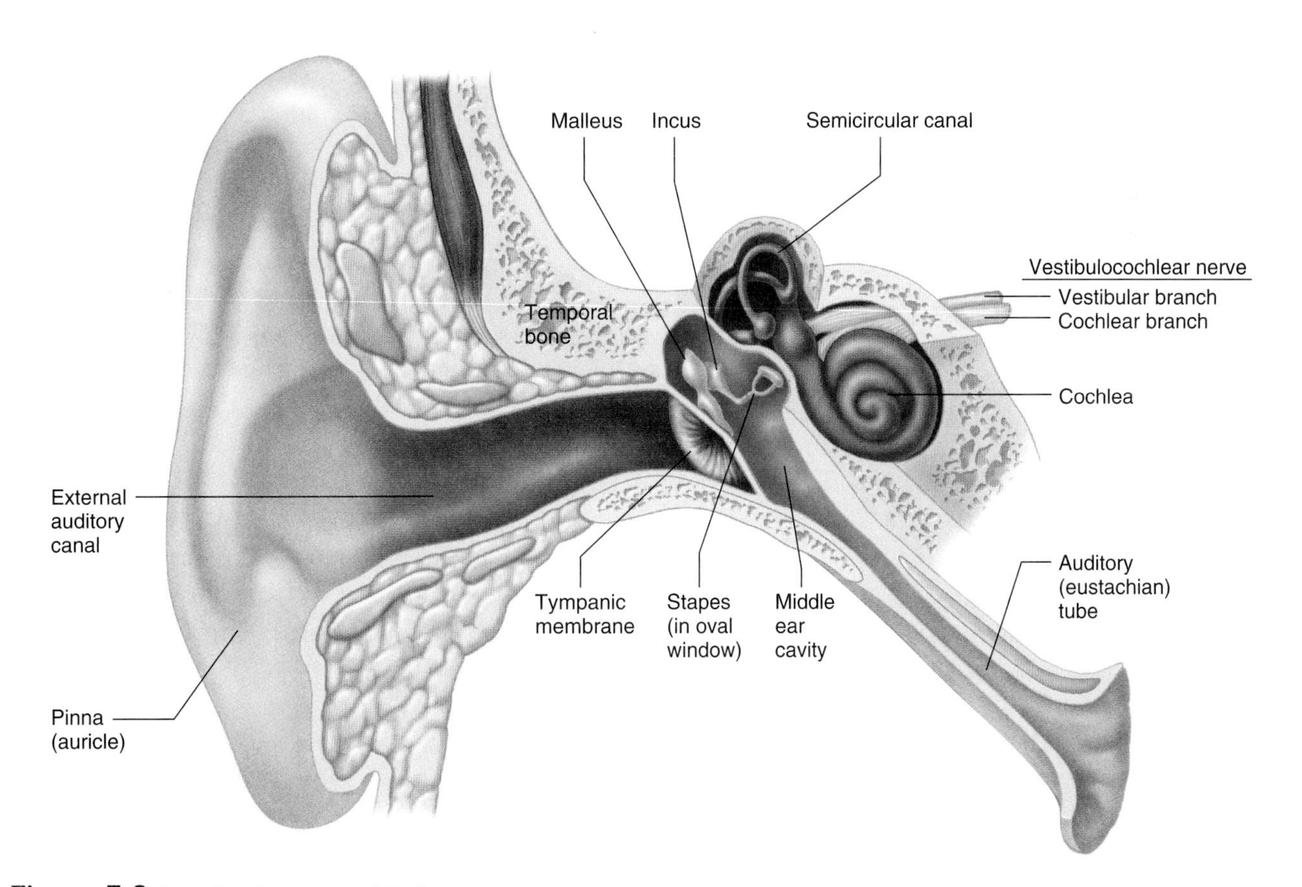

Figure 7-2 Functional anatomy of the human ear.

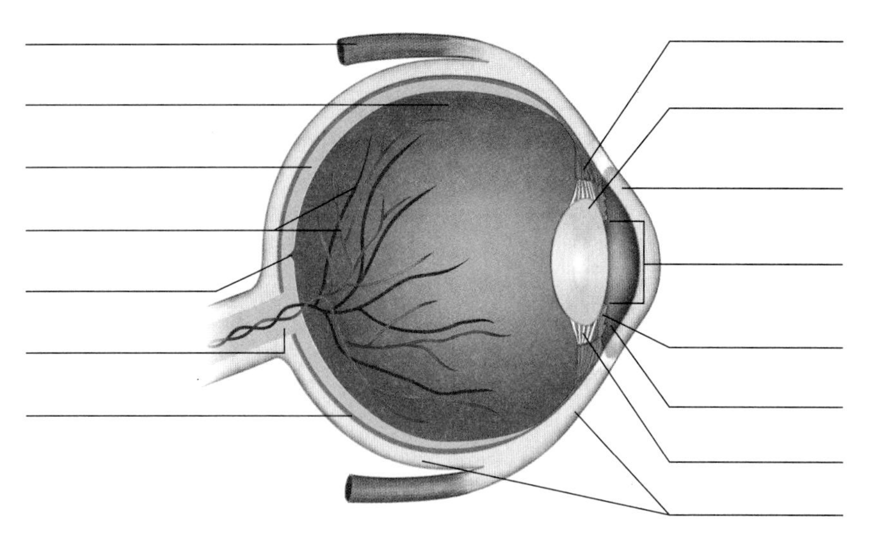

Figure 7-3 Cross section of the eye.

coarse focusing of light rays as they enter the eye. The fine focusing of light rays is performed by the transparent crystalline **lens.** Together, the cornea and lens focus light rays from external objects onto the **retina,** a delicate layer of tissue composed of specialized sensory receptors. These light receptors, called **photoreceptors,** convert light energy into a neural signal that is relayed to the central nervous system.

Two major types of photoreceptors are involved in the transduction of light energy: rods and cones. **Rods** have low light perception (it does not require many photons to excite rods) and are used in black-and-white vision. **Cones** are used in color vision and require much more light energy to become excited (this is why color vision is not sharp in dim light conditions such as those found at dusk). Humans have three different types of cones that respond maximally to particular wavelengths of light that correspond approximately to red, green, and blue light.

In humans, the area of highest visual acuity is the **fovea centralis,** an area of the retina located precisely at the eye's posterior pole. Although the fovea centralis contains no rods, it contains the highest concentration of cones. In this region, retinal structures are displaced to the sides of the fovea centralis. This allows light to pass almost directly to the photoreceptors and greatly enhances visual acuity.

At the area where the axons of all the ganglion cells of the retina exit the eye to form the optic nerve, an anatomical structure of the retina is formed, the **optic disc.** The optic disc contains no photoreceptors. Objects focused on the optic disc will not be perceived because no photoreceptors are present in this area. Thus, a **blind spot** in the visual field of both eyes exists where the optic disc occurs. Note that optic disc and blind spot are not synonymous terms.

7-6 Label the anatomical structures of the eye shown in Figure 7-3.

■

Accommodation is the process of keeping an object in focus as it is moved closer to the eye. At the **infinite distance** (about 20 ft. or 6.1 m), light rays entering the eye from an object are parallel and are naturally focused on the fovea centralis while the lens is in its relaxed state. If an object is moved closer to the eye, then the lens must change shape to refract the light rays to a point on the fovea centralis. Together, the ciliary muscles and zonular fibers induce accomodation of the eye. When the ciliary muscle contracts, the zonular fibers (also called suspensory ligaments) become loose, and the lens is allowed to bulge and become more convex (more refractive) (Figure 7-4). When the ciliary muscle relaxes, the zonular fibers become taut, and the lens is flattened and thus becomes less refractive.

7-7 What is the significance of standing 20 feet from an eye chart when you are examined for visual acuity?

■

Collectively, the cornea, lens shape, and length of the eyeball determine the point where light rays

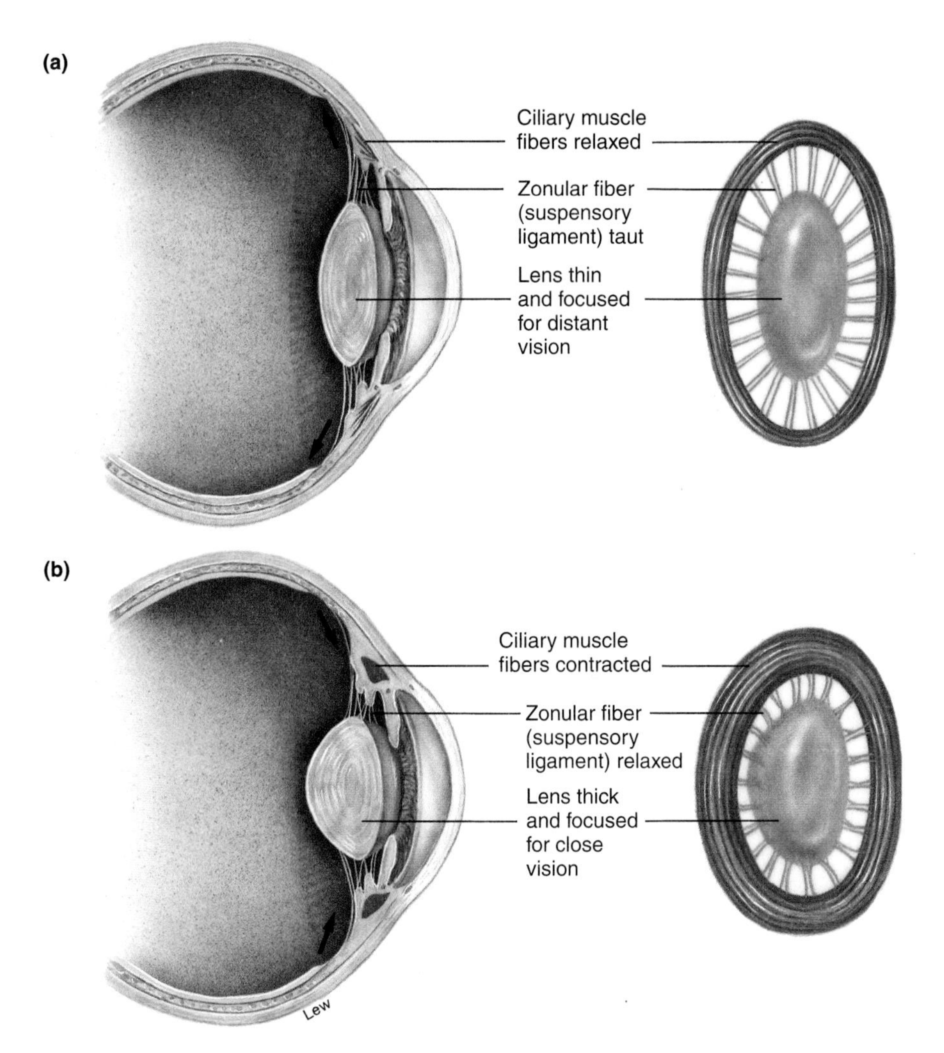

Figure 7-4 (a) An unaccommondated eye. (b) During accommodation, contraction of the ciliary muscle causes the lens to become more refractive.

converge. If the light rays converge on the retina, then the object will be perceived as a distinct, focused object. If light rays do not converge on the retina, then the image will be out of focus and the person will have difficulty seeing. There are many causes of eyesight deficits in humans, including the age-related decline in lens elasticity and/or the ability of the ciliary muscle to change the shape of the lens during accommodation. This age-related decline in the ability to accommodate for near vision is called **presbyopia.** Defects in vision may also occur if the eyeball is too long or too short in relation to the lens size. One eyesight deficit, **hyperopia** or farsightedness, is an inability to see near objects clearly. If the eyeball is too short for the lens, images of faraway objects fall on the retina, but the images of near objects focus (converge) at a point behind the retina (Figure 7-5). This visual disorder may be corrected with a convex lens, which helps the light rays to converge before they are focused by the lens of the eye. The focal point of light rays entering a convex lens occurs behind the lens. Thus, convex lenses are referred to as positive (+) lenses, and eyeglass prescriptions of hyperopic patients are designated as "+." In contrast, the focal point of a concave lens occurs in front of the lens and concave lenses are thus negative (−) lenses.

7-8 Figure 7-6 depicts the problems associated with **myopia,** or nearsightedness. Which type of corrective lens must be used to correct the problem? Draw the appropriate corrective lens and the redirected light rays in the figure.

■

Visual acuity is the ability of the lens of the eye to bend or refract light to focus an image on the retina. As discussed earlier, if the light is refracted too much or too little, the image will be focused in

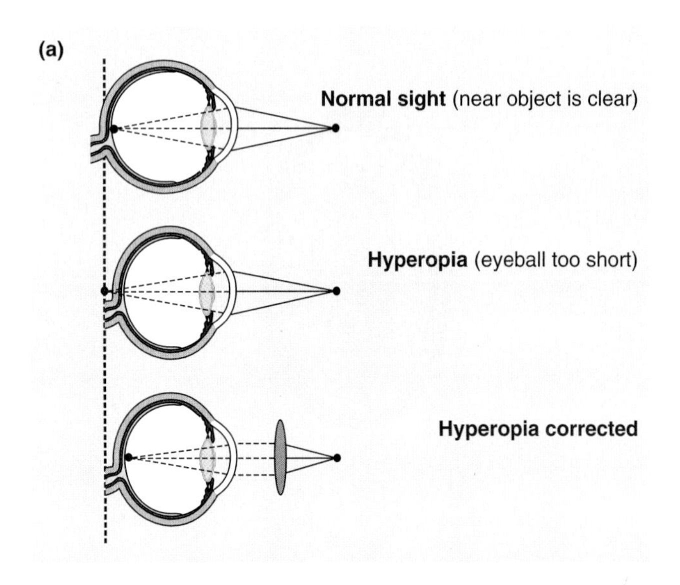

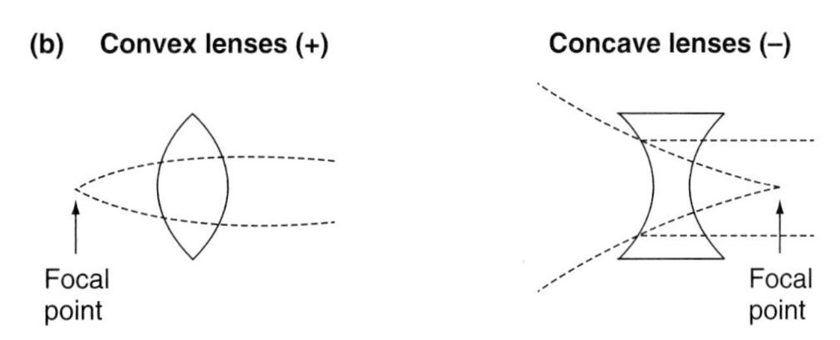

Figure 7-5 Diagram showing (a) a corrective convex lens refracting light rays prior to entering the natural lens of the hyperopic eye; (b) the focal points of convex and concave lenses.

front of or behind the retina, respectively. The refractive power or strength of the lens can be adjusted by the ciliary muscle, which causes the lens to become more or less convex. The refractive power of a lens is measured in **diopters.**

$$\text{Diopter} = \frac{1}{\text{focal length in meters}}$$

The focal length is the distance from the lens to the focal point of the light rays. The refractive power of an unaccommodated normal eye (i.e., when an image is 6.1 m or 20 feet from the eye) is 67 diopters.

To measure visual acuity, calculate the size of the retinal image by drawing lines from the top and bottom of the observed object through the nodal point of the eye (the lens) to the retina. The

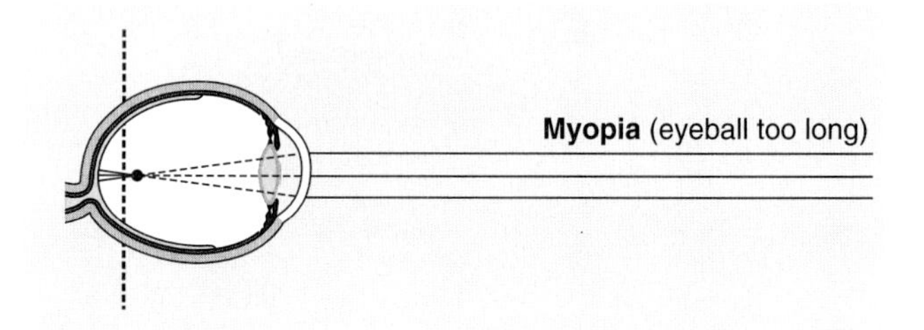

Figure 7-6 Diagram of an uncorrected myopic eye.

angle formed at the nodal point is the visual angle (Figure 7-7). As an image moves farther away from the eye and becomes smaller, the visual angle will also become smaller until the image can no longer be seen. The smallest visual angle that can be seen by the normal eye is about a one-minute arc and is a measurement of visual acuity. The Snellen Eye Chart, which is used to measure visual acuity, is based upon a visual angle of one minute (Figure 7-8).

If line 5 is the smallest line you can read, your visual acuity is expressed as $20/40 = 0.5$ and is below normal visual acuity. This means that you can see at 20 feet what a person with normal vision can see at 40 feet. Thus, you have 0.5 or half of the visual acuity of normal vision. Normal visual acuity is $20/20 = 1.0$, which means you can see line 8 of the chart from 20 feet away.

$$\text{Visual acuity} = \frac{d}{D}$$

where d = distance from the chart

D = distance at which the line can be read with normal vision

7-9 A patient standing 6.1 meters (20 feet) from a Snellen Eye Chart can distinguish the letters perfectly on line 10 but not line 11. Calculate the patient's visual acuity. How does this patient's visual acuity compare to normal vision?

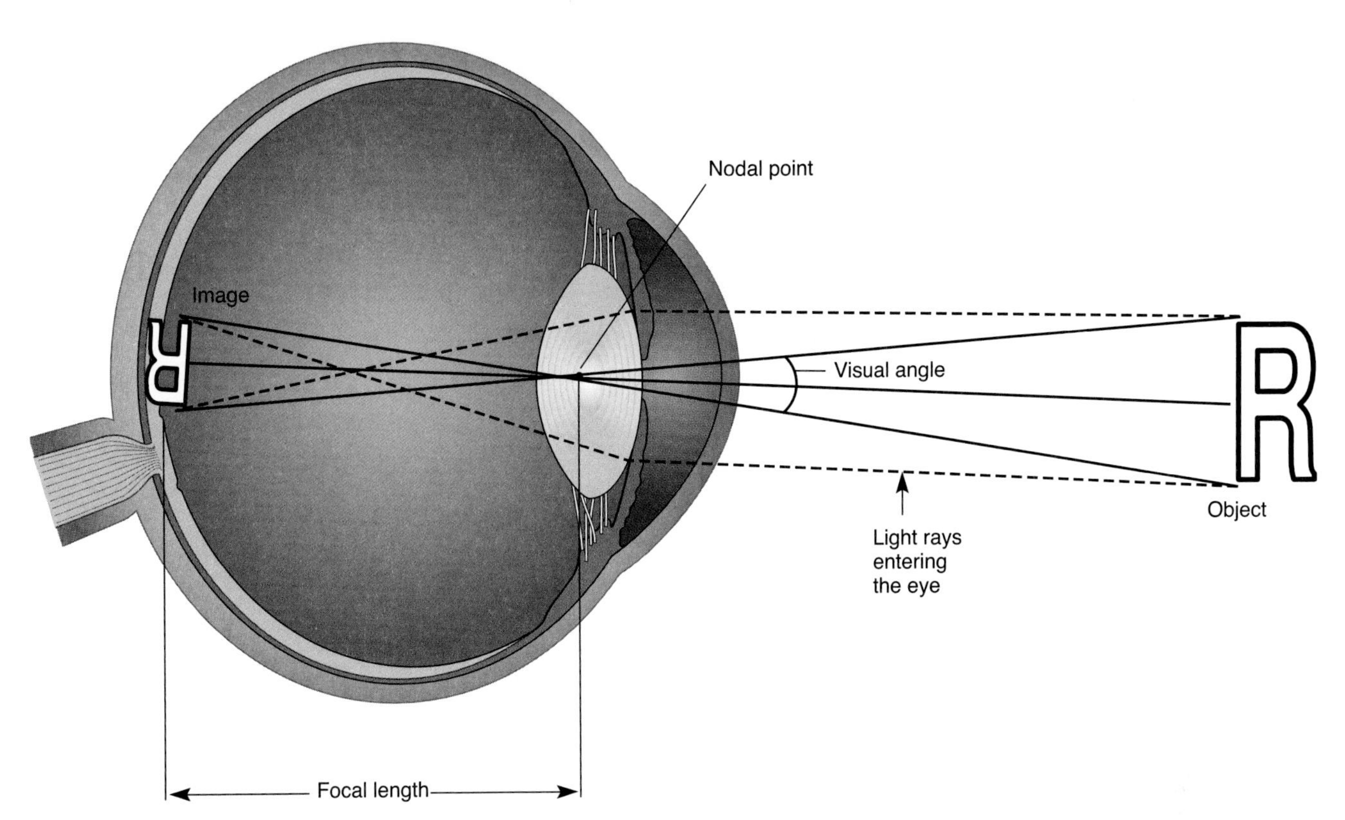

Figure 7-7 Light from a distant object being focused on the retina. Objects 6.1 m (20 ft.) from the eye produce parallel rays of light that are focused on the retina by the unaccommodated eye. As an image moves farther away from the eye and becomes smaller, the visual angle will also become smaller until the image can no longer be seen.

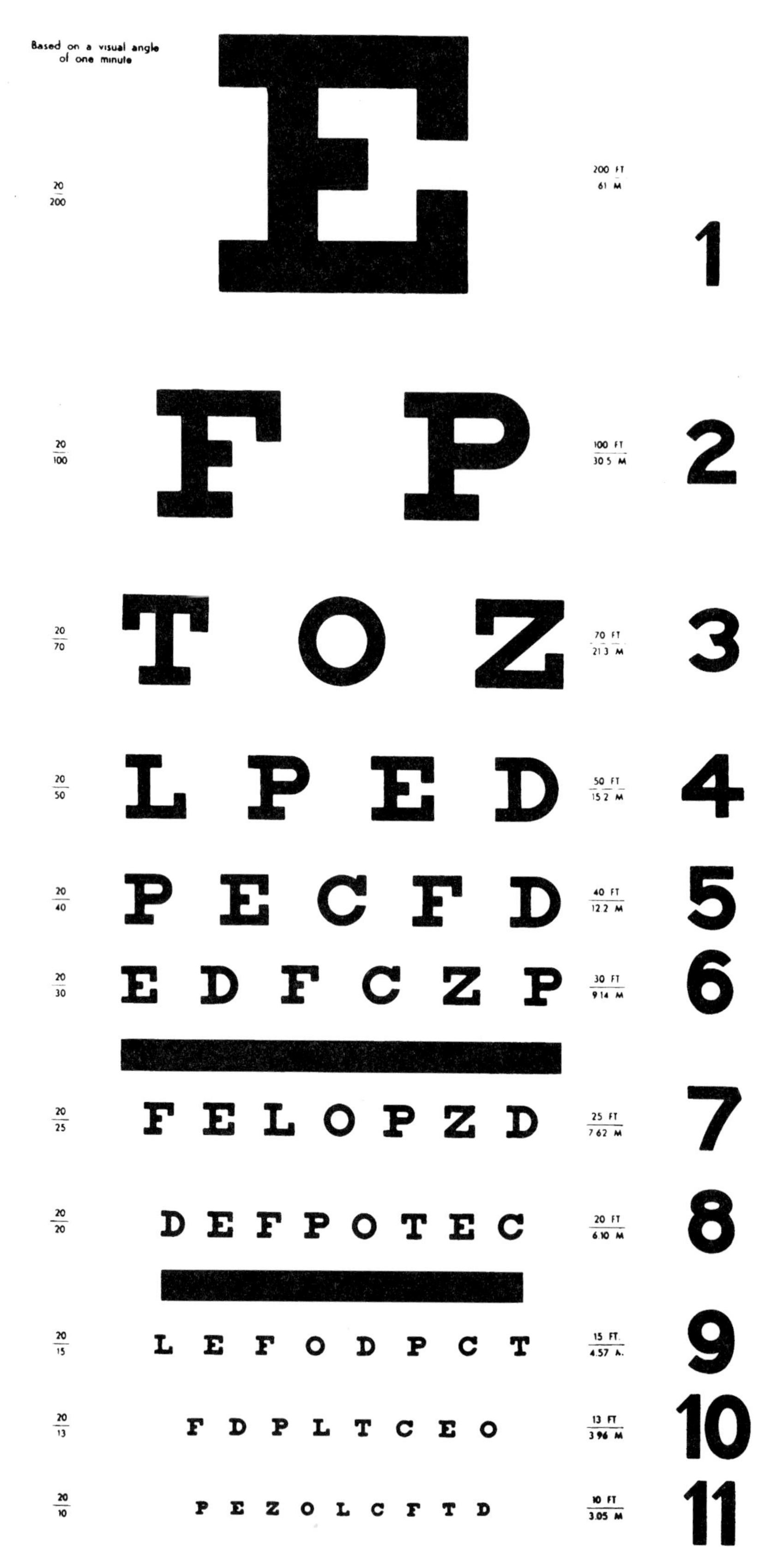

Figure 7-8 The Snellen Eye Chart.

Methods and Materials

EXPERIMENT 7.1 Two-Point Threshold Test

In this experiment you will investigate the density of touch receptors in different regions of the body and statistically determine which areas may differ significantly.

- The density of cutaneous receptors can be determined by measuring the distance between two points that touch the skin when they are no longer perceived as two distinct points. An adjustable caliper with two plastic or rubber points will be used in this experiment.
- Ask your laboratory subject to close his or her eyes. Open the calipers to 70 mm (7 cm) and lightly touch the skin on the back of the subject's hand with the two points simultaneously. Ask the subject if he or she feels one or two points. You may want to randomly touch the subject's skin with only one point to make sure your subject is unbiased.
- Close the caliper in 5 mm increments and continue testing until your subject can no longer distinguish two separate points. Record the distance between the caliper's two points as your **descending distance** in Results Table 7.1a of the Laboratory Report.
- Continue to test the back of the hand region by now closing the calipers at least 10 mm from the descending distance and repeating the experiment. This time open the calipers in 5 mm increments until your subject can once again distinguish two points. Record this second distance between the caliper's two points as the **ascending distance** in Results Table 7.1a.
- Repeat this experiment using the subject's fingertip, forearm, back of the neck, and upper lip.

EXPERIMENT 7.2 Clinical Investigation of Hearing

In this experiment you will investigate some clinical tests used to evaluate auditory function. Because sound energy must be transmitted through the middle ear into the cochlea, damage to either area can produce a hearing deficit. Sound conduction through the bones in the skull (a solid medium) can bypass the middle ear and directly cause vibrations in the fluid, or endolymph, of the cochlea. In contrast, the conduction of sound waves through air must be transmitted to the cochlea by the middle ear. The following exercises will allow you to test for **conduction deafness,** an inability to hear that results from impaired conduction of sound waves. These tests enable us to distinguish between conduction damage to the middle ear and conduction damage to the cochlea. Note that **sensorineural deafness** (or perceptive deafness) occurs when the transduction of sound waves into nerve impulses from the cochlea to the auditory cortex is impaired. Such damage would result in no perception of sound. Use Figure 7-9 to aid you in the placement of the tuning fork for the Weber's and Rinne's tests.

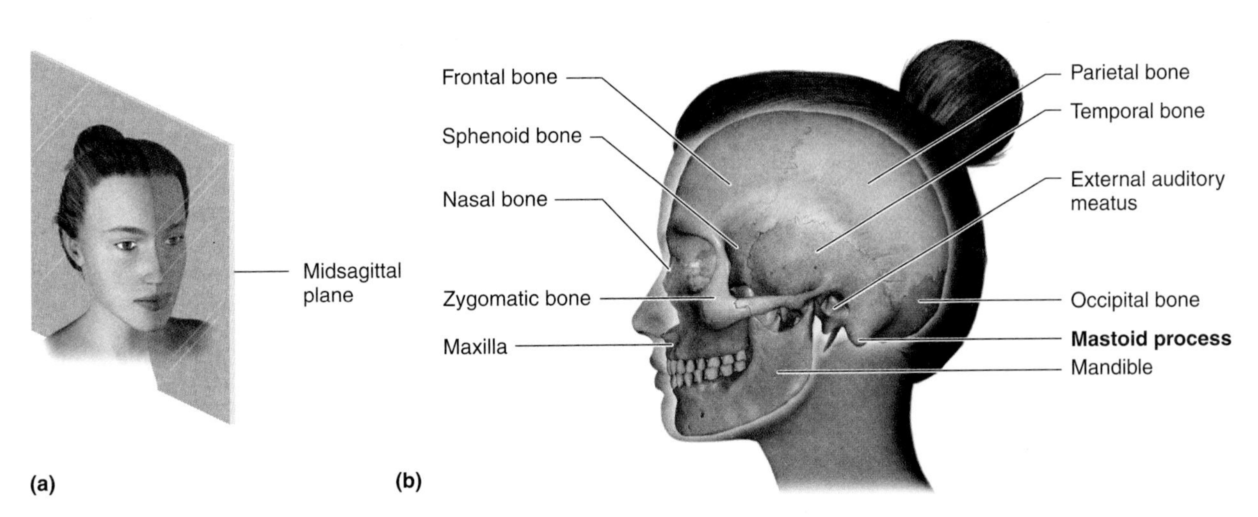

Figure 7-9 Diagram showing (a) the midsagittal plane and (b) the location of the mastoid process of the temporal bone for placement of the tuning fork.

Weber's Test for Conduction Deafness

- Use a rubber reflex mallet to strike a tuning fork. The tuning fork will begin to vibrate and produce a sound with a particular pitch (vibrations/second).
- Place the handle of the vibrating tuning fork along the midsagittal line of your lab partner's head (the handle must contact the head). Does your partner hear the sound? The sound will be perceived because of the vibrations being conducted to the cochlea through the bones of the skull as well as the air vibrations being conducted to the cochlea through the middle ear. This procedure tests both ears simultaneously and tests for conduction deafness due to either middle ear or cochlea damage.
- Is the sound perceived to be equally loud in both ears?
  - If there is conduction damage to the cochlea of one ear, the sound will seem louder in the normal ear. This occurs because the cochlea of one ear is damaged and does not vibrate as well as the healthy or normal ear, regardless of whether the sound is being conducted through bone or air.
  - Conduction deafness due to middle ear damage is indicated by the sound being louder in a damaged ear because external noise through the auditory canal is excluded. This result occurs because the damaged middle ear cannot conduct the sound waves from the surrounding noisy classroom into the cochlea. Thus, the sound perceived by the damaged ear (due to bone conduction) is not dampened and seems louder.
  - Note that Weber's test does not enable us to distinguish between conduction deafness due to cochlea damage and conduction deafness due to middle ear damage. Weber's test simply indicates potential damage to the auditory system. For example, if the sound seems louder in the left ear, it indicates either cochlea damage to the right ear or middle ear damage to the left ear. Other tests, such as the Rinne's test, are used to distinguish the cause of conduction deafness.
- Perform Weber's test again but now with an earplug or a piece of cotton placed in one ear. The sound in the plugged ear (simulating air conduction deafness due to middle ear damage) should seem louder because external noise from the classroom has been eliminated.

Rinne's Test for Conduction Deafness and Distinguishing Between Cochlea and Middle Ear Damage

- Place the handle of the vibrating tuning fork against the mastoid process of your lab partner's temporal bone.
- Ask your lab partner if he or she can hear the tuning fork. Note that an inability to perceive the sound of the tuning fork while it is held against the mastoid process would suggest conduction deafness due to cochlea damage. Allow the tuning fork to remain in this position until your lab partner can no longer perceive the sound.
- Now move the still-vibrating tuning fork in front of and close to your lab partner's external auditory canal. Can your partner now hear the sound? If the sound is once again perceived, there is no damage to the middle ear and there is no conduction deafness.
- Repeat this experiment but now with an earplug or a piece of cotton placed in the ear. When you remove the tuning fork from your lab partner's mastoid process to the front of his or her auditory canal, the sound should now not be heard. This is a simulation of conduction deafness due to middle ear damage where bone (the mastoid process) is more efficient at conducting sound waves than the middle ear.
- Test both the right and left ears.

EXPERIMENT 7.3 Clinical Investigation of Vision

In this experiment you will examine the phenomena of afterimages and the blind spot and calculate the diameter of the optic disc based upon measures of the blind spot. You will also learn how to clinically evaluate visual acuity.

Blind Spot

- Remove the blind spot figure (Figure 7-11a) from page 82.
- Place your hand over your left eye.
- With your right hand, hold the blind spot card and extend your arm. Now bring the card closer to your face and allow your right eye to focus on the black circle. Make sure the cross is on the right side of this circle.

- Continue to move the black circle closer to your right eye until the cross in your peripheral vision can no longer be seen. *If you focus on the cross, the experiment will not work.* Make sure your right eye is focused completely on the circle and the cross is only seen in your peripheral vision.

- Once the cross has disappeared from your peripheral vision, have your laboratory partner measure the distance the card is away from your right cheek with a ruler. Place the ruler under the cheekbone when making your measurement and *exercise extreme caution when placing anything near someone's eye.*

- Continue to move the card (while focusing on the circle) toward your eye until the cross reappears. Have your lab partner measure this second distance.

- Because the eye is recessed in relation to the cheekbone, add 3 cm to the two measurements (B_1 and B_2) to account for the depth of the eye or cornea from the cheekbone.

- You now have two measurements: the distance at which the cross disappeared (B_1 = ______cm) and the distance at which the cross reappeared (B_2 = ______cm).

- Now you can use the measurements from the disappearance and reappearance of the cross to calculate the diameter of the optic disc using the following information and ratios:

  A = distance between the circle and cross on the card (5 cm)

  B_1 = distance between the cornea of your eye and the card when the cross disappears from your peripheral vision.

  B_2 = distance between the cornea of your eye and the card when the cross reappears in your peripheral vision.

  C_1 = first distance between the optic disc and the fovea centralis.

  C_2 = second distance between the optic disc and the fovea centralis.

  d_{Eye} = diameter of the eye or distance from the cornea to the retina. This distance is approximately 2 cm.

  d_{OD} = diameter of the optic disc

- Calculations: Follow the three steps shown in Figure 7-10 to calculate the diameter of the optic disc. Make sure all of your measurements are in centimeters. d_{OD} = ________

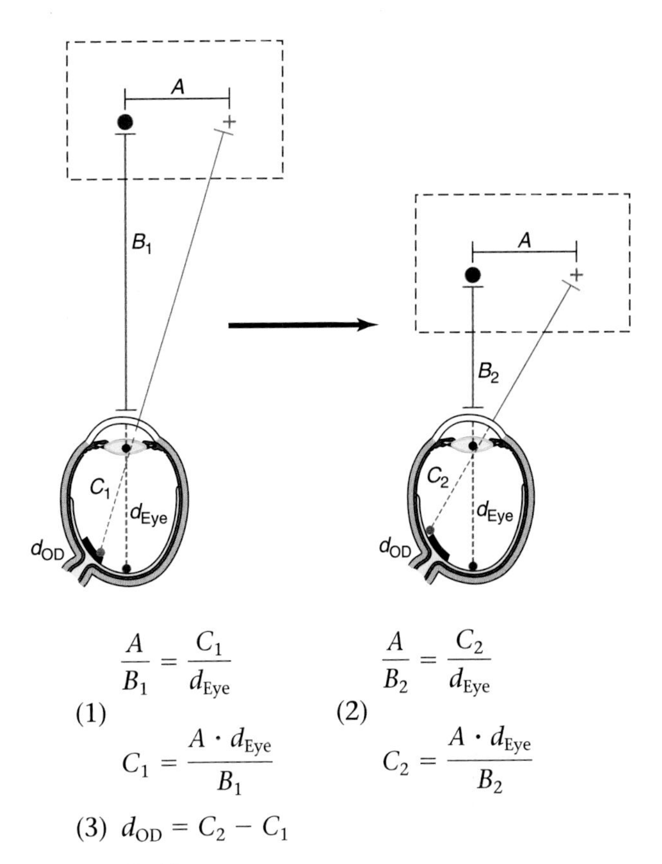

Figure 7-10 Calculations of the optic disc diameter.

- Repeat this entire procedure for the left eye. Make sure the cross is now on the left side of the circle when testing the left eye.

Afterimages

- Remove the negative afterimage figure (Figure 7-11b) from page 82. Stare at the image for 60 seconds. Make sure you give the image your full concentration and do not focus on any surroundings.

- After the 60 seconds, quickly focus your eyes on a blank piece of white paper.

- Describe what you see in your results.

- To investigate color afterimage, remove the color afterimage figures (Figure 7-12a and b) from pages 83 and 84. Stare at a colored square for 60 seconds and then focus on a blank sheet of white paper.

- Describe and report the afterimage in your results. What color(s) did you see in your afterimage?

- Repeat this procedure for all of the colored squares.

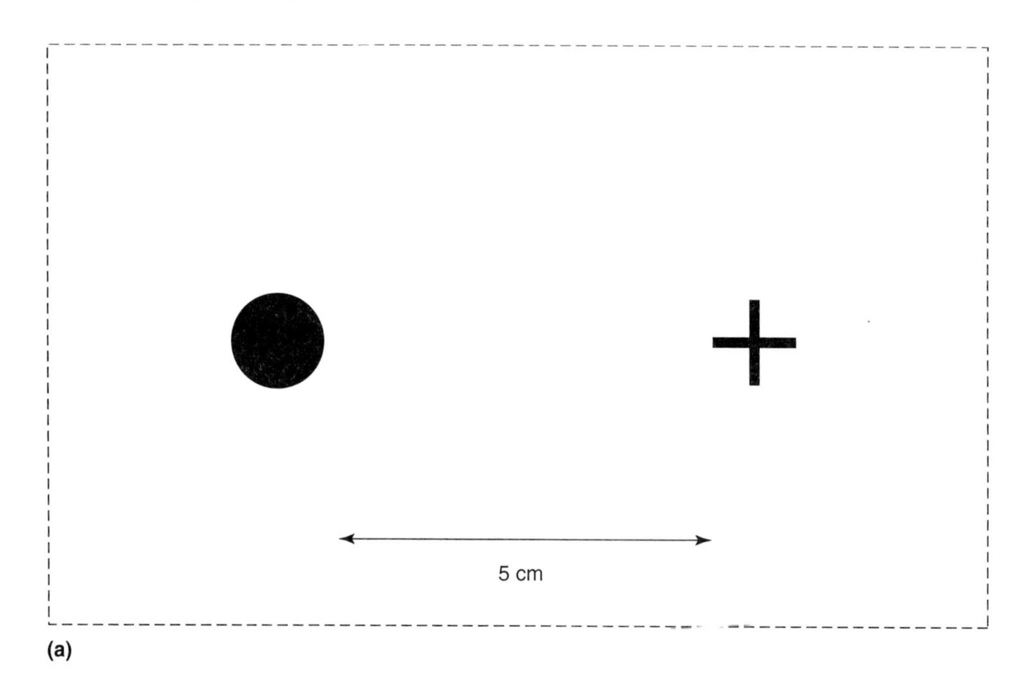

Figure 7-11 Cards for examining the (a) blind spot and (b) negative afterimage.

Figure 7-12a Cards for examining color afterimages.

Figure 7-12b Cards for examining color afterimages.

Evaluating Visual Acuity with the Snellen Eye Chart

- Stand 6.1 m (20 feet) from the Snellen Eye Chart. Cover one eye and read the letters indicated by your laboratory partner in the first line of the chart. Your lab partner will point to specific letters within the line for you to read and evaluate the number of correct responses.
- Your lab partner will continue to point to letters within lines that have smaller and smaller letters. Once your laboratory partner indicates that you have made at least two mistakes in identifying letters within a line, your visual acuity for that eye is determined using the last line read without mistakes.
- Repeat this procedure with your other eye.
- If you wear glasses, repeat the procedure without your glasses to determine your unaided visual acuity.

COMPARATIVE NOTE *Laboratory 7*

Infrared Modality: Seeing Heat

Remember that all sensory receptors decode stimuli in regard to duration, intensity, location, and modality. To perceive a stimulus, a receptor must transduce that particular stimulus into a nerve impulse. For example, you can be exposed to X rays and no matter how long, intense, or where on your body you receive these waves, you do not have the ability to perceive this stimulus. This is because you do not have a modality for X rays, meaning that there is no sensor with the ability to transduce the energy from X rays into electrical stimuli in your nervous system. The only part of the electromagnetic spectrum that we can transduce is visible light with our eyes (wavelengths of 400 to 700 nm). However, some animals can perceive other wavelengths in the electromagnetic spectrum.

Snakes in the families Boidae (*Boas*), Pythonidae (*Pythons*) and Viperidae (*Vipers*) have special receptors to process infrared wavelengths and "see" heat. These snakes have facial pits that contain the infrared receptors. These receptors are composed of branching sensory neurons that detect slight changes in tissue temperature within the pit. An increased frequency of action potentials from the receptor transduces this infrared stimulus to the snake's nervous system. Rattlesnakes can detect a temperature change as slight as 0.002°C. This is important to an ambush or sit-and-wait predator so that it may detect the body heat from a mouse. Because many rodents are nocturnal (active at night), this sensory modality that allows a pit viper to see body heat from its prey, even in the dark, may help in the viper's foraging success.

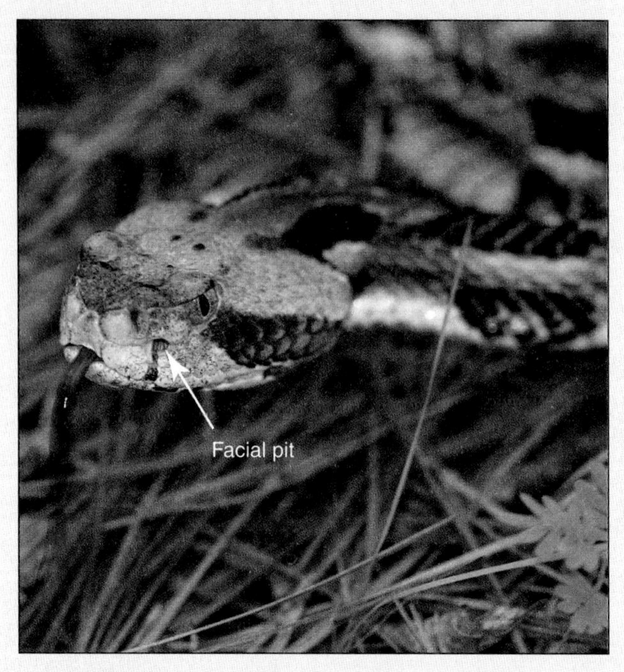

Facial pit of a timber rattlesnake (*Crotalus horridus*).

RESEARCH OF INTEREST

Chiszar, D., D. Dickman, and J. Colton. 1986. Sensitivity to thermal stimulation in prairie rattlesnakes *Crotalus viridis* after bilateral anesthetization of the facial pits. Behavioral and Neural Biology 45:143–149.

Dickman, J.D., J.S. Colton, D. Chiszar, and C.A. Colton. 1987. Trigeminal responses to thermal stimulation of the oral cavity in rattlesnakes *Crotalus viridis* before and after bilateral anesthetization of the facial pit organs. Brain Research 400:365–370.

Kardong, K.V. and H. Berkhoudt. 1999. Rattlesnake hunting behavior: correlations between plasticity of predatory performance and neuroanatomy. Brain Behavior and Evolution 53:20–28.

RESULTS AND DISCUSSION

LABORATORY REPORT 7

EXPERIMENT 7.1
Two-Point Threshold Test

Results Table 7.1a Individual Group Data for Experiment 7.1: Two-Point Threshold Test.

	Two-Point Threshold Distance (mm)					
	Back of Hand			Fingertip		
Student (Name)	Descending	Ascending	Mean	Descending	Ascending	Mean
1						
2						
3						
4						
5						
6						

	Two-Point Threshold Distance (mm)					
	Forearm			Back of Neck		
Student (Name)	Descending	Ascending	Mean	Descending	Ascending	Mean
1						
2						
3						
4						
5						
6						

Student (Name)	Two-Point Threshold Distance (mm)		
	Upper Lip		
	Descending	Ascending	Mean
1			
2			
3			
4			
5			
6			

Results Table 7.1b Class Data for Experiment 7.1: Two-Point Threshold Test.

Student (Name)	Individual Mean Two-Point Threshold Distances (mm)				
	Back of Hand	Fingertip	Forearm	Back of Neck	Upper Lip
1					
2					
3					
4					
5					
6					
7					
8					
9					
10					
11					
12					
13					
14					
15					
16					
17					
18					
19					
20					
21					
22					
23					
24					
Mean (±SE) of Class Data					

Problem Set 7.1: Two-Point Threshold Test

Using the data from Results Table 7.1b, create a graph showing the mean two-point threshold distances (with 95% CI) versus area tested. This experiment is designed so that each subject serves as his or her own control (i.e., the same subject was used to measure two-point threshold distance for the five body areas). Thus, a repeated measures ANOVA should be used to determine if there are differences in the two-point threshold test distances among the different areas.

a. State the null hypothesis.

b. Should you accept or reject the null hypothesis? Support your conclusions with the appropriate statistical results.

c. Which area has the greatest density of sensory receptors? Use the data collected on two-point threshold distances to support and thoroughly explain your answer.

d. Which area has the lowest density of sensory receptors? Use the data collected on two-point threshold distances to support and thoroughly explain your answer.

e. If fingertips had a sensory receptor density similar to that reported in question d, would people be able to train themselves to read Braille? Explain.

f. Suppose you perform an experiment similar to the two-point threshold test, but you use tiny flashlights to test the two-point threshold of the retina. How would the retina's two-point threshold distance to light stimuli compare to the finger's two-point threshold to cutaneous stimuli? Explain.

Problem Set 7.2: Clinical Investigation of Hearing

Weber's Test for Conduction Deafness

a. What were the results of the Weber's test for both the left and right ears? Was the sound equally loud in both ears? If not, in which ear was the sound louder?

b. How did using an earplug alter the results of the Weber's test? What type of damage to the ear does the earplug simulate? Explain your results.

c. While you are conducting the Weber's test, a patient reports that the sound seems louder in her right ear. What are the two possible explanations for this result? Explain.

Rinne's Test for Conduction Deafness

d. What were the results of the Rinne's test for the right ear (i.e., did air conduction of sound waves to the cochlea result in greater sensitivity to sound than bone conduction of sound waves, as is normally observed)? What were the results of the Rinne's test for the left ear?

e. How did using an earplug alter the results of the Rinne's test? What type of damage to the ear does the earplug simulate? Explain your answer.

f. In question c, you listed two possible explanations for the Weber's test result of a patient. How would you use the Rinne's test to distinguish between these two possible explanations? Be sure to explain what the possible outcomes of the Rinne's test would indicate for both ears.

Problem Set 7.3: Clinical Investigation of Vision

Blind Spot

a. What is the d_{OD} for the left eye? What is the d_{OD} for the right eye?

b. Are the optic discs in your left and right eyes similar in size?

c. How does the size of your optic discs compare to those of your lab partners? Do you think there is a relationship between optic disc size and body size?

d. What would happen to the visual field of the right eye if the optic disc of the right eye was twice the size?

e. If each of your eyes has an optic disc, then why do you not have two small portions of every scene that are not perceived?

Afterimages

f. Describe the afterimages observed after staring at the black-and-white figure and each of the colored squares for 60 seconds.

Image	Afterimage
Black-and-white figure	
Yellow square	
Red square	
Blue square	
Green square	

g. What is the relationship between the color of the square and the color seen in the afterimage?

h. For each of the afterimage results reported, explain the phenomenon of afterimages in relation to the stimulation of photoreceptors.

Evaluating Visual Acuity with the Snellen Eye Chart

i. What is the smallest line you could read with your right eye? What is the visual acuity of your right eye? What does this mean?

j. What is the smallest line you could read with your left eye? What is the visual acuity of your left eye? What does this mean?

k. Is your visual acuity normal (i.e., do you have "perfect" vision)?

l. Using your knowledge of myopia and hyperopia, describe how corrective lenses improve vision. Use the unaided visual acuity of one of your labmates who wears glasses as an example to answer this question.

m. Mike has a visual acuity of 20/10, while Tim has a visual acuity of 10/20. If an object was placed 25 feet from Mike and Tim, who would be able to clearly and accurately describe the object? How do these visual acuities lead you to your conclusions?

LABORATORY 8

Functional Anatomy of Muscle and Mechanics of Contraction

PURPOSE

This laboratory will introduce you to the structural and functional anatomy of skeletal muscle and the mechanics of skeletal muscle contraction.

Learning Objectives

- Describe the microanatomy of skeletal muscle.
- Understand the sliding-filament mechanism of muscle contraction and the role of each microstructure during muscle contraction.
- Describe the sequence of events that occurs during muscle contractions, starting with the transmission of a nerve impulse to a muscle fiber and ending with the relaxation of the muscle.
- Understand the relationship between the length of a muscle sarcomere and the tension or force of muscle contraction.
- Describe the difference between an after-loaded and direct-loaded muscle's ability to perform work during contraction.

Laboratory Materials

Experiment 8.1: Muscle Length–Tension Relationship

Hand dynamometers

Experiment 8.2: After-Loaded and Direct-Loaded Muscle Contractions

Hand or free weights (5-, 10-, 15-, 20-, 25-, 30-, 35-lb. dumbbells)

Introduction and Pre-Lab Exercises

Muscle is one of the four primary tissue types found in the body. There are three different types of muscle: skeletal, cardiac, and smooth. Skeletal muscle is the only muscle type that is under voluntary control, and thus it is called voluntary muscle (Figure 8-1). In this laboratory, you will examine the functional anatomy of muscle and the mechanics of its contraction. Because the three different types of muscles differ somewhat in their anatomy and mechanics of contraction, we will focus on skeletal muscle here and during Laboratory 9 ("Physiology of Muscle Contraction").

If you look at a longitudinal section of skeletal muscle under a microscope, you will see alternating light and dark bands along the length of the muscle fibers (Figure 8-2). Because of its striped pattern, skeletal muscle is also called **striated muscle.** These striations result from the regular arrangement of muscle proteins into cylindrical bundles called **myofibrils.** The cytoplasm of each muscle cell (also called a **muscle fiber** or **myofiber**) is filled with many myofibrils bundled together. Myofibrils are so prevalent in muscle fibers that the multiple nuclei in each skeletal muscle fiber lie on the outside of these myofibril bundles, adjacent to the plasma membrane of the muscle fiber.

Each myofibril, in turn, consists of **thick** and **thin filaments** arranged in repeating patterns (Figure 8-3). One unit of this repeating pattern (which corresponds to one unit of the alternating dark and light bands of muscle fibers examined under a microscope) is called a **sarcomere.** A sarcomere is the functional unit of muscle. A myofibril consists of many sarcomeres joined end to end by a network of interconnecting proteins called the **Z line** or Z disc. Thus, the borders of a sarcomere are defined by the Z lines on either side.

Each sarcomere of a myofibril contains two sets of thin filaments, each anchored to the Z line on

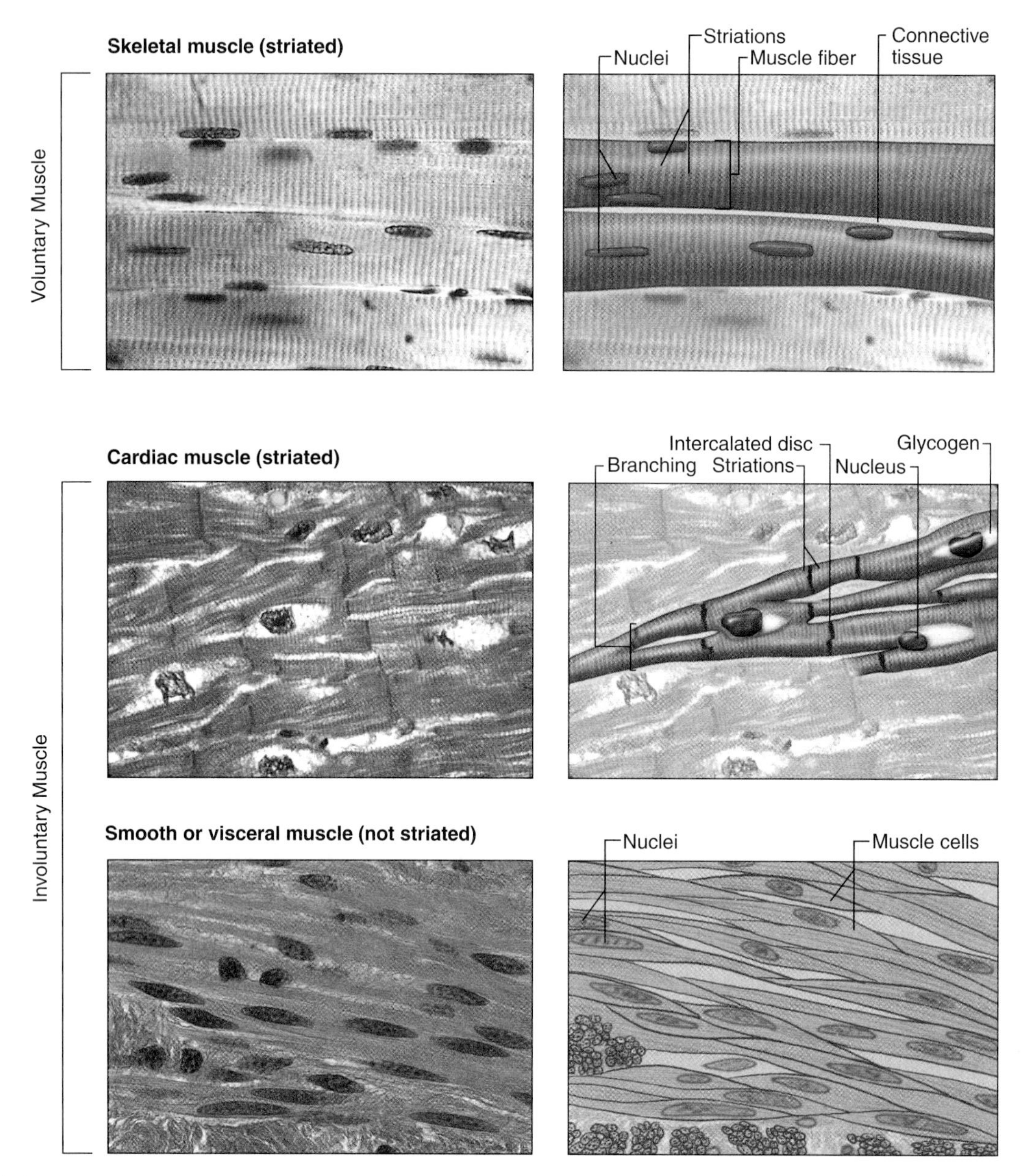

Figure 8-1 Comparison of skeletal, cardiac, and smooth muscle. Both skeletal and cardiac muscle have a striated appearance. Cardiac and smooth muscle cells tend to have a single nucleus, while skeletal muscle fibers are multinucleated.

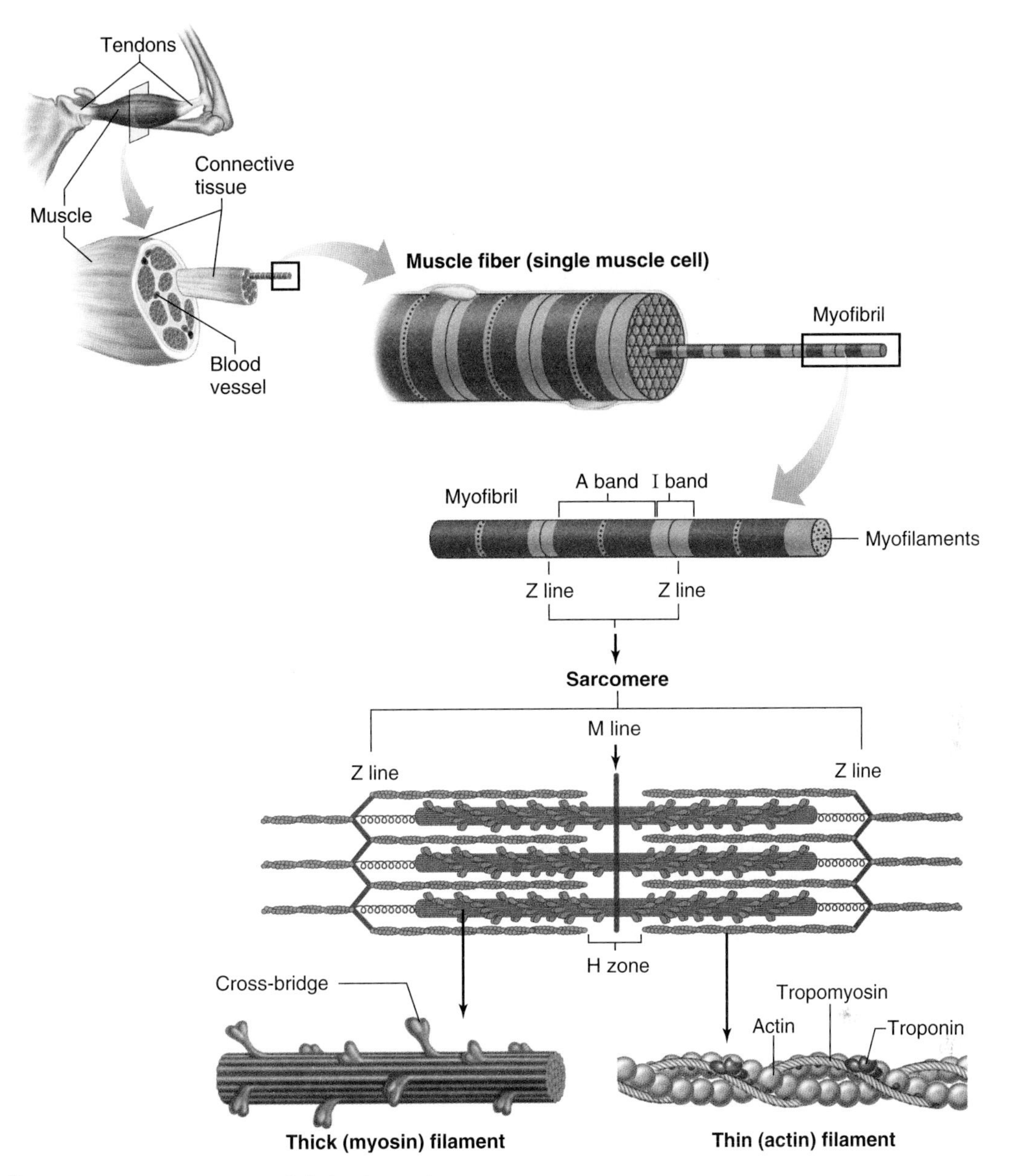

Figure 8-2 Functional microanatomy of skeletal muscle.

either side of the sarcomere. The thick filaments are located in the middle of the sarcomere and overlap slightly with a portion of each set of thin filaments. The length of the myosin filaments defines a portion of the sarcomere called the **A band;** the parallel arrangement of myosin filaments in the middle of each sarcomere is what produces the dark bands (or stripes) that are apparent when you microscopically examine a muscle fiber. The light bands, called the **I bands,** are defined as the distance from the end of one A band to the beginning of the next A band. The I bands appear lighter than the A bands because only the thin filaments are present in these regions. Note that the I band is bisected by the Z line, and thus the distance between one Z line and the A band forms one-half of the I band.

The thick filaments of myofibrils are composed almost entirely of the contractile protein **myosin.** The myosin molecule consists of a long tail and two globular heads that protrude from the molecule. These protruding globular heads, called **cross-bridges,** attach to myosin binding sites on the thin filaments during muscle activation.

The thin filaments contain three different proteins, but they are only about half the diameter of the thick filaments. Thin filaments consist of **actin** proteins associated with the proteins troponin and tropomyosin. **Tropomyosin** forms a loose coil

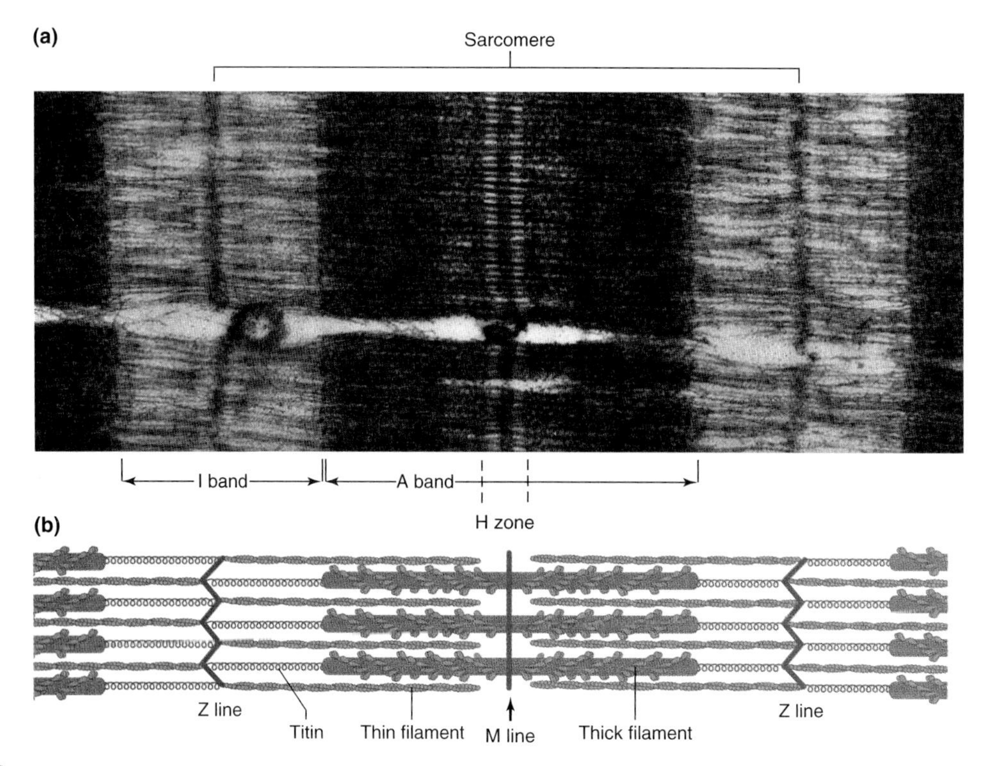

Figure 8-3 (a) High magnification of a sarcomere within a myofibril. (b) Arrangement of the thick and thin filaments in the sarcomere.

around the actin filament, while **troponin** is bound to the tropomyosin coil. The site of myosin cross-bridge attachment is located on the actin molecules. However, when a muscle is at rest, the tropomyosin and troponin protein complex physically blocks the myosin attachment sites on the actin filaments.

The Sliding-Filament Mechanism of Muscle Contraction

When a motor neuron (or somatic efferent neuron) is activated, it releases acetylcholine at the neuromuscular junction to stimulate skeletal muscle contraction. When acetylcholine binds to acetylcholine receptors on the postsynaptic **sarcolemma** (the plasma membrane of a muscle cell), an **end-plate potential** is generated in the motor end plate. This excitatory graded potential depolarizes the sarcolemma adjacent to the end plate to its threshold potential. The muscle cell in turn produces one or more action potentials. These action potentials are propagated along the sarcolemma and deep into the muscle cell along the transverse tubules.

The **transverse tubules,** or **T-tubules,** are a continuation of the sarcolemma and provide a means by which the deep myofibrils can be stimulated by action potentials (Figure 8-4). The T-tubules are closely associated with the **sarcoplasmic reticulum,** the cellular organelle (modified endoplasmic reticulum) that sequesters and stores calcium ions (Ca^{2+}). The T-tubules and sarcoplasmic reticulum are connected by structures called **junctional feet** or **foot proteins.** This junction involves two integral membrane proteins: **dihydropyridine** (DHP, a voltage-sensitive protein in the cell membrane of T-tubules) and **ryanodine** (a foot protein that forms a calcium ion channel in the membrane of the sarcoplasmic reticulum). When an action potential is propagated down the T-tubules and into the muscle cell, the change in membrane potential induces a conformational change in DHP. This conformational change in DHP ultimately induces a conformational change in the ryanodine Ca^{2+} channel, which results in its opening. While these Ca^{2+} channels are open, Ca^{2+} diffuses down its concentration gradient from the sarcoplasmic reticulum to the sarcoplasm of the muscle cell.

Once in the sarcoplasm, Ca^{2+} attaches to calcium binding sites on the troponin proteins, causing a conformational change (and hence movement) in the troponin-tropomyosin complex (Figure 8-5). When the troponin-tropomyosin complex shifts along the actin filament, the binding sites for myosin cross-bridges are exposed. Myosin then binds to the actin filament. When cross-bridge binding occurs, it causes a conformational change and movement in the myosin cross-bridges. This phenomenon is known as the power

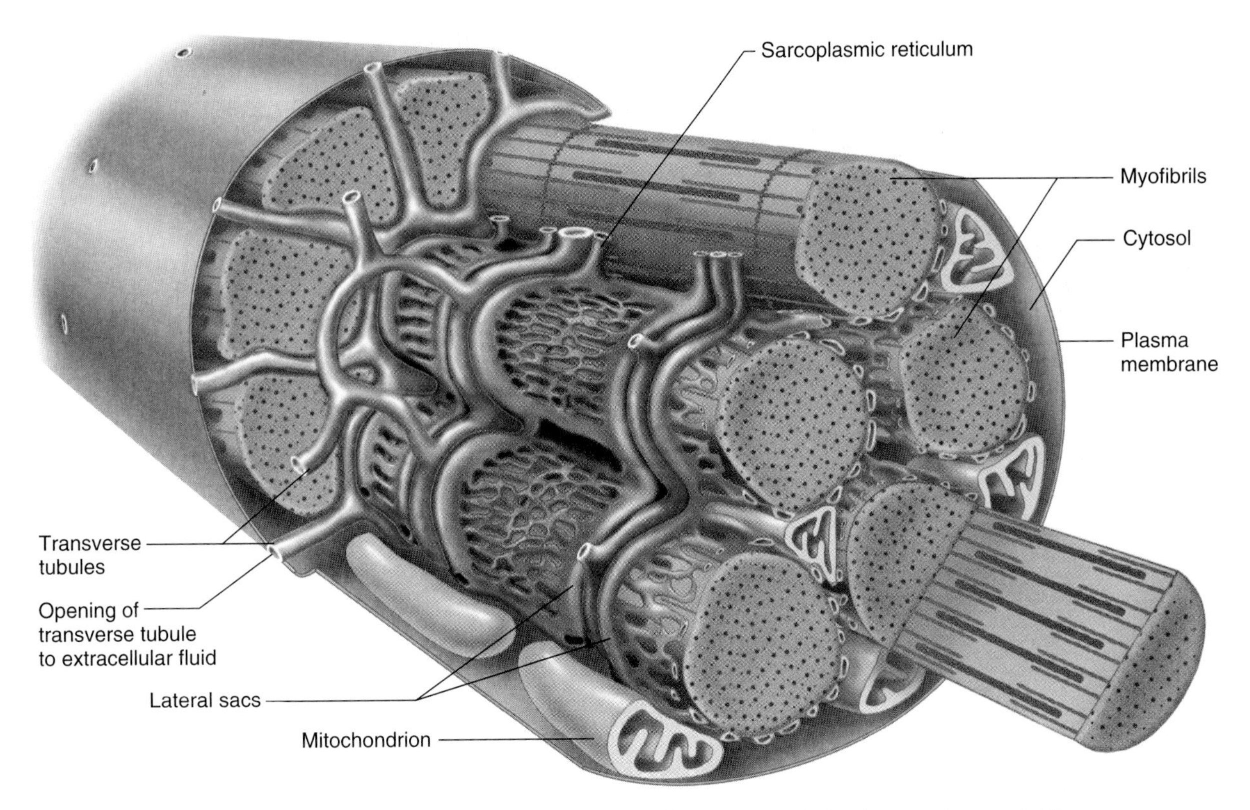

Figure 8-4 Anatomical structure of the sarcoplasmic reticulum and transverse tubules in a single muscle fiber.

stroke of cross-bridge cycling, as the moving myosin heads pull the attached thin filaments and Z lines toward the center of the sarcomere. Thus, the entire sarcomere shortens during muscle contraction as thick and thin filaments slide past one another. Note that the length of neither the thick nor the thin filaments changes during muscle shortening.

8-1 Does the length of the A band change during sarcomere shortening? Why or why not? ■

8-2 Does the length of the I band change during sarcomere shortening? Why or why not? ■

Release of the cross-bridges, and therefore relaxation of the muscle, requires the binding of ATP to the myosin cross-bridges. ATP acts as an allosteric modulator, lowering the affinity of myosin for actin and resulting in the detachment of cross-bridges from actin. When ATP is hydrolyzed by the ATP-ase enzyme activity present in myosin's ATP binding site, the energy released is used to reset or reenergize the myosin cross-bridge. As long as Ca^{2+} remains bound to troponin, the myosin binding sites on actin will be exposed, and myosin and actin will continue the cross-bridge cycle (provided that ATP is available).

8-3 If the supply of ATP to a shortening muscle is completely depleted, what will happen to the cross-bridge cycle? What is the name of the phenomenon that results from complete ATP depletion? ■

8-4 Order the following 10 events that summarize muscle contraction. Remember that a muscle contraction begins with the transmission of a nerve impulse to a muscle fiber.

______ The power stroke

______ Opening of ryanodine Ca^{2+} channels in the sarcoplasmic reticulum

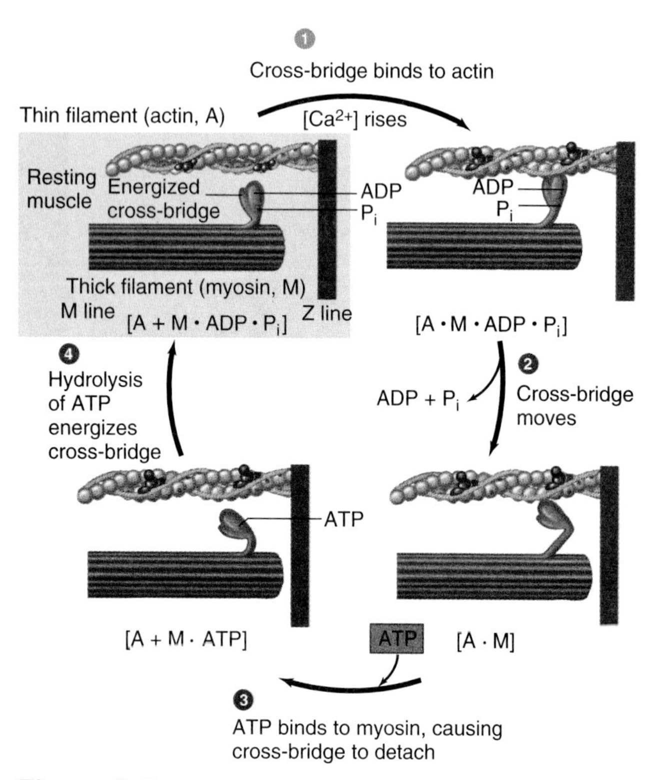

Figure 8-5 Interactions between actin and myosin during the cross-bridge cycle.

_____ Binding of myosin cross-bridges to actin

_____ Release of acetylcholine into the neuromuscular junction

_____ Conformational change in tropomyosin-troponin complex

_____ Binding of Ca^{2+} to troponin

_____ Binding of ATP to myosin cross-bridges

_____ Depolarization of the muscle fiber

_____ Hydrolysis of ATP bound to myosin cross-bridges

_____ Action potential propagation along the transverse tubules ■

Now that you have been introduced to the cellular structure and microanatomy of muscles, we can investigate the influence of these microanatomical properties on muscle contraction. Because the myosin cross-bridges literally pull the thin filaments toward the center of the sarcomere, the amount of muscle shortening that can be generated is influenced by the sarcomere's resting length. Each muscle sarcomere has an optimal length that will produce an optimal muscle force. In the first experiment, you will investigate how the stretching of the sarcomere affects the alignment of muscle proteins and thus the force of muscle contraction. The second experiment will demonstrate how the work performed by a contracting muscle varies with its initial length. Use the drawings that you create in the spaces that follow to help you understand the results of your experiments and draw conclusions about the influence of muscle stretch on maximal force production.

8-5 Sketch a figure of a sarcomere from a muscle that is overstretched. Be sure to label the sarcomere components.

■

8-6 Sketch a figure of a sarcomere from a muscle that is understretched or compressed. Be sure to label the sarcomere components.

■

Methods and Materials

EXPERIMENT 8.1 Muscle Length–Tension Relationship

In this experiment you will investigate how the stretching of the sarcomere affects the alignment of muscle proteins and thus the force of muscle contraction.

- Use a hand dynamometer to perform this experiment.
- Position your hand and wrist in each of the three positions (+90, 0, and −90°) shown in Figure 8-6. Squeeze the hand dynamometer and record the force that your digital flexor muscles are able to exert on the dynamometer. Be careful not to move your wrist out of position or use muscles other than your digital flexors when squeezing the dynamometer. Record three measurements of force for each of the three wrist positions in Results Table 8.1a of the Laboratory Report.

EXPERIMENT 8.2 After-Loaded and Direct-Loaded Muscle Contractions

In this experiment you will examine how initial muscle length, determined by after-loaded and direct-loaded conditions, will affect the work performed by a contracting muscle. Use your nondominant hand for this experiment. Thus, if you are right-handed you will use your left hand and arm. Extend your arm and supinated hand (palm facing upward) and have your laboratory partner measure and record the distance from the inside of your elbow (on the anterior arm) to the joints between your metacarpals and proximal phalanges (these are your large knuckles). This distance is the radius of a circle = _________ cm that will be used for future calculations of muscle work (Figure 8-7).

Direct-Loaded Muscle Contractions

- In a sitting position and with your nondominant hand and arm, hold the 5-pound dumbbell at the side of your body with your arm fully extended. This position will represent the **direct-loaded** muscle position because the weight is stretching the biceps brachii muscle of your upper arm prior to contraction.
- Contract and shorten your biceps brachii muscle, moving the dumbbell an angular distance through an arc until it reaches your shoulder. This simple weightlifting exercise is referred to as a curl. It is very important to use only your biceps brachii muscle while lifting the weight slowly and at a constant speed. *Do not try to swing the weight upward or use your back to lift the weight.*
- To calculate the distance that the dumbbell was moved through the arc or circle segment, estimate the angle between the long axis of the body and forearm when the weight is lifted (Figure 8-7). Use rough estimates of 0° (arm fully extended; dumbbell not moved), 30°, 60°, 90° (arm flexed with the forearm at a right angle to the upper arm), 120°, 150°, and 180° (arm fully flexed with the dumbbell lifted the full distance of the arc to the shoulder). When you are only able to complete a partial

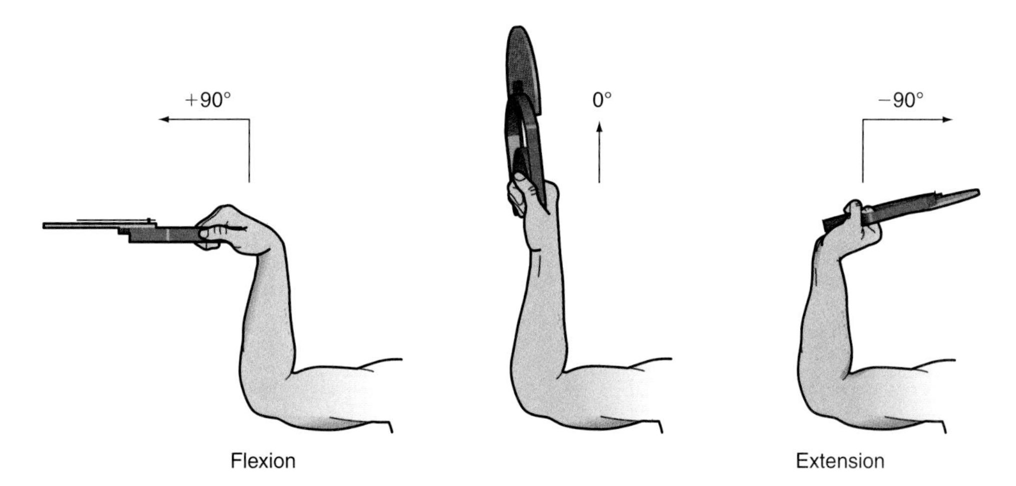

Figure 8-6 Illustration showing the wrist positions for testing the length–tension relationship in the digital flexors.

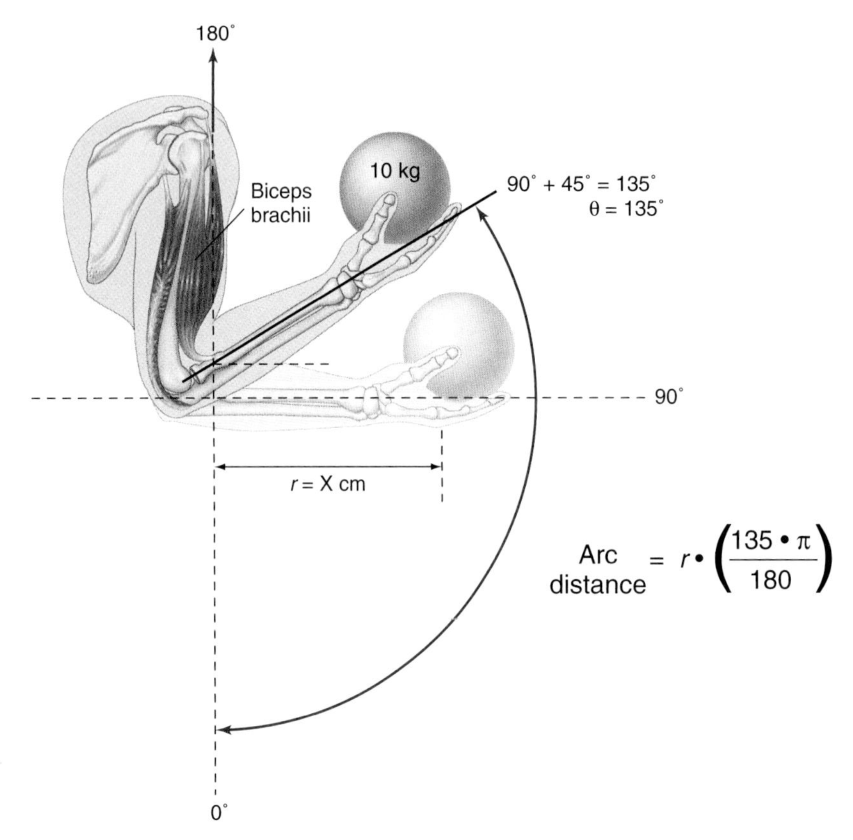

Figure 8-7 Illustration of how to calculate arc distance.

curl and are unable to move through the entire arc distance due to the increase in load, have your laboratory partner estimate the angle as shown in Figure 8-7.

- Using the angle estimated in the previous step, calculate the actual distance the weight was moved with the following formula. You will calculate an arc distance because this is the true circular distance or segment the weight was moved.

$$\text{Segment of a circle or arc distance} = r \cdot \left(\frac{\theta \cdot \pi}{180}\right)$$

Where: r = radius of a circle (the distance from your elbow to the joints between your metacarpals and proximal phalanges)

θ = the estimated angle between the long axis of the body and forearm.

$\pi = 3.14$

- Record this calculated arc distance (cm) in Results Table 8.2.
- Repeat the preceding steps for the 10-, 15-, 20-, 25-, 30-, and 35-pound dumbbells.

After-Loaded Muscle Contractions

- In a sitting position and with your nondominant side, position your forearm at a 90° angle to your upper arm. Have your lab partner support the weight of the 5-pound dumbbell in your hand so that your biceps brachii muscle is at rest. This position will represent the **after-loaded** muscle position because the weight is applied to the biceps brachii muscle after it has been partially contracted.
- Contract and shorten your biceps brachii muscle, moving the dumbbell an angular distance through an arc until it reaches your shoulder. As in the direct-loaded experiment, it is very important that you lift only with your biceps brachii muscle and do not attempt to swing the weight upward or use your back for lifting. Lift the weight slowly and at a constant speed.

- Calculate the distance the dumbbell was moved by estimating the angle between the long axis of the body and forearm when the weight is lifted. Because your arm is already at 90°, estimate the distance the weight was moved as 90° (arm unable to flex and lift the weight), 120°, 150°, or 180° (arm able to fully flex and move the dumbbell to the shoulder).

- Using the angle estimated in the previous step, calculate the actual distance the weight was moved with the following equation. Note that this calculation is slightly different from the direct-loaded calculations because your arm was already positioned at 90°. The following equation subtracts 90° from the estimated angle in calculating the total arc distance for an after-loaded muscle.

$$\text{Arc distance} = r \cdot \left(\frac{(\theta - 90) \cdot \pi}{180}\right)$$

- Record this calculated arc distance (cm) in Results Table 8.2.

- Repeat the preceding steps for the 10-, 15-, 20-, 25-, 30-, and 35-pound dumbbells.

Calculations of Muscle Work

- Calculate the amount of work (in joules) performed by the after- and direct-loaded muscles for each dumbbell mass by using the arc distance the dumbbell was moved.

- Work (W), measured in joules (J), is calculated by multiplying a known force (F) by the distance (d) through which that force is applied to move an object. Thus, if a force is applied but the object does not move, no work is done.

$$\boldsymbol{W = F \cdot d}$$

If distance = 0 because the object is not moved, then $\boldsymbol{W = F \cdot d = 0}$, no matter how great the force.

- Force is equal to mass (m) multiplied by acceleration (a). Thus, the object's mass, acceleration, and the distance the object was moved are needed to calculate the amount of work done by a muscle.

$$\boldsymbol{W = (m \cdot a) \cdot d}$$

Mass (kg): Convert the mass of each dumbbell from pounds to kilograms. There are 2.2 pounds in 1 kg. Thus, the 20-pound dumbbell has a mass of 9.1 kilograms.

$$20 \text{ pounds} \cdot \frac{1 \text{ kilogram}}{2.2 \text{ pounds}} = 9.1 \text{ kilograms}$$

Acceleration (m/s^2): In this experiment, you are lifting the mass of the dumbbell against gravity. Thus, we can use the acceleration due to gravity (9.82 m/s^2) to calculate the force (in Newtons) exerted to lift the dumbbell. For simplicity, we will assume that the acceleration of the dumbbell through the arc of the curl remains constant. We will also assume that all students lift the dumbbell at the same rate.

Distance (m): Use the arc distances you calculated and recorded in Results Table 8.2 as the distance each dumbbell was moved. Don't forget to convert your arc distances from centimeters to meters before calculating work.

8-7 If someone was able to move the 20-pound dumbbell to an angle of $\theta = 120°$ in the direct-loaded experiment, what is the amount of work done in joules? The distance from the person's elbow to the metacarpal-phalanges joint is 30 cm. Show all calculations.

Did you get an answer of 56.1 joules? If not, check your calculations. ■

8-8 What is the amount of work done in joules if the preceding observation was made for the after-loaded condition? Show all your calculations.

Did you get an answer of 14.0 joules? If not, check your calculations. ■

COMPARATIVE NOTE *Laboratory 8*

Behavioral Natural History and Muscle Type: You Are What You Do

You may not have given much thought to the types of muscle fibers you eat when you enjoy a slice of turkey. However, the characteristics of different muscle types shown in the following table can be easily illustrated with the different types of meat we eat. For instance, turkeys move about using mostly their leg muscles. Although these birds do not typically fly great distances, they are capable of very quick, strong bursts of flight. The muscles responsible for these fast, powerful bursts of flight are primarily fast glycolytic muscle fibers (white meat) located in the breast. The leg muscles of turkeys are primarily oxidative muscle fibers (dark meat) and are very resistant to fatigue. Migratory ducks, in contrast, have very dark breast meat. Can you explain this muscle characteristic?

The characteristics of muscle fibers and their predominant locations in animals are directly related to the activity levels of the muscles in that part of the body. Think about the different types of fish we eat. Tuna and salmon are mostly dark red meat, while flounder are white meat. What might these differences in muscle morphology have to do with the natural history of these fishes? Tuna are highly active, ram-ventilating fish that swim great distances through the open oceans. Salmon migrate upstream hundreds of miles to return to spawning areas. Thus, these fishes require red oxidative muscle tissue with the characteristics to resist muscle fatigue. A flounder, however, has mostly glycolytic muscle fibers. Flounder burrow into the sand of the ocean floor to sit and wait for prey. A flounder then ambushes its prey with a very quick burst of movement, which is permitted by the fast twitch rate of its white muscle fibers. Although the flounder is fast, don't ask it to swim a marathon with the tuna.

As you can see, muscle morphology can tell us a great deal about an animal's behavior and natural history, including modes of locomotion and activity. From this information and discussion, what can you infer about the controversy expressed by animal activists regarding veal?

	MUSCLE TYPES		
Characteristics	**Red** Slow Oxidative Type I	**Intermediate** Fast Oxidative Type IIA	**White** Fast Glycolytic Type IIB
Fiber diameter	Small	Intermediate	Large
Oxidative capacity	High	High	Low
Twitch rate	Slow	Fast	Fast
Rate of fatigue	Slow	Intermediate	Fast
Mitochondria	Many	Many	Few
Capillaries	Many	Many	Few
Glycogen content	Low	Intermediate	High
Myoglobin content	High	High	Low

RESEARCH OF INTEREST

Gamperl, A.K. and E.D. Stevens. 1991. Sprint-training effects on trout *Oncorhynchus mykiss* white muscle structure. Canadian Journal of Zoology 69:2786–2790.

Meyer-Rochow, V.B. and J.R. Ingram. 1993. Red-white muscle distribution and fibre growth dynamics: a comparison between lacustrine and riverine populations of the Southern smelt *Retropinna retropinna* Richardson. Proceedings of the Royal Society of London Series B Biological Sciences 252:85–92.

Meyers, R.A. and J.W. Hermanson. 1994. Pectoralis muscle morphology in the little brown bat, *Myotis lucifugus:* a non-convergence with birds. Journal of Morphology 219:269–274.

RESULTS AND DISCUSSION

LABORATORY REPORT 8

EXPERIMENT 8.1
Muscle Length–Tension Relationship

Results Table 8.1a Individual Group Data for Experiment 8.1: Muscle Length–Tension Relationship.

Student (Name)	Muscle Force											
	Flexion Position 1 (+90°)				Position 2 (0°)				Extension Position 3 (−90°)			
	Trial 1	Trial 2	Trial 3	Mean	Trial 1	Trial 2	Trial 3	Mean	Trial 1	Trial 2	Trial 3	Mean
1												
2												
3												
4												
5												
6												

Results Table 8.1b Class Data for Experiment 8.1: Muscle Length–Tension Relationship.

Student (Name)	Mean Muscle Force		
	Position 1 (+90°)	Position 2 (0°)	Position 3 (−90°)
1			
2			
3			
4			
5			
6			
7			
8			
9			
10			
11			
12			
13			
14			
15			
16			
17			
18			
19			
20			
21			
22			
23			
24			
Mean of Class Data (±SE)			

Problem Set 8.1: Muscle Length–Tension Relationship

Using the data from Results Table 8.1b, create a bar graph showing the mean muscle force produced at each of the three wrist positions. Be sure to include 95% CI in your graph. Using the data from Results Table 8.1b, perform a repeated measures ANOVA to examine possible differences in muscle force among the three wrist positions and answer the following questions.

a. State the null hypothesis.

b. Should you accept or reject the null hypothesis? Support your conclusions with the appropriate statistical results.

c. How does wrist position affect the force produced by the digital flexor muscles? Which position produces the optimal (i.e., maximal) muscle force?

d. Using your knowledge of muscle microanatomy, explain the mechanism underlying the effects of wrist position on the production of muscle force by the digital flexors.

e. Based upon these results, what is the functional importance of tendons (besides connecting muscles to bones)?

f. With regard to sarcomere length, what is the functional importance of monosynaptic stretch reflexes?

EXPERIMENT 8.2

After-Loaded and Direct-Loaded Muscle Contractions

Results Table 8.2 Individual Group Data for Experiment 8.2: After-Loaded and Direct-Loaded Muscle Contractions.

The distance from your elbow to the joint between your metacarpals and proximal phalanges is **r** = ________ **cm** (this is the radius of an arc made by the movement of the weight).

Weight (lb)	Mass (kg)	Direct-Loaded Muscle			After-Loaded Muscle		
		θ (°)	Arc Distance (cm)	Muscle Work (J)	θ (°)	Arc Distance (cm)	Muscle Work (J)
5							
10							
15							
20							
25							
30							
35							

Problem Set 8.2: After-Loaded and Direct-Loaded Muscle Contractions

Using the data from Results Table 8.2, create a graph showing the relationship between load (mass in kg) and work for both after-loaded and direct-loaded muscles. Examine Figure 8-8 before preparing your graph.

a. Compare your graph to Figure 8-8, a theoretical plot for the after- and direct-loaded gastrocnemius muscle of a frog. Does your graph look similar?

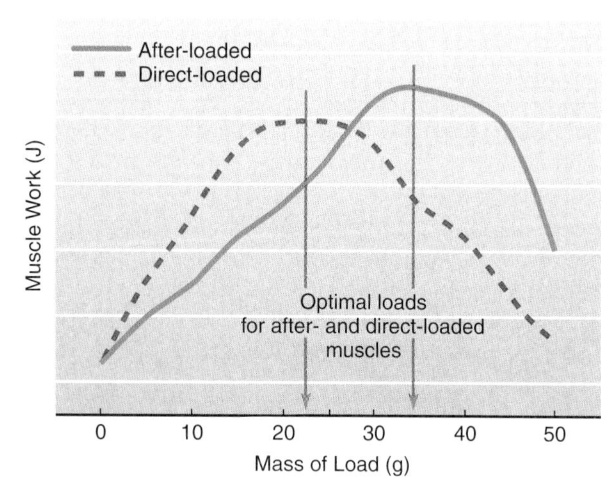

Figure 8-8 This graph shows the relationship between work and load for a frog gastrocnemius muscle and indicates the optimal load for after- and direct-loaded muscle conditions. The **optimal load** is equal to the load at which the maximum amount of work is performed by a muscle. Your relationships for after- and direct-loaded muscles should demonstrate a similar relationship.

b. Mathematically demonstrate why work is a function of load and the distance a muscle can move that load.

c. How does after-loading versus direct-loading affect the production of muscle force? Relate your results to the length–tension relationship of muscles.

d. Explain why there is a trade-off between after-loaded and direct-loaded muscles in relation to the amount of work they can perform at a given load.

LABORATORY 9

Physiology of Muscle Contraction

PURPOSE

This laboratory will introduce you to the different physiological phenomena of skeletal muscle contractions with the help of computer simulation.

Learning Objectives

- Review the properties of the length–tension relationship you investigated during Laboratory 8 (*Functional Anatomy of Muscle and Mechanics of Contraction*) using a computer simulation of a frog gastrocnemius muscle.
- Define the terms *threshold* and *maximal stimulus voltages* and understand how they relate to the generation of muscle contractions.
- Understand why increased stimulus strength increases the amount of force produced by a muscle via motor unit recruitment.
- Describe the effects of increased stimulus frequency on cross-bridge cycling and the generation of force in a muscle.
- Define the terms *wave summation* and *tetanus*.
- Describe the differences between wave summation and treppe.

Laboratory Materials

Experiment 9.1: Muscle Length–Tension Relationship

Physiology Interactive Lab Simulations (Ph.I.L.S.)

Experiment 9.2: Motor Unit Recruitment

Ph.I.L.S.

Experiment 9.3: Wave Summation and Tetanus

Ph.I.L.S.

Introduction and Pre-Lab Exercises

Like neurons, muscle fibers are subject to the threshold and maximal stimulus strength properties you learned about during Laboratory 5, *Action Potentials*. These characteristics of skeletal muscle contractions result from the neural innervation of skeletal muscle fibers. When acetylcholine is released from a motor neuron (or somatic efferent neuron) at the neuromuscular junction, it binds to acetylcholine receptors on the sarcolemma. The acetylcholine receptor is itself an ion channel; when acetylcholine binds to its receptor, it causes a conformational change in the protein receptor that opens the ion channel. Cations such as Na^+ and K^+ are then permitted to diffuse through the channel, which generates a local depolarization of the motor end plate, called an end-plate potential. (Although both Na^+ and K^+ diffusion occur through these channels, more Na^+ diffuses into the cell than K^+ out of the cell because of the differences in the electrochemical gradients across the membrane.) If the sarcolemma adjacent to the motor end plate is depolarized to the threshold voltage, voltage-gated Na^+ channels in the sarcolemma open, and the muscle generates an action potential. Action potential propagation along the muscle cell initiates the cross-bridge cycle via release of Ca^{2+} from the sarcoplasmic reticulum.

9-1 Curare is a poison that binds to acetylcholine receptors on the neuromuscular junction but does not activate the acetylcholine receptor (an antagonist). What effect does curare have on skeletal muscle contractions?

How might the effects of curare be useful in a surgical setting?

What is the cause of death in curare poisoning? Why?

■

When a large muscle like the biceps brachii muscle of the upper arm shortens, it is usually a result of a portion of the muscle fibers (muscle cells) within the muscle being stimulated. Because the proportion of muscle fibers responding to a stimulus can be modulated, each whole muscle can respond to stimuli with varying degrees of force. This process of regulating the number of muscle fibers that respond to a stimulus is called **motor unit recruitment**. A **motor unit** is defined as one somatic motor neuron and all the muscle fibers it innervates. Each motor neuron may innervate a few or many muscle fibers. Thus, the stimulation of a motor neuron results in the activation and shortening of a few or many muscle fibers. In this way, the amount of force produced by a muscle can be regulated by the number of motor units activated (Figure 9-1).

In vivo, the nervous system regulates the force of muscle contraction by altering the number of activated motor units. We can experimentally investigate the process of motor unit recruitment by directly stimulating an excised muscle with electrodes. In such an in vitro muscle preparation, the stimulus is applied to the entire muscle, and the number of muscle fibers responding to the stimulus determines the force of muscle contraction.

Recall that a stimulus must depolarize the sarcolemma to the threshold voltage to produce an action potential in the muscle fiber and initiate muscle shortening. Subthreshold stimuli will not, of course, produce any muscle shortening. Weak stimuli that are slightly greater than the threshold voltage will typically activate small numbers of muscle fibers. The activation of small numbers of muscle fibers will in turn produce small muscle forces. As the stimulus increases in strength, more and more muscle fibers will be recruited, and the force produced by the whole muscle will increase. Obviously, there is a fixed number of muscle fibers in each of your muscles. Thus, if the stimulus is strong enough to recruit all of the available muscle fibers, then the muscle will produce its maximal force. Increasing the stimulus beyond this maximal stimulus strength will not increase the force produced by the muscle because you have already recruited all of the muscle fibers.

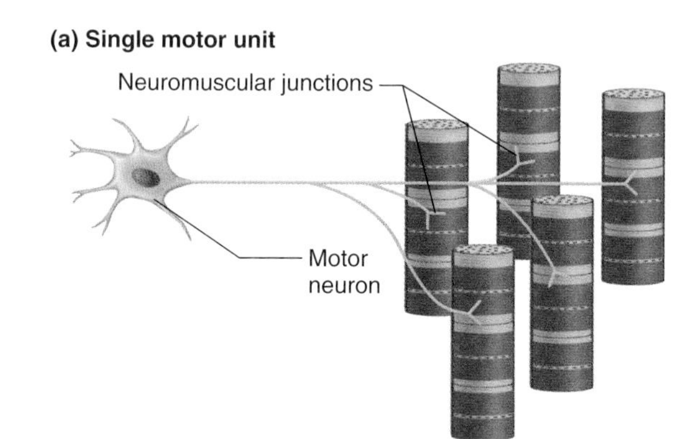

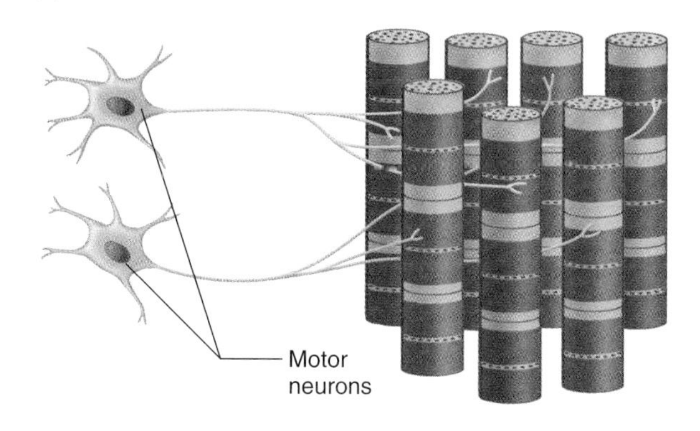

Figure 9-1 (a) Single motor unit consisting of one motor neuron and the muscle fibers it innervates. (b) Two motor units and their intermingled fibers in a muscle.

Again, it is important to make the distinction between experimentally investigating motor unit recruitment in muscle preparations and motor unit recruitment under normal physiological conditions. In these experiments, you will investigate the recruitment of muscle fibers by increasing stimulus voltage. In vivo, the nervous system regulates the number of muscle fibers activated during a contraction by regulating the number of motor units recruited. You experience the phenomenon of motor unit recruitment when you perform everyday tasks. Picking your pencil up from your desk requires only a few muscle fibers to be activated; it wouldn't be energetically conservative to activate all of your motor units to pick up such a light object. In contrast, moving your desk across the room requires the activation of many more motor units to perform the task.

9-2 The maximal stimulus voltage delivered to a muscle produces a maximal force of contraction. What happens to muscle force if the muscle is stimulated by a voltage above the maximal stimulus? Explain your reasoning and the meaning of the term *maximal stimulus*.

■

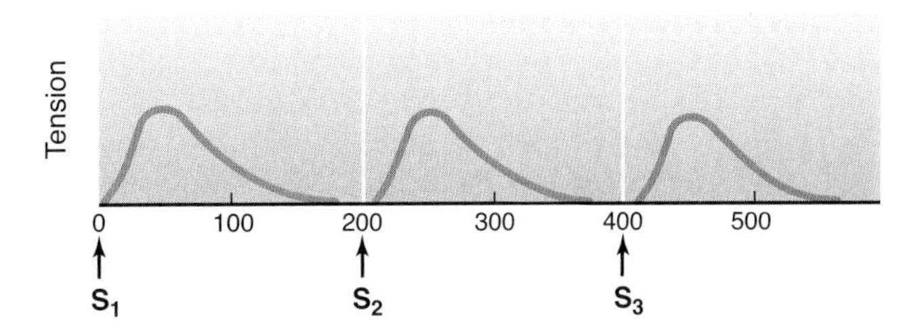

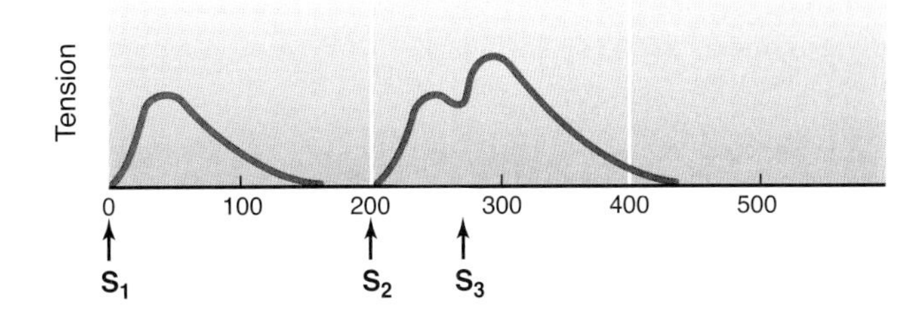

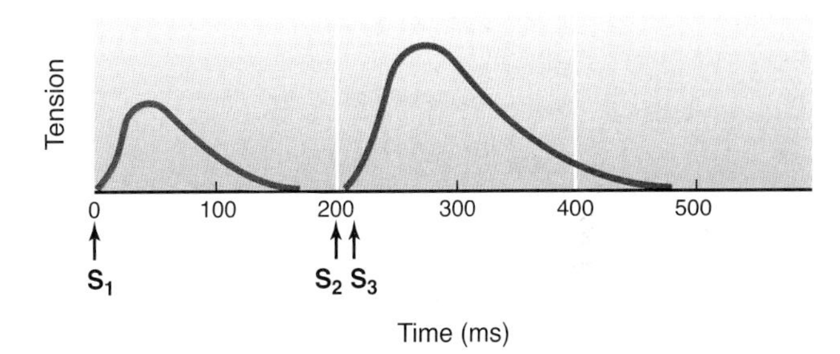

Figure 9-2 Wave summation of muscle contractions produced by increasing stimulus frequency, thus shortening the time between stimuli S_2 and S_3. Muscle tension was measured by keeping the length of the muscle constant (an isometric contraction).

The production of force by a muscle is also regulated by the frequency at which the muscle is stimulated. Remember that Ca^{2+} is released from the sarcoplasmic reticulum in response to an action potential in the muscle fiber. Calcium then binds to troponin and results in the exposure of the myosin binding sites on actin so that the cross-bridge cycle can begin. At all times, Ca^{2+} is pumped against its concentration gradient into the sarcoplasmic reticulum by Ca^{2+} pumps located in the membrane of the sarcoplasmic reticulum. These Ca^{2+}-ATPases consume large quantities of ATP for the active transport of Ca^{2+} into the sarcoplasmic reticulum. When most of the Ca^{2+} released into the cytoplasm of the muscle fiber during an action potential is returned to the sarcoplasmic reticulum, the troponin-tropomyosin complex once again blocks the myosin attachment site on actin. Thus, the cross-bridge cycle stops and the muscle returns to its relaxed state. The Ca^{2+} released from the sarcoplasmic reticulum during a single muscle action potential is almost immediately returned to the sarcoplasmic reticulum. This produces brief periods of muscle shortening, called a muscle twitch. A muscle twitch is so brief that actin and myosin filaments do not slide completely past one another, and the sarcomere is only partially shortened.

When a muscle fiber receives multiple stimuli at a high frequency, however, the Ca^{2+} pumps do not have adequate time to pump all of the Ca^{2+} back into the sarcoplasmic reticulum before the second stimulus is received. Thus, there is residual Ca^{2+} in the muscle fiber when the second stimulus induces Ca^{2+} channels to open. As a result, there is now more Ca^{2+} present in the cytoplasm than what is normally present following a single stimulus. Thus, it will take more time for the Ca^{2+} pumps to restore the cytoplasmic Ca^{2+} levels to resting concentrations. Because cytoplasmic Ca^{2+} levels remain elevated for a longer duration, there is now more time for cross-bridge cycling. Thus, when stimuli occur at a high frequency, each sarcomere will shorten more (i.e., produce more force) than if a single stimulus is applied. This increase in muscle force in response to high-frequency stimuli is called **wave summation.** Characteristics of wave summation include the presence of distinct muscle twitches, each having a greater force than the muscle twitch before it, without complete muscle fiber relaxation between the twitches (Figure 9-2).

A sustained contraction in response to repetitive stimulation is called **tetanus. Unfused tetanus** occurs at lower frequencies, when the muscle fibers only partially relax between stimuli. When stimulus frequency is so high that absolutely no muscle relaxation takes place between stimuli, a smooth, fluid muscle contraction occurs. This is called **fused tetanus.** Tetanic contractions are the types of contractions you use in everyday tasks to perform work. They are sustained, fluid contractions that have maximal muscle force (Figure 9-3).

9-3 _____ True or False? Fatigue is defined as a decrease in the maximal force produced by contracting skeletal muscles despite a continuous high-frequency stimulus. ■

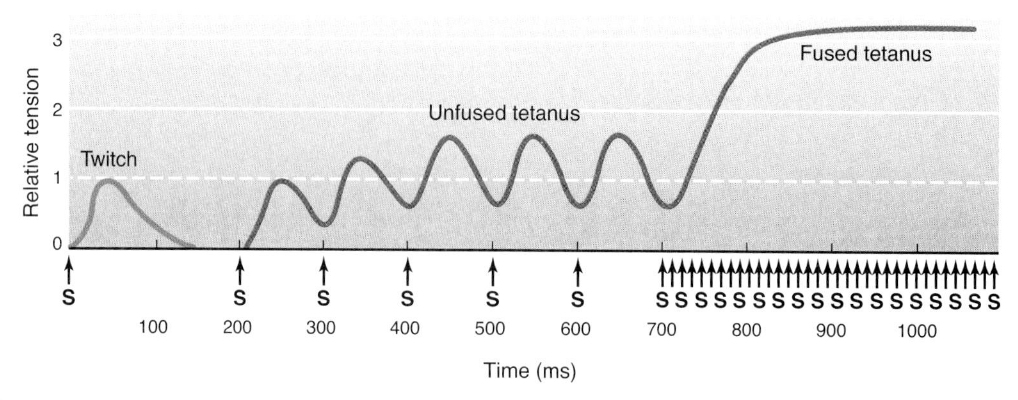

Figure 9-3 Muscle contractions produced by a single stimulus (twitch) versus multiple stimuli at frequencies of 10 stimuli per second (unfused tetanus) and 100 stimuli per second (fused tetanus). (S = stimulus)

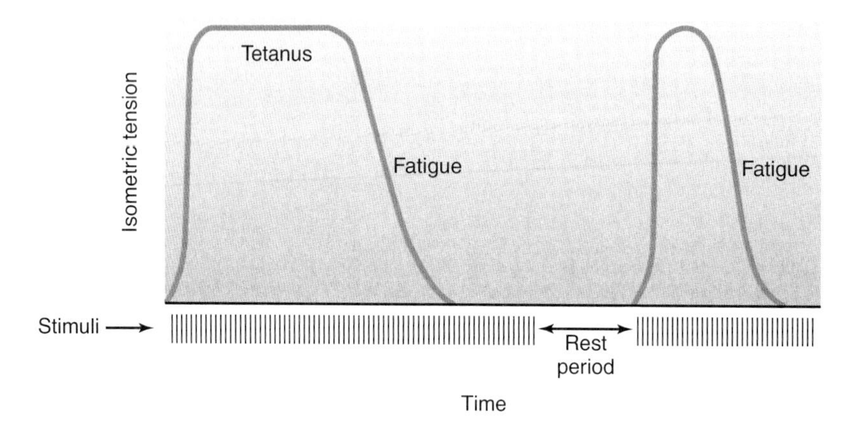

Figure 9-4 Muscle fatigue during a maintained isometric tetanus and recovery following a period of rest.

9-4 Describe what causes the condition of fatigue illustrated in Figure 9-4.

The experiments in this laboratory will introduce a computer simulation for investigating the physiology of skeletal muscle contraction. The first experiment will review the principles of the length–tension relationship you investigated in Laboratory 8. The next two experiments will introduce the principles of motor unit recruitment and wave summation, tetanus, and fatigue.

Methods and Materials

EXPERIMENT 9.1 Muscle Length—Tension Relationship

In this exercise you will again examine how different muscle lengths affect muscle force.

- Launch the Physiology Interactive Lab Simulations (Ph.I.L.S.). In the section entitled ***Skeletal Muscle Function,*** select Experiment **5.** ***The Length-Tension Relationship.***
- Read the Objectives of this simulation.
- Click the **Continue** button in this window to continue reviewing the introductory material. Be sure to click on and view all informational links presented within the *Introduction* window of this simulation. These links present helpful information and animations to assist you in the understanding of the introductory exercise.
- Once you have completed your review of the introductory material, the **Continue** button will open the next tab entitled *Pre-Lab Quiz*. You may also click the *Pre-Lab Quiz* tab directly to enter this section of the simulation. Answer all questions in the Pre-Lab Quiz to test your comprehension and understanding of the introductory material. *Note:* The printed Lab Report will indicate your correct and incorrect responses on the *Pre-Lab Quiz* and report the number of correct responses out of the total number of questions possible. The simulation will only allow one attempt at the quiz, so answer each question carefully.
- Once you have completed the *Pre-Lab Quiz*, the **Continue** button within this window will open the next tab entitled *Wet Lab*. You may also click the *Wet Lab* tab directly to enter this section of the simulation. Read the material presented and be sure to click on and view all informational links presented within the *Wet Lab* window of this simulation. These links provide valuable information and videos to introduce the apparatus and procedures necessary to conduct this experiment. *Note:* The printed Lab Report will indicate the number of video clips you viewed in the *Wet Lab* section.
- Once you have completed the *Objectives & Introduction*, *Pre-Lab Quiz*, and *Wet Lab* sections of the simulation, you are now ready to conduct the experiment. Click the **Continue** button to open the next tab entitled *Laboratory Exercise*, or click on the *Laboratory Exercise* tab directly to enter this section of the simulation. All of the experimental procedures are organized into 13 steps presented at the bottom of the simulation window. The simulation will indicate which step you are currently conducting and provide links for information about the apparatus. You may also click on **View All** to view the entire experimental procedure.

Setting Up the Apparatus

- Click on the *Data Acquisition Unit* link to learn about data acquisition and processing in this experiment.
- Click on the **Power Switch** of the Data Acquisition Unit.
- Notice that the black electrical plug (called a patch cord) is connected to the force transducer that will measure the force produced by the contracting muscle. This patch cord should be connected where on the data acquisition unit (the stimulator output or the recording input)? ______________________
- Connect the positive and negative patch cords of the stimulating electrodes to the stimulator outputs. Which color is positive? __________ Which color is negative? _________
- Prior to conducting the experiment you must click the **Zoom** button beneath the ruler. This view will remain open for the rest of the simulation.

Conducting the Experiment

- On the **Control Panel,** increase the stimulus voltage to 2.0 volts. This is the highest voltage the stimulator can deliver.
- Click the **Shock** button in the control panel to stimulate the muscle and generate a contraction.
- Record the maximum amplitude of this muscle trace by clicking on the peak of the voltage trace. Notice a black inverted triangle has marked where you have clicked. You may adjust this pointer to the highest peak of the voltage trace.
- Once you have marked the peak of the voltage trace, now click on the zero baseline of the voltage trace to determine the peak's amplitude which is a measure of muscle tension in grams (g). This measured amplitude now appears in the Data box after *AMP*.
- Now click on the **Journal** icon to record this data point in your Journal data sheet. Notice you have just collected data for muscle tension (g) produced at a muscle length of 26.0 mm.
- Close the journal window and erase the previous muscle trace by clicking on the **Erase** button.
- Lengthen the frog gastrocnemius muscle to 26.5 mm by clicking the **Lengthen Arrow** on the femur clamp apparatus.
- Click the **Shock** button to stimulate the muscle and generate a contraction. Once again, use the black pointer to obtain the data and write these data into the Journal.
- Repeat these steps and increase muscle tension in 0.5 mm increments until you complete the data sheet for each of the muscle lengths from 26.0 to 30.0 mm.
- Click on the *Post-Lab Quiz* and *Lab Report* tab. Answer the Post-Lab questions.
- Once you have completed the *Post-Lab Quiz*, a *Conclusion* will be presented for this experiment. Read the conclusion carefully and click the **Finish Lab** button.
- The *Print Your Laboratory Report* window will allow you to enter your name and course details. Click the **Print Lab** button at the bottom of the window. Once you have successfully printed your laboratory report, you have completed the laboratory simulation.

EXPERIMENT 9.2 Motor Unit Recruitment

In this exercise you will examine how stimulus strength influences the amount of force produced by a muscle and the duration of muscle contraction.

- Launch the Physiology Interactive Lab Simulations (Ph.I.L.S.). In the section entitled ***Skeletal Muscle Function,*** select Experiment **4.** ***Stimulus-Dependent Force Generation.***

Repeat the same steps found in Experiment 9.1 until the section for Setting up the Apparatus.

Setting Up the Apparatus

- In this experiment you will set up the apparatus as you did in Experiment 9.1.
- On the **Control Panel,** increase the stimulus voltage to 1.6 volts (not 1.0 volt as indicated by the instructions at the bottom of the simulation panel). Note that in this experiment the maximum voltage the simulator can deliver is 1.6 volts.
- Click the **Shock** button to stimulate the muscle and generate a contraction.
- Record the maximum amplitude of this muscle trace by clicking on the peak of the voltage trace. Notice a black inverted triangle has marked where you have clicked. You may adjust this pointer to the highest peak of the voltage trace.
- Once you have marked the peak of the voltage trace, now click on the zero baseline of the voltage trace to determine the peak's amplitude which is a measure of muscle tension in grams (g). This measured amplitude now appears in the Data box after *AMP*.
- Record the amplitude of the muscle contraction in the journal data sheet by clicking on the **Journal** icon. This will write the data into the Journal. Notice you have just collected data for muscle tension (g) produced at a stimulus amplitude of 1.6 volts.
- Close the journal window and erase the previous muscle trace by clicking on the **Erase** button.
- Decrease the stimulus amplitude in 0.1 volt-increments until you complete the data sheet. Record these data in Results Table 9.2 of the Laboratory Report which correspond with data for the duration of muscle contraction at each stimulus amplitude.'
- Click on the *Post-Lab Quiz* and *Lab Report* tab. Answer the Post-Lab questions.
- Once you have completed the *Post-Lab Quiz,* a *Conclusion* will be presented for this experiment. Read the conclusion carefully and click the **Finish Lab** button.
- The *Print Your Laboratory Report* window will allow you to enter your name and course details. Click the **Print Lab** button at the bottom of the window. Once you have successfully printed your laboratory report, you have completed the laboratory simulation.

EXPERIMENT 9.3 Wave Summation and Tetanus

In this exercise you will observe how stimulus frequency influences the amount of force produced by a muscle.

1. Launch the Physiology Interactive Lab Simulations (Ph.I.L.S.). In the section entitled ***Skeletal Muscle Function,*** select Experiment **6. *Principles of Summation and Tetanus.***

   Repeat the same steps found in Experiment 9.1 until the section for Setting up the Apparatus.

Setting Up the Apparatus

- Set up the apparatus as you did in Experiment 9.1.
- On the **Control Panel,** increase the stimulus voltage to 2.0 volts. The stimulator will continually administer a stimulus every 500 ms to the muscle when you start the simulation.
- Review the information presented within the *Summation link* located at the bottom of the simulation screen.
- Click the **Start** button to stimulate the muscle and generate muscle contractions. Notice that a stimulus is applied every 500 ms to the muscle. Each stimulus produces a distinct and separate muscle twitch.
- Decrease the interval between each stimulus thus stimulating the muscle at a higher frequency until you observe summation.
- Click the **Journal** icon and record the stimulus interval at which you first observe summation as presented in the *Summation* link. This is incomplete relaxation of the muscle between stimuli. You may wish to print the muscle summation trace by clicking the Print Button (**P**) located in the bottom left corner of the virtual monitor.
- Record the stimulus interval at which the muscle first demonstrates summation. Interval = ______
- Now review the information presented within the *Incomplete Tetanus* link located at the bottom of the simulation screen.
- Click the **Start** button to stimulate the muscle and generate muscle contractions.
- Again decrease the interval between the each stimulus thus stimulating the muscle at a higher frequency until you observe incomplete tetanus.
- Click the **Journal** icon and record the stimulus interval at which you first observe incomplete tetanus as presented in the *Incomplete Tetanus* link. You may wish to print the incomplete tetanus trace by clicking the Print Button (**P**) located in the bottom left corner of the virtual monitor.
- Record the stimulus interval at which the muscle first demonstrates incomplete tetanus. Interval = ______
- Now review the information presented within the *Complete Tetanus* link located at the bottom of the simulation screen.
- Click the **Start** button to stimulate the muscle and generate muscle contractions.
- Again decrease the interval between the each stimulus thus stimulating the muscle at a higher frequency until you observe complete tetanus.
- Click the **Journal** icon and record the stimulus interval at which you first observe incomplete tetanus as presented in the *Complete Tetanus* link. You may wish to print the incomplete tetanus trace by clicking the Print Button (**P**) located in the bottom left corner of the virtual monitor.
- Record the stimulus interval at which the muscle first demonstrates complete tetanus. Interval = ________
- Click on the *Post-Lab Quiz* and *Lab Report* tab. Answer the Post-Lab questions.
- Once you have completed the *Post-Lab Quiz,* a *Conclusion* will be presented for this experiment. Read the conclusion carefully and click the **Finish Lab** button.
- The *Print Your Laboratory Report* window will allow you to enter your name and course details. Click the **Print Lab** button at the bottom of the window. Once you have successfully printed your laboratory report, you have completed the laboratory simulation.

RESULTS AND DISCUSSION

LABORATORY REPORT 9

Problem Set 9.1: Muscle Length–Tension Relationship

How do the results from the length–tension simulation exercise in a frog gastrocnemius muscle compare to the results you collected in the length–tension relationship of the digital flexors during Laboratory 8? Which frog gastrocnemius muscle lengths correspond to the different wrist positions?

EXPERIMENT 9.2

Motor Unit Recruitment

Results Table 9.2 Individual Group Data for Experiment 9.2: Motor Unit Recruitment

		Calculation of Contraction Duration		
Stimulus Voltage (V)	Muscle Tension (g)	Ending Time (sec)	Beginning Time (sec)	Contraction Duration (sec)
0.0		0.00	0.00	
0.1		0.00	0.00	
0.2		0.00	0.00	
0.3		0.00	0.00	
0.4		0.78	0.11	
0.5		0.80	0.12	
0.6		0.84	0.11	
0.7		0.89	0.11	
0.8		0.89	0.12	
0.9		0.99	0.12	
1.0		1.04	0.11	
1.1		1.16	0.12	
1.2		1.23	0.12	
1.3		1.34	0.11	
1.4		1.44	0.12	
1.5		1.47	0.12	
1.6		1.49	0.11	

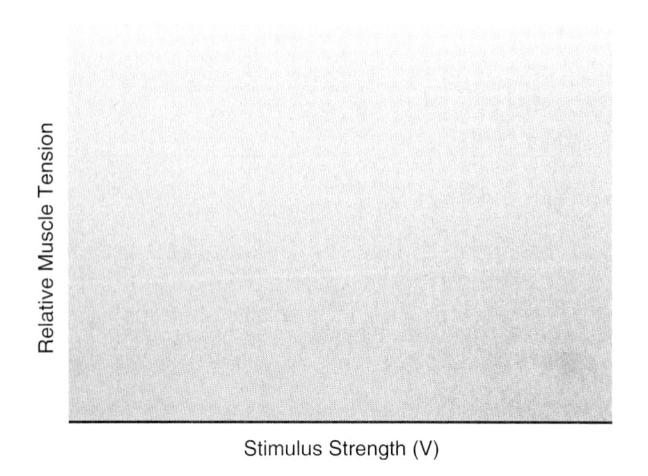

Figure 9-5 Relationship between muscle tension or force and stimulus strength (mV).

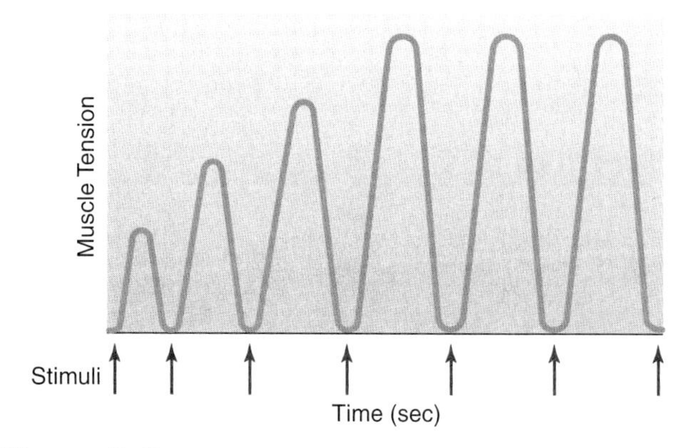

Figure 9-6 A computer trace of muscle tension over time. Arrows indicate stimulus application; stimulus strength held constant.

Problem Set 9.2: Motor Unit Recruitment

Using the data from Results Table 9.2, create a graph showing the relationship between stimulus voltage (on the x-axis) and the duration of muscle contraction (on the y-axis). Using these data, perform the appropriate statistical test to examine the relationship between stimulus voltage and muscle contraction duration and answer the following questions:

a. State the null hypothesis.

b. Should you accept or reject the null hypothesis? Support your conclusions with the appropriate statistical results.

c. Use your knowledge of cross-bridge cycling in a muscle contraction to explain why we observed an increase in the duration of muscle contraction with an increase in stimulus voltage.

d. In question 3 of the simulation, you are asked to address the effects of high stimulus voltage on the contraction force of a muscle. What values of force would you predict for a voltage of 1.7 to 2.0 volts? Make a sketch in Figure 9-5 of what these data would look like and explain your answer.

Problem Set 9.3: Wave Summation and Tetanus

If a muscle is repetitively stimulated with a maximum voltage so that each stimulus elicits an individual muscle twitch, each successive twitch will produce a greater muscle force than the previous twitch (up to a maximal muscle force). This phenomenon is called **treppe.** The increase in muscle force during treppe is due to residual Ca^{2+} in the muscle fiber and is known as the staircase or warm-up effect. Because muscle contractions use large amounts of ATP, muscle fibers contribute significantly to the production of body heat. As the muscle fibers shorten, they produce small increases in heat, which in turn helps enzymes to function more efficiently (recall the principles of enzyme kinetics you learned during Laboratory 4).

Is the trace of muscle contractions shown in Figure 9-6 an example of treppe or wave summation? Support your answer by describing the differences between treppe and wave summation muscle traces.

COMPARATIVE NOTE *Laboratory 9*

Muscles in Flight

Many flying insects, such as dragonflies, moths, butterflies, and grasshoppers, have relatively low wing beat frequencies. In such insects, each muscle contraction occurs in response to a single nerve impulse or action potential. This type of muscle, in which each contraction is synchronized with a single nerve impulse, is called synchronous muscle. In insects with very high wing beat frequencies (bees, flies, and mosquitoes), wing beats occur as fast as 100 to more than 1000 beats per second. This frequency is much too high for each muscle contraction to be controlled by a single nerve impulse, and so a different mechanism is employed. High wing beat frequencies are produced by asynchronous muscles, in which the frequency of nerve impulses is slower than the frequency of muscle contractions. In asynchronous muscles, each nerve impulse corresponds to many muscle contractions, in some instances as many as 40 contractions per action potential.

High wing beat frequencies are produced by two sets of muscles, neither of which actually attach to the wings. The first set, called the vertical muscles, run vertically from the dorsal surface of the thorax to the ventral surface of the thorax. The second set of muscles, the longitudinal or horizontal muscles, run longitudinally through the thorax from anterior to posterior. When these muscles contract, the primary effect of muscle contraction is distortion of the elastic thorax. Contraction of the vertical muscles, for example, causes distortion of the thorax so that the wings are moved upward (the upstroke of the wing beat cycle). Distortion of the thorax in turn causes stretching of the longitudinal muscles, which contract in response to the stimulus of being stretched. Contraction of the longitudinal muscles also distorts the thorax, pushing the wings into the downstroke as well as stretching the vertical muscles. The vertical muscles contract in response to being stretched, and a new wing beat cycle is begun. Although many wing beat cycles may occur in response to one action potential, nerve impulses are needed to initiate and maintain continued flight.

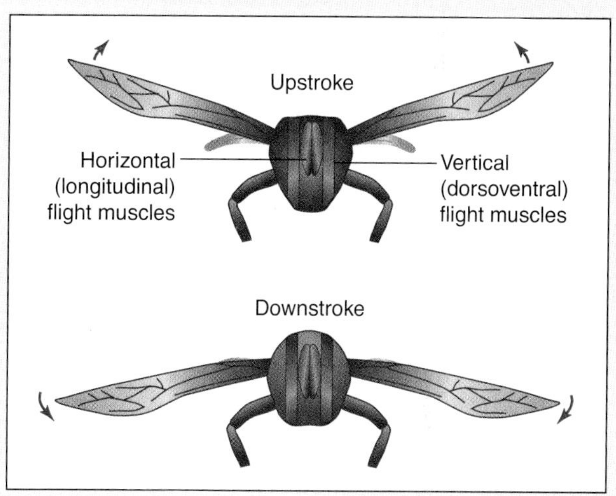

The flight muscles and elastic thorax from an oscillating system allow for faster wing beat frequency than nerve impulse frequency.

RESEARCH OF INTEREST

Agianian, B., U. Krzic, F. Qiu, W.A. Linke, K. Leonard, and B. Bullard. 2004. A troponin switch that regulates muscle contraction by stretch instead of calcium. European Molecular Biology Organization Journal 23:772–779.

Nongthomba, U., S. Clark, M. Cummins, M. Ansari, M. Stark, and J.C. Sparrow. 2004. Troponin I is required for myofibrillogenesis and sarcomere formation in *Drosophila* flight muscle. Journal of Cell Science 117:1795–1805.

Nongthomba, U., M. Cummins, S. Clark, J.O. Vigoreaux, and J.C. Sparrow. 2003. Suppression of muscle hypercontraction by mutations in the myosin heavy chain gene of *Drosophila melanogaster.* Genetics 164:209–222.

Pringle, J.W.S. 1949. The excitation and contraction of the flight muscles of insects. Journal of Physiology 108:226–232.

Schilder, R.J. and J.H. Marden. 2004. A hierarchical analysis of the scaling of force and power production by dragonfly flight motors. Journal of Experimental Biology 207:767–776.

LABORATORY 10

Endocrine Physiology

PURPOSE

The purpose of this laboratory is to introduce you to the principles of the endocrine system. The influence of thyroxine on ventilation rate in goldfish will be investigated experimentally. The principles of the negative feedback systems that regulate the hormones produced by the hypothalamus and pituitary will also be discussed.

Learning Objectives

- Define the term *hormone.*
- Understand how hormones exert effects on their target sites.
- Know the major functions of the thyroid hormones.
- Understand why thyroxine influences the ventilation rate in goldfish.
- Learn the different hormones that are produced by the hypothalamus and pituitary gland.
- Understand how these hypothalamic and pituitary hormones are regulated by negative feedback systems.
- Learn about traditional hormone manipulation experiments.

Laboratory Materials

Experiment 10.1: The Influence of Thyroxine on Ventilation Rate: Tonicity and Osmosis Experiment

1. 12 goldfish treated with thyroxine for 6 days (0.023 mg thyroxine per liter of aquarium water)
2. 12 goldfish treated with propylthiouracil for 6 days (30 mg propylthiouracil per liter of aquarium water)
3. 12 control goldfish not treated with either drug
4. Testing chamber (e.g., 600 ml beaker)

Experiment 10.2: Experimental Hormone Manipulations in Rats

Hormone manipulation cards showing rat necropsy data

Introduction and Pre-Lab Exercises

The endocrine system is extremely important to the maintenance of homeostasis in the body. Like the nervous system, the endocrine system functions in communication by sending chemical messengers (i.e., hormones) to all cells in the body. Because the nervous and endocrine systems have similar functions and are intricately related, it is likely that these two systems co-evolved. Even organisms with the simplest nervous systems (like the jellyfish with a "nerve net") also have endocrine systems. In addition, some neurotransmitters can function as hormones and some hormones can function as neurotransmitters. Such experimental observations have fueled the rapidly expanding field of neuroendocrinology.

An **endocrine** gland is defined as a ductless gland whose secretory products enter the circulatory system after being secreted into the extracellular fluid. This definition was developed to emphasize the differences between the endocrine system and the **exocrine** system, which consists of glands that release their secretions into a duct. An example of an exocrine gland is the pancreas, which releases its exocrine secretions (digestive enzymes) into the pancreatic duct. The pancreatic duct then carries these enzymes to the beginning of the small intestine. (Note that the pancreas is also an endocrine gland that produces the hormones insulin and glucagon.)

Traditionally, a hormone has been defined as a chemical messenger that enters the blood and travels to a distant target site. However, because of the advances of neuroendocrinology, this traditional definition is no longer precise. For example, some hormones, when released from their endocrine glands, do not travel to a distant target site but act directly on their own endocrine glands to regulate their secretory activity. In such instances, a hormone secreted into the extracellular fluid may act upon the cells that secreted it (**autocrine stimulation**) or upon other cells near the secretion site (**paracrine stimulation**).

A hormone exerts its actions on its target cell by binding to a specific receptor located either on the plasma membrane of the cell or inside the cell. The ability of a cell to respond to a hormone depends on whether or not that cell has receptors for the hormone. For example, calcitonin is a hormone released by the C-cells of the thyroid gland and functions in maintaining normal levels of blood calcium. When blood calcium levels are too high, calcitonin is released. Calcitonin inhibits the activity of osteoclasts, specialized cells that resorb the calcified matrix of bone. Thus, calcitonin favors bone deposition, and its net effect is to decrease blood calcium to homeostatic levels. Osteoclasts respond to calcitonin because they have receptors on their cell membranes that are specific for calcitonin. In contrast, osteoblasts (bone-producing cells) are not affected by calcitonin because they lack calcitonin receptors.

Most hormones influence a variety of tissues in the body, and most cells contain many different hormone receptors. Thus, the ultimate response of a cell or tissue to a hormone often depends on the presence or absence of other hormones. All of the inputs to a cell are integrated into the final response of the cell.

10-1 Complete the following table of hormone target sites and actions. Be sure to fill in the table carefully and accurately because you will use this table as a resource for Experiment 10.2 (Experimental Hormone Manipulations in Rats). Note that some glands (e.g., the pancreas) produce more than one hormone, and therefore more than one correct answer is possible.

Hormone	Secreted by	Action
	Pancreas	
Atrial natriuretic peptide (ANP)		
		Stimulates the anterior pituitary to secrete thyroid-stimulating hormone (TSH)
Thyroid hormones: Thyroxine (T_4) and Triiodothyronine (T_3)		Thyroid hormones influence most physiological processes and are often required for (or are permissive to) the actions of other hormones. For example, thyroid hormones are required both for the production of growth hormone and for its systemic actions. Among many other physiological actions, thyroid hormones indirectly stimulate mitochondrial oxidative phosphorylation and therefore regulate basal metabolic rate. Typically, the higher the concentration of thyroid hormones, the higher a person's metabolic rate. Hypo- and hyperthyroidism are common diseases that are directly related to having too little or too much thyroid hormone, respectively. Thyroid hormones also exert feedback inhibition on both pituitary TSH and hypothalamic TRH secretion.
Adrenocorticotropic hormone (ACTH)		
Cortisol		Cortisol stimulates the breakdown of proteins and lipids and helps the body adapt to stress. Cortisol can also function as an anti-inflammatory drug and can cause a decrease in immune system function. Thus, cortisol can affect organs that are necessary for immune system function. The thymus, which is the site of T-lymphocyte differentiation, is one example. Cortisol regulates ACTH release through negative feedback inhibition.
Luteinizing hormone (LH)		
	Leydig cells	This hormone is responsible for the development of the secondary sex characteristics in males and functions to maintain the male reproductive system, especially the seminal vesicles and prostate. This hormone also inhibits the production of LH through negative feedback inhibition.
Estradiol		
	Pineal gland	
Erythropoietin		
Vasopressin or antidiuretic hormone (ADH)	Posterior pituitary	

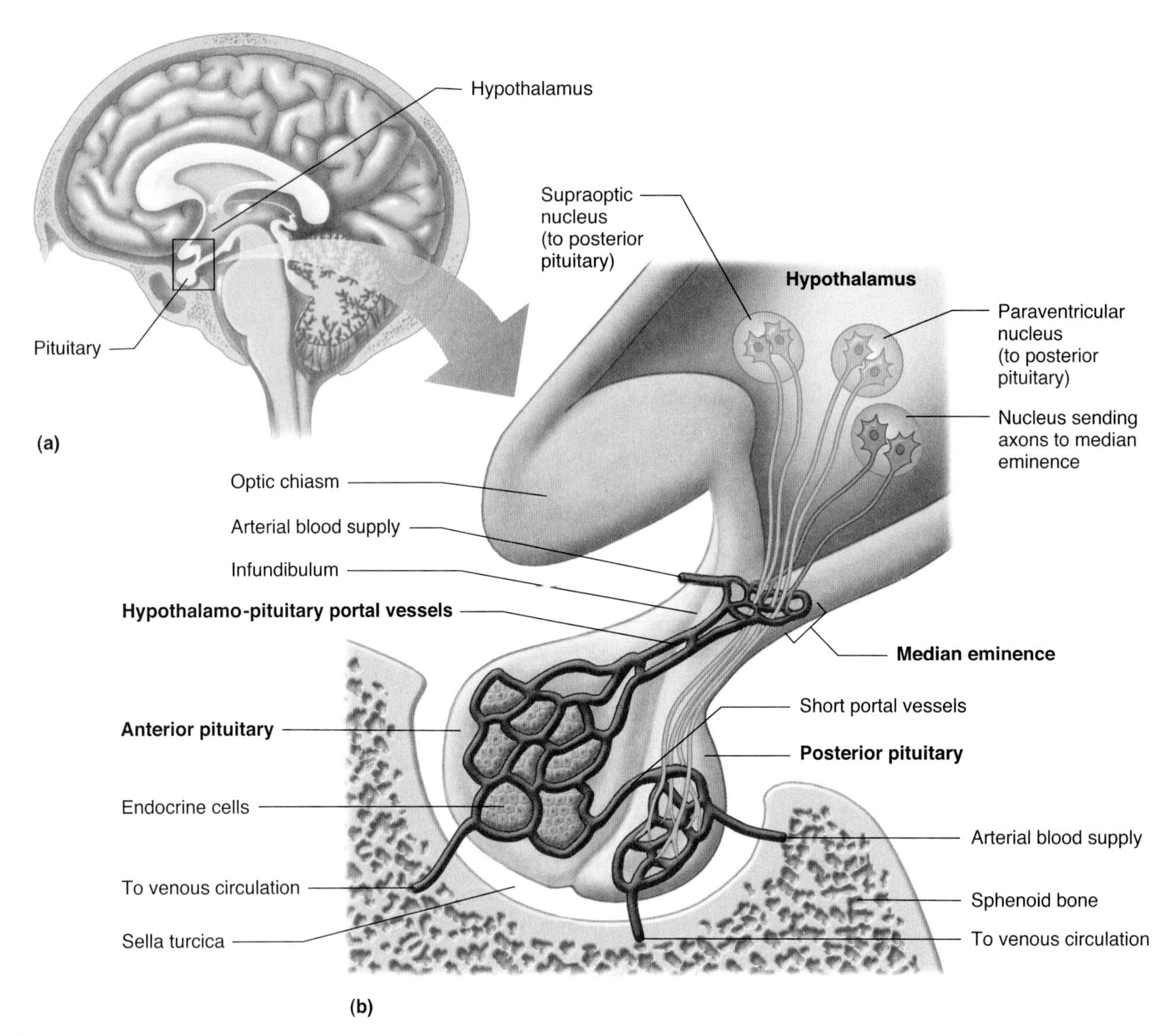

Figure 10-1 Diagram showing (a) the location of the hypothalamus and pituitary gland in the brain and (b) the neural and vascular connections between the hypothalamus and pituitary gland.

The endocrine system contributes to the maintenance of homeostasis mainly through the use of antagonistic systems or **negative feedback loops.** In a traditional negative feedback loop, the final product of a system feeds back on the prior system components to inhibit their activity or secretion. In this manner, most hormones are maintained at fairly constant concentrations. Note, however, that many hormones, such as growth hormone and gonadotropin-releasing hormone, are secreted in short bursts, with little or no hormone being secreted between bursts. In addition, the secretion of some hormones (cortisol and melatonin) shows distinct 24-hour (circadian) cycles.

A series of well-described negative feedback loops regulates the secretion of hormones from the hypothalamus and pituitary gland in the brain (Figure 10-1). For example, the hypothalamus releases thyrotropin-releasing hormone (TRH), which stimulates the anterior pituitary gland to release thyroid-stimulating hormone (TSH) into the blood. TSH then stimulates the thyroid gland to produce thyroid hormones (thyroxine, T_4, and triiodothyronine, T_3). When thyroid hormone concentrations in the blood become elevated, thyroid hormones will negatively regulate their own production by inhibiting the secretion of TRH and TSH from the hypothalamus and pituitary gland, respectively (Figure 10-2). Again, remember that to respond to thyroid hormones, both the hypothalamus and pituitary gland must have thyroid hormone receptors.

 **10-2** Negative feedback loops maintain hormone levels within homeostatic boundaries.

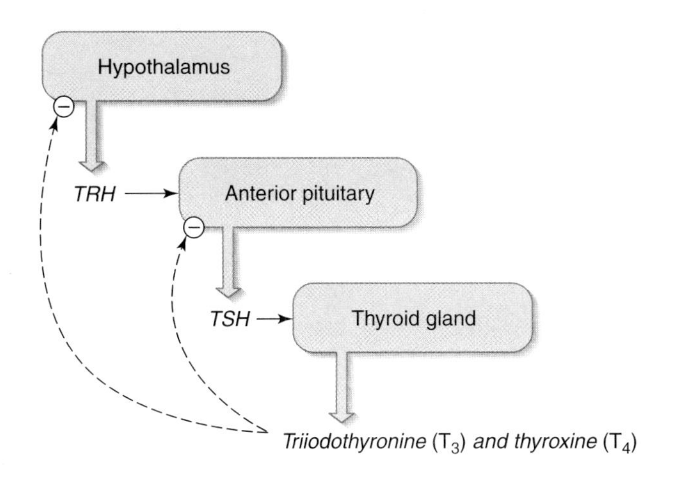

Figure 10-2 Negative feedback loop regulating the production of thyroid hormones by the thyroid gland.

Many diseases and/or dietary deficiencies can upset the balance of negative feedback systems, producing abnormally high (hyper-) or low (hypo-) hormone levels. One form of abnormally low thyroid hormone levels, or **hypothyroidism,** results from a deficiency of iodine, an element that is necessary for the synthesis of thyroid hormones. Advanced stages of hypothyroidism result in the growth (**hypertrophy**) of the thyroid gland, producing large goiters (Figure 10-3). Using the negative feedback loop for thyroid hormones, explain why iodine deficiency results in goiters.

■

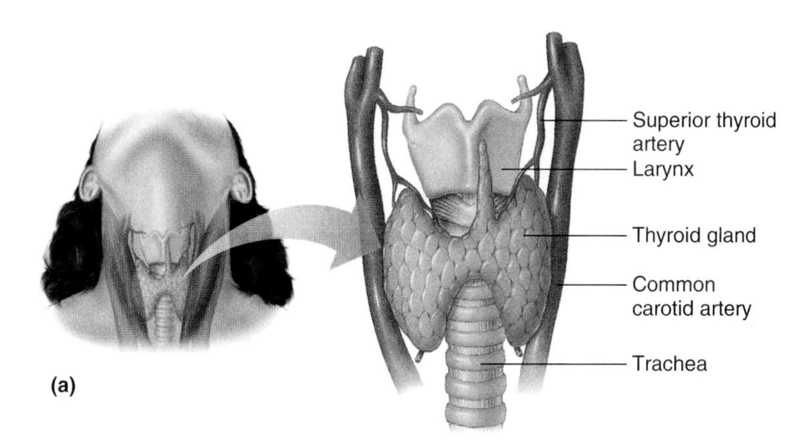

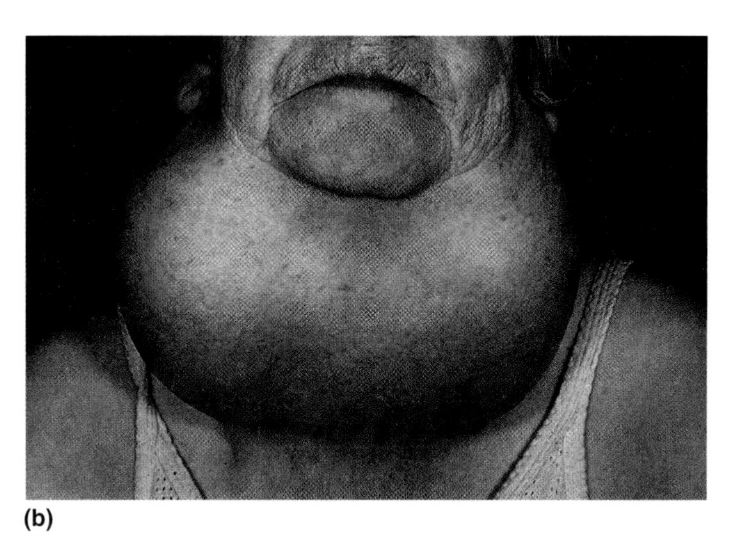

Figure 10-3 Gross anatomy of the thyroid gland (a) and the hypertrophy of the gland forming an advanced-stage goiter (b).

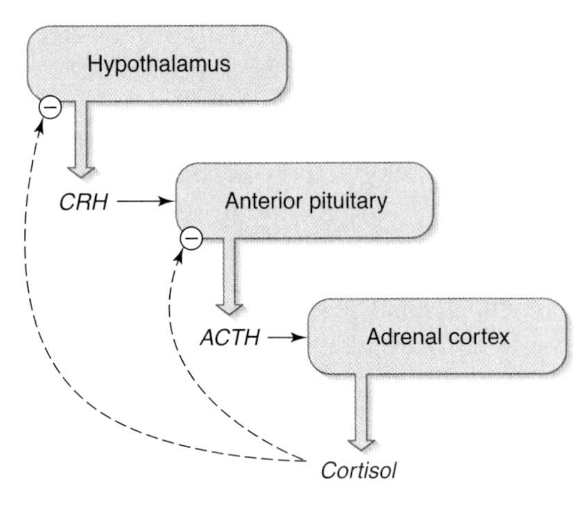

Figure 10-4 Negative feedback loop regulating the production of cortisol by the adrenal cortex.

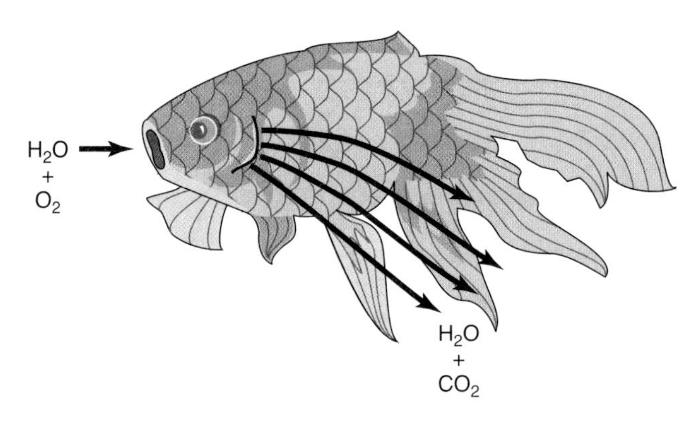

Figure 10-5 Rate of opercular opening and closing as a measure of metabolic rate in the goldfish.

A second example of an antagonistic system is the negative feedback loop that regulates the production of cortisol by the adrenal glands. The hypothalamus produces corticotropin-releasing hormone (CRH), which stimulates the pituitary gland to release adrenocorticotropic hormone (ACTH) into the blood. ACTH then stimulates the adrenal cortex to produce cortisol. The hormone cortisol modifies rates of protein and lipid metabolism to mobilize energy stores. When cortisol concentrations in the blood become elevated, cortisol will negatively regulate its own production by inhibiting the secretion of CRH and ACTH from the hypothalamus and pituitary gland, respectively (Figure 10-4).

10-3 Sketch the negative feedback loop that controls the secretion of growth hormone and insulin-like growth factor.

In this laboratory, you will examine the influence of thyroxine and propylthiouracil on metabolism in goldfish (*Carassius auratus*). Propylthiouracil inhibits the production of thyroid hormones by the thyroid gland. This fish species is an excellent experimental model because it is easily accessible, inexpensive, and morphologically suitable for investigating respiration and metabolism by direct observation of its ventilation rate.

Most bony fishes such as goldfish ventilate a continuous flow of water through their gills by a synchronous expansion and contraction of their buccal and opercular cavities (Figure 10-5). The rate at which this continuous flow of water is passed through their gills is dependent on the ventilation rate, or the number of buccal and opercular cycles of expansion and contraction per minute. As you know, metabolic demand for oxygen will determine respiratory rate. From Laboratory 2 ("Homeostasis"), recall how exercise increased the metabolic demand for oxygen and thus increased both heart and breathing rates. Similarly, we can estimate the goldfish's metabolism, or metabolic demand for oxygen, by directly observing the number of ventilations made per minute.

Methods and Materials

EXPERIMENT 10.1 The Influence of Thyroxine on Ventilation Rate

In this experiment you will examine the influence of thyroxine and propylthiouracil on the ventilation rate of goldfish. Ventilation rate is directly related to metabolic rate. You will collect three measurements of ventilation rate for each of two control-treated goldfish.

- Fill your testing chamber (for example, a 600 ml beaker) with water from the control treatment aquarium.
- Obtain a control-treated goldfish from this treatment aquarium and place the fish in your testing chamber.
- Allow the fish to habituate to its new container for five minutes.
- Record the number of ventilation cycles (closing and opening of the mouth) that occur during one minute.
- Repeat the previous step to obtain two additional trials of ventilation rate. Record these data in Results Table 10.1a of your Laboratory Report.
- When you have finished measuring the ventilation rate of the first goldfish, place the fish in a designated post-experiment holding aquarium for control-treated fish (this will avoid repeated sampling of the same fish).
- Obtain a second control-treated goldfish, allow the fish to habituate for five minutes, and record three measures of ventilation rate.
- When you have finished measuring the ventilation rate of this second goldfish, again place the fish in the designated post-experiment holding aquarium for control-treated fish. Empty your testing chamber into the sink and rinse it thoroughly.

You will now collect three measurements of ventilation rate for each of two thyroxine-treated goldfish.

- Follow the steps listed previously to obtain similar measurements of ventilation rate for thyroxine-treated fish.
- After experimentation, be sure to place these fish in the designated post-experiment holding aquarium for thyroxine-treated fish (this will avoid repeated sampling of the same fish).

Now collect three measurements of ventilation rate for each of two propylthiouracil-treated goldfish.

- Follow the steps listed previously to obtain similar measurements of ventilation rate for propylthiouracil-treated fish.
- After experimentation, be sure to place these fish in the designated post-experiment holding aquarium for propylthiouracil-treated fish (this will avoid repeated sampling of the same fish).

EXPERIMENT 10.2 Experimental Hormone Manipulations in Rats

In this puzzle-like exercise, you will examine a series of data sets collected from hormone manipulation experiments in rats (from Odenweller *et al.* 1997).

- You have been conducting an experiment for 10 weeks to investigate the effects of different hormones on the production of neurotransmitters in the brain.
- During this 10-week experiment, you have treated groups of male rats with hormones for approximately two weeks. For each of the six different hormones you are investigating, you also treated a group of castrated rats to investigate whether an intact reproductive system influences the results. Castration is the surgical removal of the testes to eliminate virtually all androgen (e.g., testosterone) production.
- On the day you have scheduled the necropsies for one of your experimental trials, you enter the Animal Facilities Laboratory to find that a technician has washed the cages and has accidentally mixed up the treatment identification cards.
- You perform the necropsies as planned, but you must now try to identify which pair of rats was treated with which hormone. This can be done by examining how these hormones influenced the body mass of rats and the size of different endocrine glands
- Using the experimental data listed on the hormone manipulation cards, match each rat pair with the appropriate hormone treatment. Recall that continuous stimulation of a tissue will cause it to **hypertrophy** (enlarge), and a lack of stimulation will cause **atrophy** (decrease in size).

- The six possible hormone treatments are:
  - testosterone
  - luteinizing hormone
  - thyrotropin-releasing hormone
  - thyroid-stimulating hormone
  - cortisol
  - adrenocorticotropic hormone
- Differences between control rats and hormone-treated rats greater than 20% should be considered significantly different.

  Formula for Calculating Percent Change:

$$\frac{\text{Experimental Value} - \text{Control Value}}{\text{Control Value}} \times 100$$

- Fill in Results Table 10.2 in the Laboratory Report to help you identify which rats were treated with which hormone.

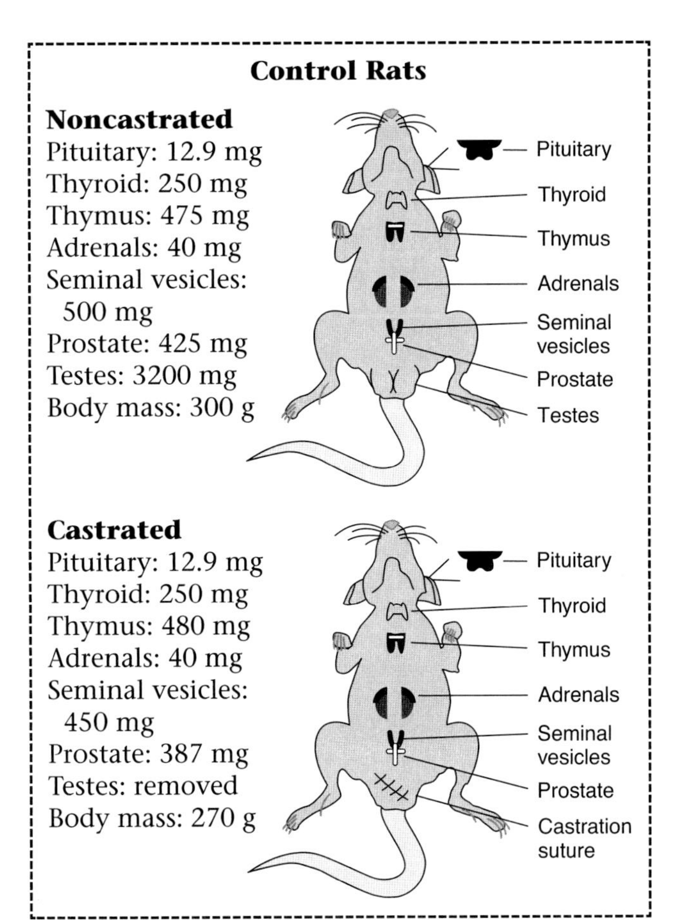
Control Rats
Noncastrated
Pituitary: 12.9 mg
Thyroid: 250 mg
Thymus: 475 mg
Adrenals: 40 mg
Seminal vesicles: 500 mg
Prostate: 425 mg
Testes: 3200 mg
Body mass: 300 g
Pituitary
Thyroid
Thymus
Adrenals
Seminal vesicles
Prostate
Testes
Castrated
Pituitary: 12.9 mg
Thyroid: 250 mg
Thymus: 480 mg
Adrenals: 40 mg
Seminal vesicles: 450 mg
Prostate: 387 mg
Testes: removed
Body mass: 270 g
Pituitary
Thyroid
Thymus
Adrenals
Seminal vesicles
Prostate
Castration suture

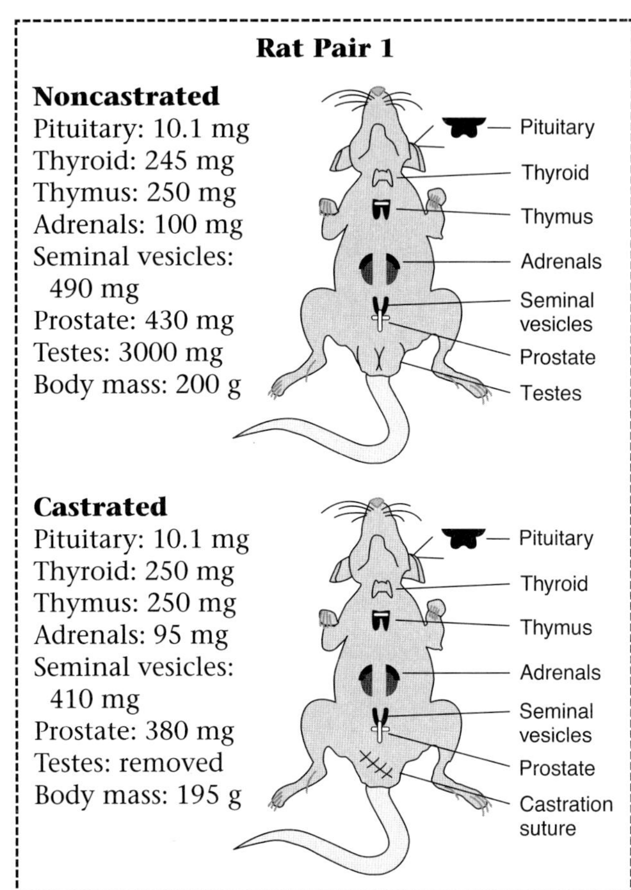
Rat Pair 1
Noncastrated
Pituitary: 10.1 mg
Thyroid: 245 mg
Thymus: 250 mg
Adrenals: 100 mg
Seminal vesicles: 490 mg
Prostate: 430 mg
Testes: 3000 mg
Body mass: 200 g
Pituitary
Thyroid
Thymus
Adrenals
Seminal vesicles
Prostate
Testes
Castrated
Pituitary: 10.1 mg
Thyroid: 250 mg
Thymus: 250 mg
Adrenals: 95 mg
Seminal vesicles: 410 mg
Prostate: 380 mg
Testes: removed
Body mass: 195 g
Pituitary
Thyroid
Thymus
Adrenals
Seminal vesicles
Prostate
Castration suture

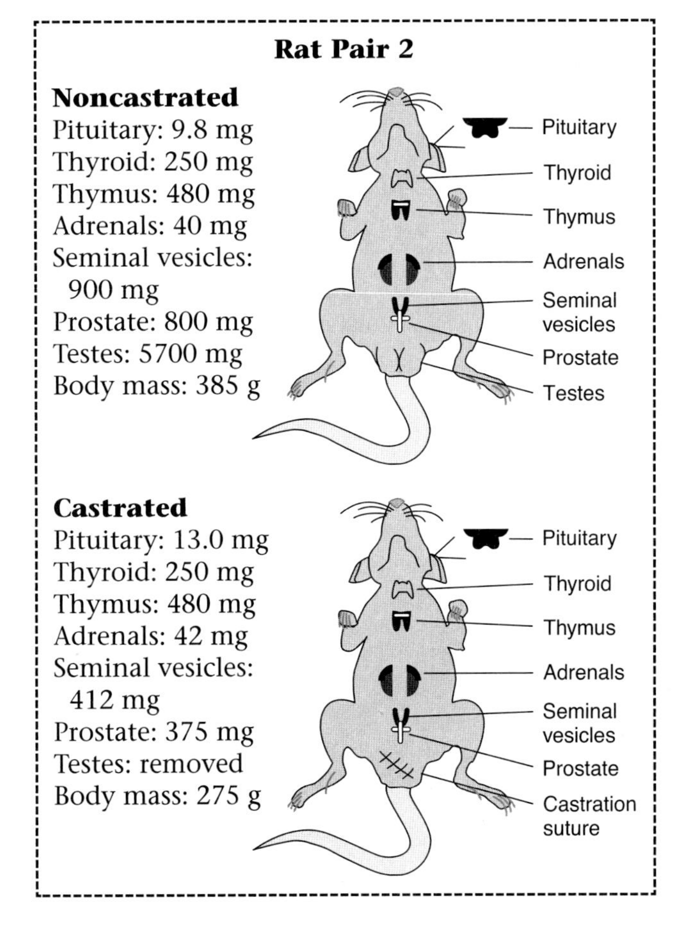
Rat Pair 2
Noncastrated
Pituitary: 9.8 mg
Thyroid: 250 mg
Thymus: 480 mg
Adrenals: 40 mg
Seminal vesicles: 900 mg
Prostate: 800 mg
Testes: 5700 mg
Body mass: 385 g
Pituitary
Thyroid
Thymus
Adrenals
Seminal vesicles
Prostate
Testes
Castrated
Pituitary: 13.0 mg
Thyroid: 250 mg
Thymus: 480 mg
Adrenals: 42 mg
Seminal vesicles: 412 mg
Prostate: 375 mg
Testes: removed
Body mass: 275 g
Pituitary
Thyroid
Thymus
Adrenals
Seminal vesicles
Prostate
Castration suture

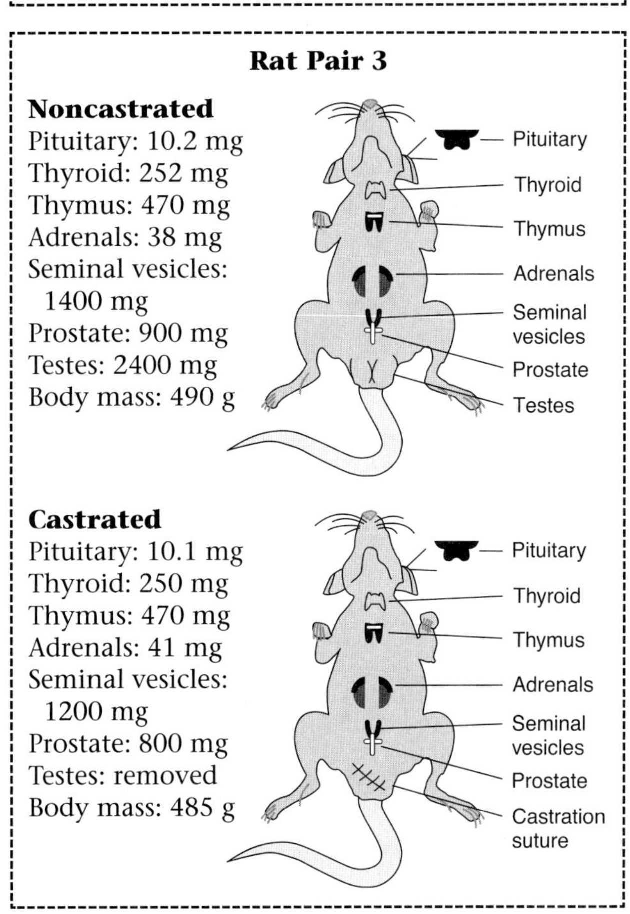
Rat Pair 3
Noncastrated
Pituitary: 10.2 mg
Thyroid: 252 mg
Thymus: 470 mg
Adrenals: 38 mg
Seminal vesicles: 1400 mg
Prostate: 900 mg
Testes: 2400 mg
Body mass: 490 g
Pituitary
Thyroid
Thymus
Adrenals
Seminal vesicles
Prostate
Testes
Castrated
Pituitary: 10.1 mg
Thyroid: 250 mg
Thymus: 470 mg
Adrenals: 41 mg
Seminal vesicles: 1200 mg
Prostate: 800 mg
Testes: removed
Body mass: 485 g
Pituitary
Thyroid
Thymus
Adrenals
Seminal vesicles
Prostate
Castration suture

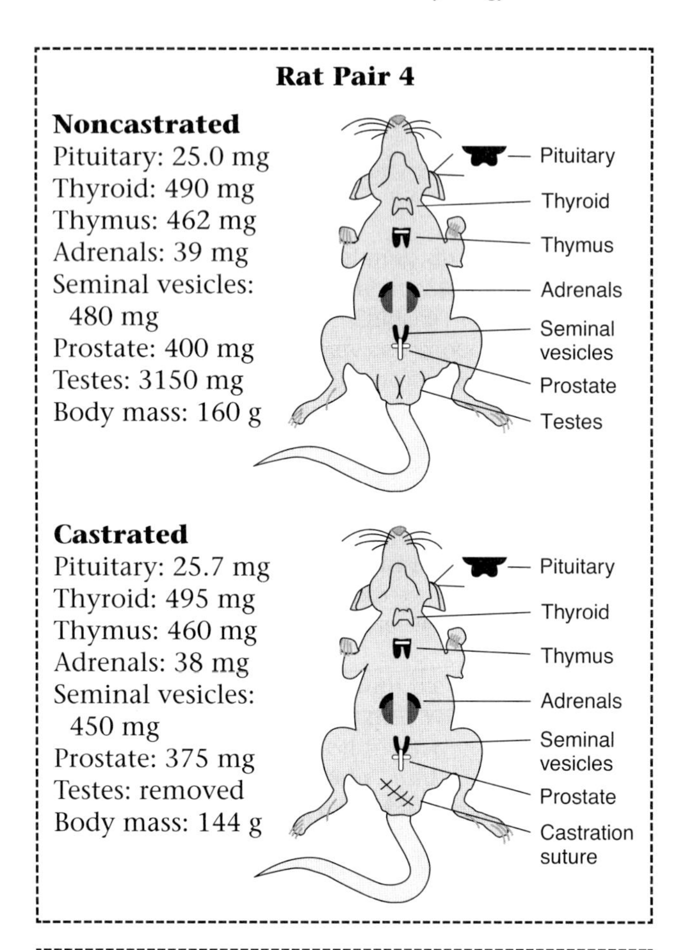
Rat Pair 4
Noncastrated
Pituitary: 25.0 mg
Thyroid: 490 mg
Thymus: 462 mg
Adrenals: 39 mg
Seminal vesicles: 480 mg
Prostate: 400 mg
Testes: 3150 mg
Body mass: 160 g
Pituitary
Thyroid
Thymus
Adrenals
Seminal vesicles
Prostate
Testes
Castrated
Pituitary: 25.7 mg
Thyroid: 495 mg
Thymus: 460 mg
Adrenals: 38 mg
Seminal vesicles: 450 mg
Prostate: 375 mg
Testes: removed
Body mass: 144 g
Pituitary
Thyroid
Thymus
Adrenals
Seminal vesicles
Prostate
Castration suture

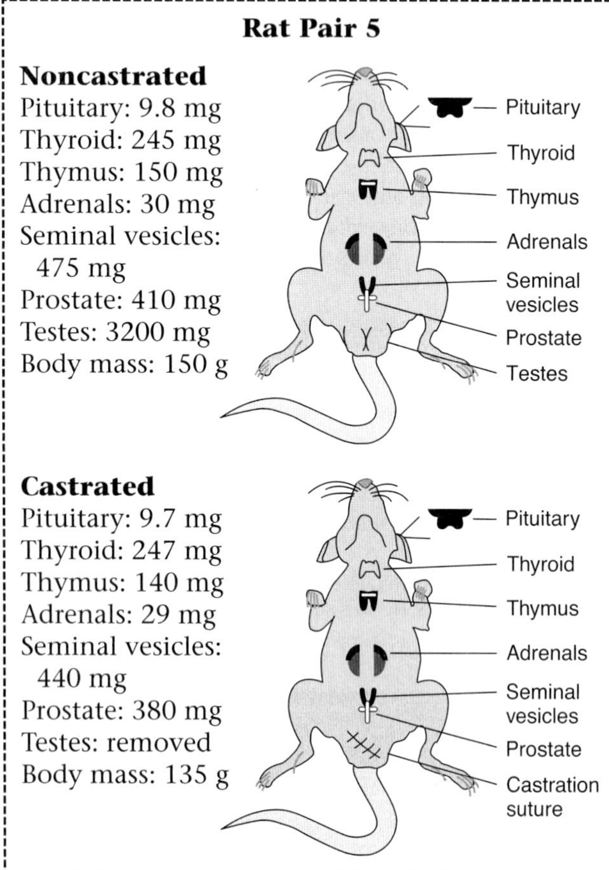
Rat Pair 5
Noncastrated
Pituitary: 9.8 mg
Thyroid: 245 mg
Thymus: 150 mg
Adrenals: 30 mg
Seminal vesicles: 475 mg
Prostate: 410 mg
Testes: 3200 mg
Body mass: 150 g
Pituitary
Thyroid
Thymus
Adrenals
Seminal vesicles
Prostate
Testes
Castrated
Pituitary: 9.7 mg
Thyroid: 247 mg
Thymus: 140 mg
Adrenals: 29 mg
Seminal vesicles: 440 mg
Prostate: 380 mg
Testes: removed
Body mass: 135 g
Pituitary
Thyroid
Thymus
Adrenals
Seminal vesicles
Prostate
Castration suture

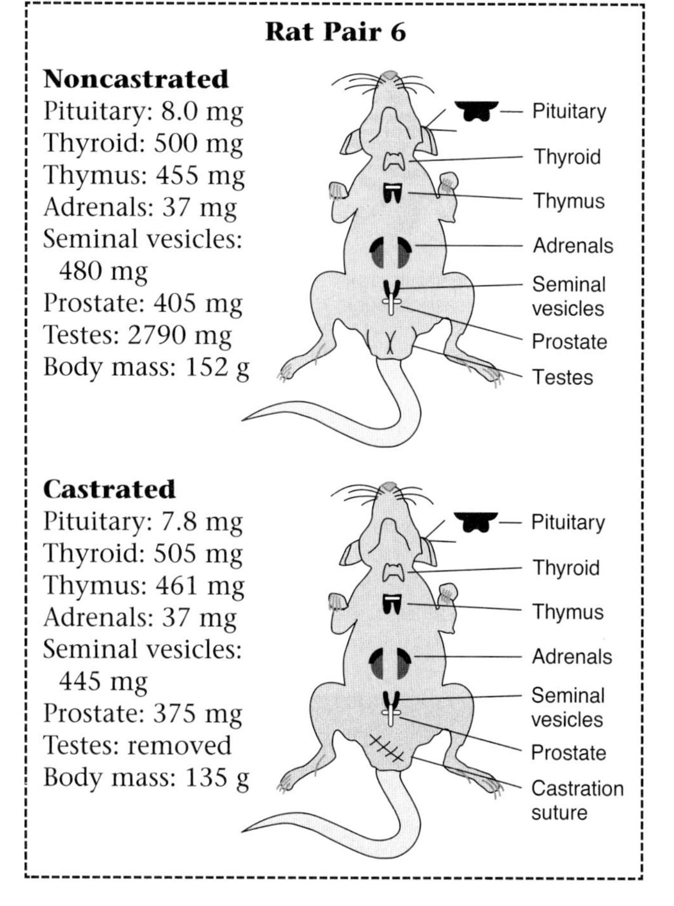
Rat Pair 6
Noncastrated
Pituitary: 8.0 mg
Thyroid: 500 mg
Thymus: 455 mg
Adrenals: 37 mg
Seminal vesicles: 480 mg
Prostate: 405 mg
Testes: 2790 mg
Body mass: 152 g
Pituitary
Thyroid
Thymus
Adrenals
Seminal vesicles
Prostate
Testes
Castrated
Pituitary: 7.8 mg
Thyroid: 505 mg
Thymus: 461 mg
Adrenals: 37 mg
Seminal vesicles: 445 mg
Prostate: 375 mg
Testes: removed
Body mass: 135 g
Pituitary
Thyroid
Thymus
Adrenals
Seminal vesicles
Prostate
Castration suture

RESULTS AND DISCUSSION

LABORATORY REPORT 10

EXPERIMENT 10.1

The Influence of Thyroxine on Ventilation Rate

Results Table 10.1a Individual Group Data for Experiment 10.1: The Influence of Thyroxine on Ventilation Rate.

Fish	Treatment	Ventilation Rate (cycles per minute)				Behavioral Observations
		Trial 1	Trial 2	Trial 3	**Mean**	
1	Control					
2	Control					
1	Thyroxine					
2	Thyroxine					
1	Propylthiouracil					
2	Propylthiouracil					

Results Table 10.1b Class Data for Experiment 10.1: The Influence of Thyroxine on Ventilation Rate.

Fish	Ventilation Rate (cycles per minute)		
	Control	Thyroxine	Propylthiouracil
1			
2			
3			
4			
5			
6			
7			
8			
9			
10			
11			
12			
Mean			
(SE)			

Problem Set 10.1: The Influence of Thyroxine on Ventilation Rate

Using the data from Results Table 10.1b, create a bar graph showing the mean ventilation rate for each of the three treatment groups. Be sure to include 95% CI. Using these data, perform the appropriate statistical test to examine possible differences in mean ventilation rate among the three different treatment groups.

a. State the null hypothesis.

b. Should you accept or reject the null hypothesis? Support your conclusions with the appropriate statistical results.

c. What is the mechanism by which thyroxine influences ventilation rate in goldfish? Remember that ventilation rate is an indirect measure of metabolic rate.

d. What is the mechanism by which propylthiouracil influences ventilation rate in goldfish? Remember that ventilation rate is an indirect measure of metabolic rate.

e. Why was an extended period of thyroxine treatment (six days) necessary to examine the effects of thyroid hormones on metabolic rate? *Hint:* What is the mode of action of thyroid hormones on target cells?

EXPERIMENT 10.2

Experimental Hormone Manipulations in Rats

Results Table 10.2 Data for Experiment 10.2: Experimental Hormone Manipulations in Rats.
In the following table, enter your observations for each of the hormone manipulations *relative* to the control group. For example, if the hormone-treated rat had an increase in body mass relative to the control rat, enter a (+) in the appropriate box. Enter a (−) for a decrease and a (0) for no change. NC = noncastrated; C = castrated.

	Rat Pair 1		Rat Pair 2		Rat Pair 3		Rat Pair 4		Rat Pair 5		Rat Pair 6	
Mass of the:	NC	C	NC	C	NC	C	NC	C	NC	C	NC	C
Pituitary gland												
Thyroid gland												
Thymus												
Adrenal glands												
Seminal vesicles												
Prostate gland												
Testes												
Body mass												
Hormone treatment:												

Problem Set 10.2: Experimental Hormone Manipulations in Rats

a. What hormone did rat pair 1 receive? Explain your reasoning.

b. What hormone did rat pair 2 receive? Explain your reasoning.

c. What hormone did rat pair 3 receive? Explain your reasoning.

d. What hormone did rat pair 4 receive? Explain your reasoning.

e. What hormone did rat pair 5 receive? Explain your reasoning.

f. What hormone did rat pair 6 receive? Explain your reasoning.

COMPARATIVE NOTE *Laboratory 10*

Sex-Changing Fish: Nemo's Dad Becomes Mommy!

In most animals, sex is determined genetically and is therefore fixed after development. Some animals, however, possess the ability to change sex in response to endocrine cues mediated by social and environmental stimuli. One such species is the tropical coral-reef fish, the bluehead wrasse (*Thalassoma bifasciatum*). These colorful fish live on coral reefs in Bermuda, the Bahamas, the Caribbean, the Gulf of Mexico, and the northern coast of South America. There are three different sexes of bluehead wrasse: the terminal-phase males, which have the distinctive blue head, initial-phase males, and initial-phase females, both of which are yellow. Typically only one or two of the largest fish on a particular reef are terminal-phase, blue-headed males. These terminal-phase males dominate the best breeding sites on the reef and may mate up to 40 times a day. If the dominant terminal-phase male dies or is experimentally removed from the reef, the largest wrasse in the group (usually a female but occasionally an initial-phase male) undergoes a series of rapid physiological changes in which it becomes the new terminal-phase male. The yellow female's transformation into the blue-headed male is complete in three to five days, and fully functional testes develop from the ovaries in only one week. The bluehead wrasse is an example of a protogynous ("female-first") sequential hermaphrodite. In contrast, some animals are protandrous ("male-first") sequential hermaphrodites, in which males change into females. The species of clownfish (*Amphiprion ocellaris*) portrayed in the movie *Finding Nemo* is an example of such a sex-changing fish. If the larger female of a monogamous mating pair dies, the male will change into a female, and a new, nearby male (usually a juvenile fish that develops into a male) will join the newly transformed female to form a new monogamous mating pair.

Current research suggests that this intriguing biological phenomenon is regulated not by the sex steroid hormones (e.g., testosterone, estrogen), but by an increased production of the hormone arginine vasotocin (AVT) in the brain. Arginine vasotocin is secreted by the neural tissue of the posterior pituitary, as well as in other locations in the brain, and is similar to the mammalian homolog arginine vasopressin (AVP, also called ADH for antidiuretic hormone). In mammals, the primary effect of vasopressin is absorption of water from the collecting ducts in the kidney and, secondarily, constriction of blood vessel smooth muscle to increase blood pressure (an effect from which the name vasopressin is derived: *vaso* = vascular, *pressin* = pressure).

(a)

(b)

Examples of hermaphroditic fishes: (a) Bluehead wrasse *(Thalassoma bifasciatum)* and (b) Clownfish *(Amphiprion ocellaris)*.

Although the hormones are similar in structure, you can see that arginine vasotocin has a very different function in sex-changing fish. This example demonstrates the importance of investigating the comparative physiology of hormone action to completely understand the role that hormones play in regulating physiology and behavior.

RESEARCH OF INTEREST

Semsar, K. and J. Godwin. 2004. Multiple mechanisms of phenotype development in the bluehead wrasse. *Hormones and Behavior* 45:345–353.

Semsar, K., F. Kandel, and J. Godwin. 2001. Manipulations of the AVT system shift social status and related courtship and aggressive behavior in the bluehead wrasse. *Hormones and Behavior* 40:21–31.

LABORATORY 11

Cardiovascular Physiology

PURPOSE

This laboratory will introduce you to both the electrical and the mechanical events of the cardiac cycle. Two experiments will demonstrate how heart rate is affected by the drug caffeine and how body position influences blood pressure. Calculations of cardiovascular variables from heart rate and systolic and diastolic pressures will also be introduced.

Learning Objectives

- Describe the electrocardiogram and how it corresponds to the electrical events of the cardiac cycle.
- Describe the heart sounds and how they correspond to the mechanical events of the cardiac cycle.
- Describe the effects that different drugs have on heart rate and explain and understand their modes of action.
- Demonstrate your ability to measure blood pressure using a stethoscope and a sphygmomanometer.
- Understand and calculate pulse pressure, mean arterial pressure, and total peripheral resistance from measures of heart rate and systolic and diastolic pressures.
- Describe and calculate cardiac output and understand how it relates to body mass.

Laboratory Materials

Experiment 11.1: The Effect of Caffeine on Heart Rate

1. Diet beverage and caffeine-free diet beverage
2. Stethoscope
3. Stopwatch

Experiment 11.2: The Effect of Body Position on Blood Pressure

1. Sphygmomanometer (manual and/or digital)
2. Stethoscope

Experiment 11.3: Calculation of Cardiovascular Variables

Data from Experiments 11.1 and 11.2

NOTE TO THE STUDENT

To obtain the best results for Experiment 11.1, volunteers should not consume any caffeine the day of their laboratory session and should not eat or drink for at least two hours prior to laboratory.

Introduction and Pre-Lab Exercises

Cardiac muscle is striated, involuntary muscle. Although there are some similarities between the microanatomy of skeletal and cardiac muscles, cardiac muscle has several characteristics that allow the heart to function as an effective pump. For example, both skeletal and cardiac muscles appear striated because of the arrangement of actin and myosin filaments within sarcomeres. Unlike skeletal muscle cells, however, cardiac muscle cells are branched and contain numerous gap junctions and intercalated discs. **Intercalated discs** are structures that contain desmosomes and join cardiac muscle fibers end to end. Adjacent to the intercalated discs are **gap junctions,** cell-to-cell contacts that allow direct and rapid communication between the cytoplasm of cells. These characteristics allow cardiac muscle to function as an efficient, rhythmic pump (Figure 11-1).

11-1 Cardiac muscle cells have a very long (about 0.3 second) refractory period during which no additional muscle contraction can be

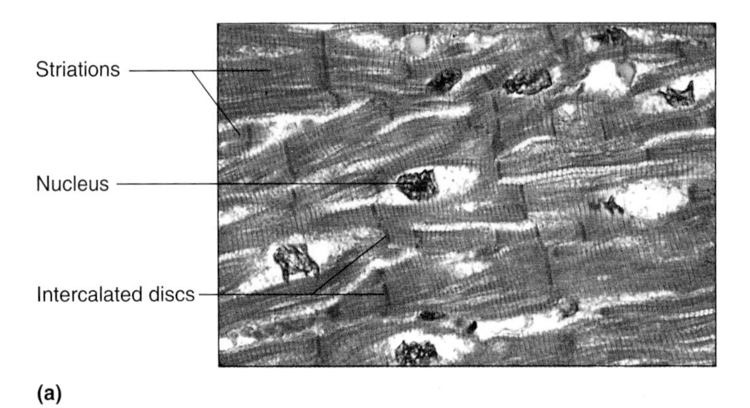

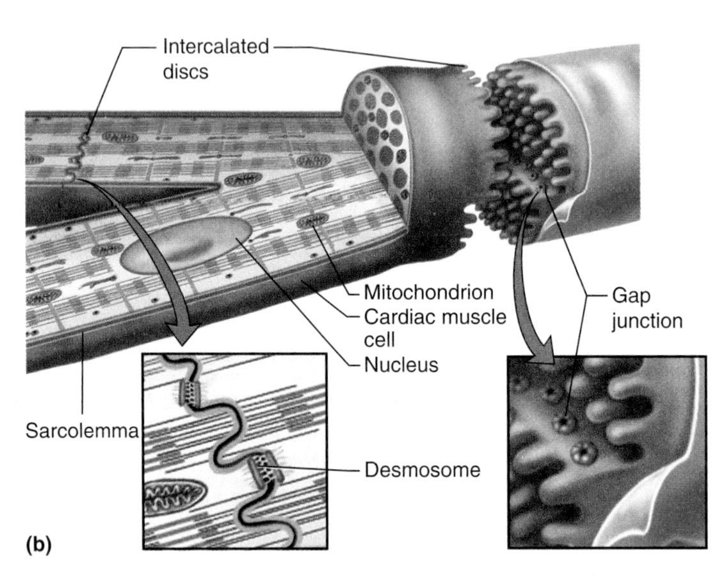

Figure 11-1 Cardiac muscle. (a) Light micrograph. (b) Cardiac myocytes and intercalated discs.

generated. What physiological phenomenon does this long refractory period prevent? What is the functional significance of this cardiac muscle feature? (*Hint:* What happens to contraction force when skeletal muscle cells receive a second stimulus before the muscle cells completely relax?)

■

The contractions of cardiac muscle are stimulated by action potentials that traverse the heart through a conducting system. Action potentials originate in the pacemaker cells of the sinoatrial node **(SA node),** which is located in the upper right atrium (Figure 11-2). These action potentials are then propagated to adjacent myocardial cells of the right and left atria through gap junctions as well as to the atrioventricular node **(AV node),** which is located on the inferior portion of the interatrial septum. Because the atria and ventricles are divided by a fibrous skeleton that does not conduct action potentials, action potentials cannot be propagated to the ventricles directly from the atria. Instead, action potentials travel inferiorly from the AV node via the **bundle of His,** a group of modified myocardial cells that lie along the interventricular septum between the left and right ventricles. At the superior interventricular septum, the bundle of His divides into left and right bundle branches, which are continuous with the **Purkinje fibers** that innervate the ventricular walls. This conduction system allows action potentials to travel quickly throughout the myocardium, stimulating rhythmic, coordinated contractions for efficiently pumping blood.

The mechanical events of the cardiac cycle may be divided into two distinct phases: diastole and systole. **Diastole** refers to the relaxation (filling) phase, while **systole** refers to the contraction (pumping) phase. Both the atria and the ventricles undergo diastole and systole. In response to the pacemaker action potentials that occur in the SA node, both the right and left atria undergo contrac-

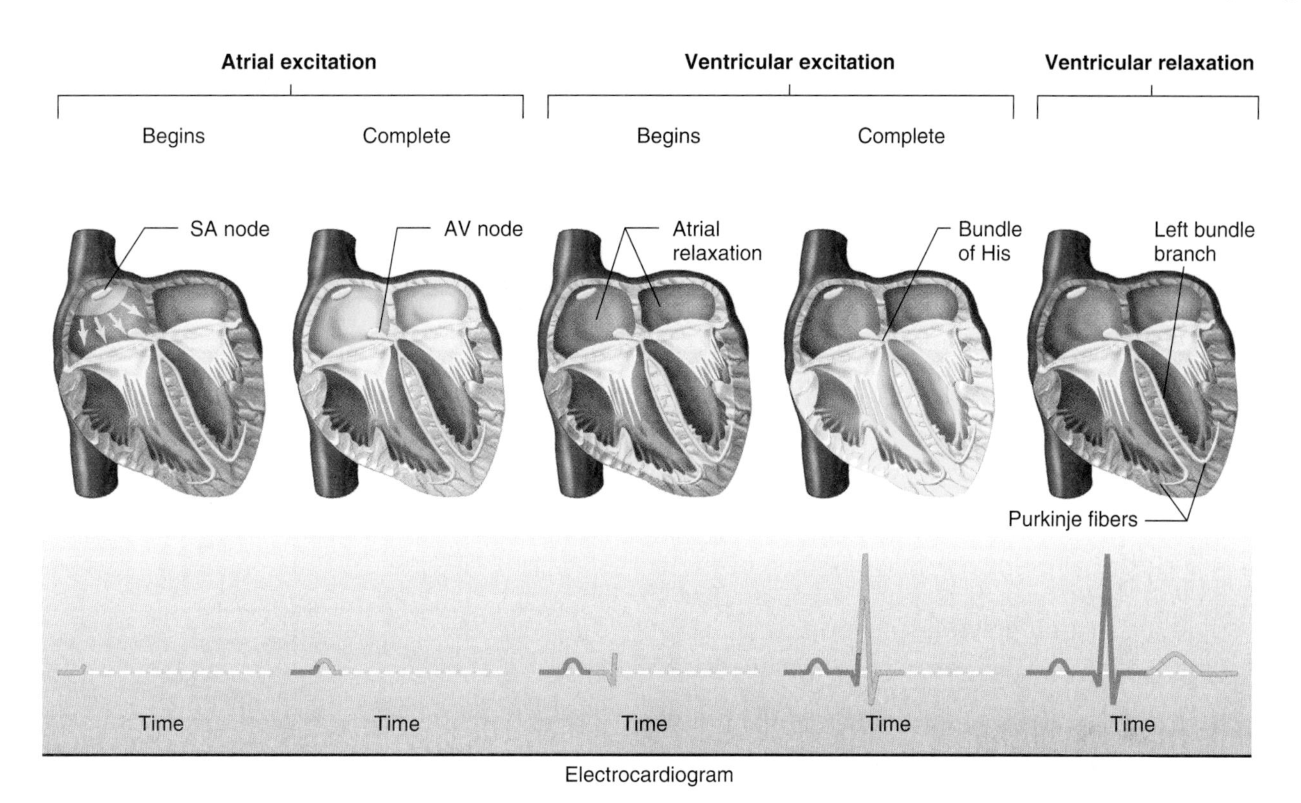

Figure 11-2 Sequence of cardiac excitation. The yellow color denotes areas that are depolarized. The impulse spreads from right atrium to left atrium via the atrial muscle cells where the atria share a wall. The electrocardiogram monitors the spread of the signal. *Adapted from Rushmer.*

tion and emptying. Note that 80% of the blood flow from the atria to the ventricles is passive, due to gravitational force and a higher pressure in the atria than in the ventricles. Thus, the contraction of the atria merely forces the last 20% of blood into the ventricles. In fact, because the majority of blood flow into the ventricles is passive, a person can continue to function normally at moderate exertion levels without a functioning SA node or atria.

In contrast, the ventricles must pump blood superiorly into the pulmonary artery (right ventricle) and the aorta (left ventricle). To function most efficiently, the ventricles should contract from the bottom upward so that the blood is forced out of the ventricles and into the superior arteries. In this elegant system, action potentials originating in the SA node are propagated to the apex of the heart along the fibrous interventricular septum via the left and right bundle branches. Thus, the apex of the heart is the first portion of the ventricles to receive action potential stimuli and to undergo muscle contraction. Action potentials are then propagated along the Purkinje fibers to further stimulate the depolarization and contraction of ventricular myocardial cells. Blood is therefore pumped out of the ventricles from the bottom up.

The force generated by the ventricles during systole is proportional to the volume of blood in the ventricles when they begin to contract (this volume is called the end diastolic volume because it is the volume of blood contained in the ventricles at the end of diastole or the relaxation phase). The relationship between ventricular force and end diastolic volume was described by the physiologists Otto Frank and Ernest Starling and is known as the **Frank-Starling law.**

11-2 When the atria contract to force the last 20% of atrial blood into the ventricles, the end diastolic volume is increased and the walls of the ventricles are stretched. This stretching of the ventricular walls in turn results in more force produced during ventricular systole. Using your knowledge of the length–tension relationship that you examined during Laboratory 8 ("Functional Anatomy of Muscle and Mechanics of Contraction"), explain why ventricular force increases in response to increases in the end diastolic volume.

■

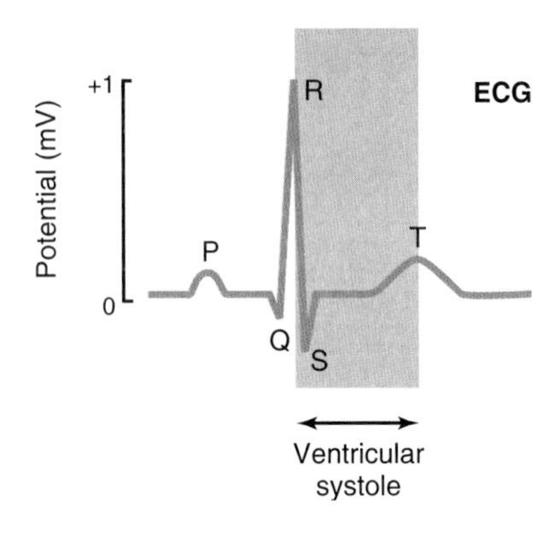

Figure 11-3 A normal ECG trace from a healthy patient.

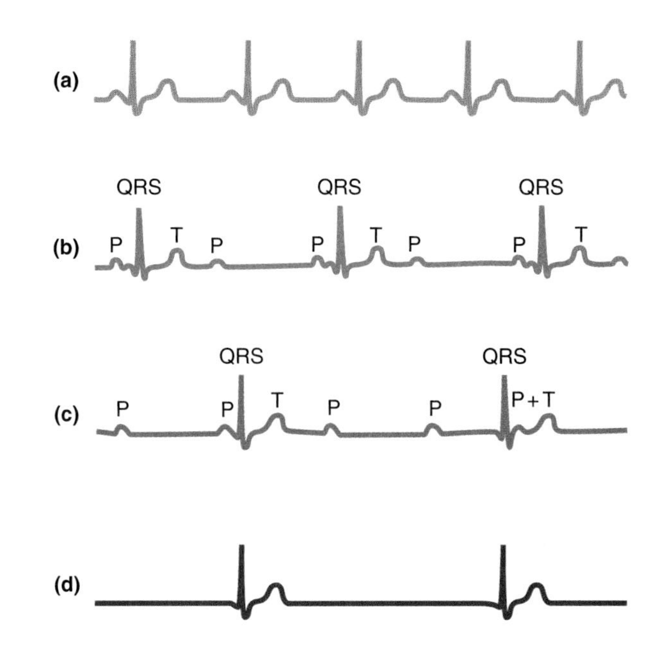

Figure 11-4 ECG traces for clinical inspection.

The mechanical events (systole, diastole) of the cardiac cycle result from the electrical events that occur during an action potential (depolarization, repolarization). Thus, as myocardial cells depolarize, calcium release from the sarcoplasmic reticulum initiates cross-bridge cycling, and the myocardial cells shorten, resulting in systole. As the cells are repolarized, calcium is returned to the sarcoplasmic reticulum, cross-bridge cycling ceases, and the myocardial cells undergo relaxation. Because the atria and ventricles undergo a cycle of depolarization and repolarization, you can observe the cardiac cycle by measuring the electrical events that occur in the heart. An **electrocardiogram** (**ECG** or **EKG**) measures these electrical events. A typical ECG is depicted in Figure 11-3. Your instructor may choose to demonstrate the recording of an ECG during this laboratory.

The P-wave corresponds to the depolarization of the atria. The QRS-wave represents depolarization of the ventricles, and the T-wave corresponds to ventricular repolarization. Repolarization of the atria is masked by the very large ventricular depolarization (i.e., atrial repolarization occurs during the large QRS-wave).

11-3 Label the P, QRS, and T-waves on the ECG tracing shown in Figure 11-4a. Is this ECG trace normal?

■

11-4 Is the ECG trace shown in Figure 11-4b from a healthy patient? Explain the clinical condition that may be causing this trace.

■

11-5 Is the ECG trace shown in Figure 11-4c from a healthy patient? Explain the clinical condition that may be causing this trace.

■

11-6 Identify and label the electrical events shown in the ECG trace in Figure 11-4d. Is this a healthy patient? Explain the clinical condition that may be causing this trace.

What is now controlling the ventricular systole and diastole depicted in Figure 11-4d?

■

11-7 Which pacemaker (the SA or the AV node) is faster at generating action potentials? Use the ECG tracings in Figure 11-4 to support your answer.

■

Although the SA node is considered the pacemaker of the heart, the rate of the cardiac cycle is regulated by many other factors. A large number of parasympathetic and sympathetic postganglionic fibers innervate the SA node and regulate the production of pacemaker potentials. For example, postganglionic neurons of the parasympathetic nervous system innervate the heart via the **vagus nerve** and decrease the heart rate. This influence of the parasympathetic nervous system is accomplished by the neurotransmitter **acetylcholine,** which binds to **muscarinic acetylcholine receptors** on myocardial cells. In contrast, the stimulatory influence of the sympathetic nervous system is accomplished by the binding of the neurotransmitter **norepinephrine** to **β-adrenergic receptors** in the SA node.

A variety of drugs and other endogenous factors (such as hormones and plasma electrolyte concentrations) also exert an influence on both heart rate and the contractility of cardiac muscle. Drugs that mimic the effects of neurotransmitters or hormones on target cells by binding to and activating their receptors are called **agonists. Antagonists** are drugs that compete with neurotransmitters or hormones for a receptor but do not activate the receptor upon binding. Thus, antagonists block the effects of neurotransmitters and hormones on target cells by preventing their binding and thus preventing the activation of receptors. Many pharmacological treatment therapies use drugs that agonize or antagonize neurotransmitters of the sympathetic and parasympathetic nervous systems. In Experiment 11.1, you will conduct an experiment to investigate the effect of caffeine on human heart rate.

11-8 Fill in the following table to describe the effects of these drugs and endogenous factors on heart rate. Explain the physiological effects of each factor by describing which receptor types are affected and why.

Drug Added to Heart	Origin	Mechanism of Action	Effect on Heart Rate	Explanation of Physiological Effect
Acetylcholine	Endogenous neurotransmitter	Binds to and activates muscarinic acetylcholine receptors	↓	*Acetylcholine decreases the heart rate by binding to and activating muscarinic acetylcholine receptors on pacemaker cells of the SA node.*
Atropine	Drug found in the nightshade plant *Atropa belladonna*	Muscarinic acetylcholine receptor antagonist		
Epinephrine	Endogenous hormone released by the adrenal medulla during the "fight-or-flight" response	Binds to and activates β-adrenergic receptors		
Caffeine	Drug found in plants (coffee beans, tea, cocoa)	Inhibits the activity of phosphodiesterase; cyclic AMP is the second messenger activated by β-adrenergic receptors		
Potassium	Electrolyte	Some K^+ channels ("leak channels") are open when cells are at resting potential.		

The mechanical events that occur during the cardiac cycle produce a series of **heart sounds,** all resulting from tissue vibrations. Although there are four heart sounds, only the first two sounds are audible with a stethoscope. These first two sounds are commonly represented as "lub-dup." The first heart sound, the "lub," is created when the atrioventricular (AV) valves close. As the atria contract and empty, the pressure in the ventricles increases. When ventricular pressure exceeds atrial pressure, the AV valves, located between the atria and the ventricles, close, thus preventing the backflow of blood into the atria. It is the vibrations created by the closing of these valves that produce the first heart sound.

The second heart sound, or the "dup," is created by the closing of the aortic and pulmonary semilunar valves. When the ventricles contract, the pressure in the aorta and pulmonary artery increases. When the arterial pressure exceeds ventricular pressure, the aortic and pulmonary semilunar valves close to prevent the backflow of blood into the ventricles. The vibrations created by the closing of the semilunar valves produce the second heart sound. The third heart sound is produced by the turbulence of blood flow in the aorta, and the fourth heart sound is produced by the contraction of the atria. In this laboratory, you may wish to listen to your own heart sounds during the cardiac cycle. Using a stethoscope, listen for two distinct sounds without extraneous noise; many valve abnormalities are first detected because of abnormal heart sounds.

11-9 A physician examines an ECG recording from a patient with a heart murmur and finds the recording to be normal. Explain.

As the ventricles undergo systole and diastole, the pressure within the blood vessels oscillates between maximum and minimum values (Figure 11-5). Every time the ventricles contract, there is an increase in the pressure in the arteries. During diastole and ventricular relaxation, pressure decreases because the blood volume drains into the capillary beds.

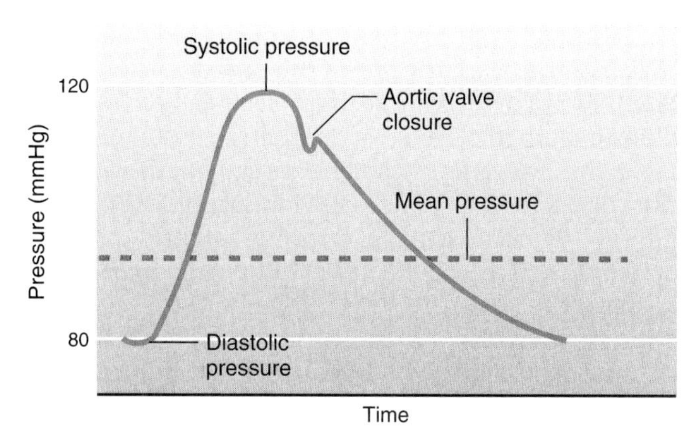

Figure 11-5 Trace of arterial pressure during a cardiac cycle.

Maximum arterial pressure occurs during ventricular systole; this is called **systolic blood pressure.** The minimum arterial pressure occurs during ventricular diastole or ventricular relaxation and is called **diastolic blood pressure.** Systolic and diastolic arterial pressures can be measured using a sphygmomanometer and a stethoscope. In sphygmomanometry, a cuff is used to increase the pressure exerted on the blood vessels in the arm. When the pressure in the cuff is increased to a point above systolic pressure, the blood vessels in the arm will be completely constricted, and blood flow through the vessels will cease. As the cuff pressure is decreased to systolic pressure, blood flow once again occurs through the partially constricted vessels. The flow of blood through partially constricted vessels is turbulent, and this turbulent blood flow can be heard through the stethoscope. The pressure at which this turbulent blood flow is first heard is used as a measure of systolic blood pressure. As the pressure in the cuff is decreased, the blood vessels will become less and less constricted. Blood flow will continue to sound turbulent until the blood vessels are no longer constricted. The cuff pressure at which the turbulence of blood flow can no longer be heard (i.e., the cuff pressure at which the blood vessels are no longer constricted) indicates diastolic blood pressure. In Experiment 11.2, you will measure both systolic and diastolic blood pressures and investigate the influence of body position on blood pressure. In Experiment 11.3, you will use these values to calculate a mean arterial pressure as well as other cardiovascular variables.

Methods and Materials

EXPERIMENT 11.1 Effect of Caffeine on Heart Rate

In this experiment, you will investigate the effect of caffeine on the human heart rate.

- Randomly determine which people in your lab group will participate in the control and treatment groups.

  Control Beverage Group—ingestion of diet caffeine-free beverage

  Treatment Beverage Group—ingestion of diet caffeinated beverage (Example: Diet Mountain Dew® or Diet Coca-Cola®).

- For both groups: Relax and sit quietly in your chair for about five minutes (do not make large movements, including standing or walking).
- After the five-minute relaxation period, take your resting heart rate while in a sitting position by recording the number of heartbeats that occur during one minute. For accuracy, do not record your heart rate over shorter periods of time and extrapolate to obtain the heart rate.
- Once you have obtained your resting heart rate, drink 355 ml (12 fluid ounces) of your designated beverage. Try to do this as quickly as possible. Most people can consume 355 ml within two minutes.
- Remain at rest in your chair after drinking the beverage. Fifteen minutes following ingestion, record your heart rate as before.
- Record your data and the class data in Results Table 11.1 in the results section of your Laboratory Report.

EXPERIMENT 11.2 The Effect of Body Position on Blood Pressure

In this experiment you will examine the influence of two different body positions on blood pressure using a pressure cuff and sphygmomanometer. As in Experiment 11.1, this experiment is designed so that each subject serves as his or her own control (i.e., each subject measures blood pressure at both of the body positions). The residual effects of caffeine on heart rate and blood pressure in those subjects who participate in Experiment 11.1 may be disregarded because of this experimental design.

- Before you begin the procedure for measuring blood pressure, inspect Figure 11-6 to help you understand the steps of this clinical procedure. Practice measuring blood pressures before collecting the experimental data.
- Ask one of your lab partners to lie down with both arms resting at his or her sides.
- Place the pressure cuff around your lab partner's upper arm just above the elbow.
- Place the bell of your stethoscope on his or her inner arm (just below the cuff) where you can feel the arterial pulse.
- Close the screw valve on the pressure cuff and squeeze the bulb to increase pressure within the cuff.
- Quickly increase the pressure within the cuff to about 200 mmHg. This is usually a sufficient cuff pressure to constrict the blood vessels and stop the flow of blood to the lower arm. At this point, you should no longer hear the sound of blood flowing through the blood vessels.
- Now slowly open the screw valve so that the pressure in the cuff decreases at a rate of 2 to 3 mmHg per second.
- When the pressure in the cuff decreases to a point that is equal to the highest pressure in the vessel (i.e., the pressure during ventricular systole), blood will once again flow through the partially constricted blood vessels. Record the pressure reading on the sphygmomanometer at which blood flow is first heard; this is your systolic blood pressure.
- Allow the pressure in the cuff to continue to decrease until you no longer hear any sounds. At this point, the cuff pressure is equal to the lowest pressure in the vessels (i.e., the pressure during ventricular diastole) and the vessel is no longer constricted. Thus, blood flow is laminar and you will no longer hear the turbulence of blood flowing through a constricted vessel. Record this pressure in your data sheet as the diastolic pressure.
- **Pulse pressure** results from the increase in pressure produced by the contraction of the ventricles followed by a sudden fall in pressure produced by ventricular relaxation. Thus, pulse pressure is calculated as the difference

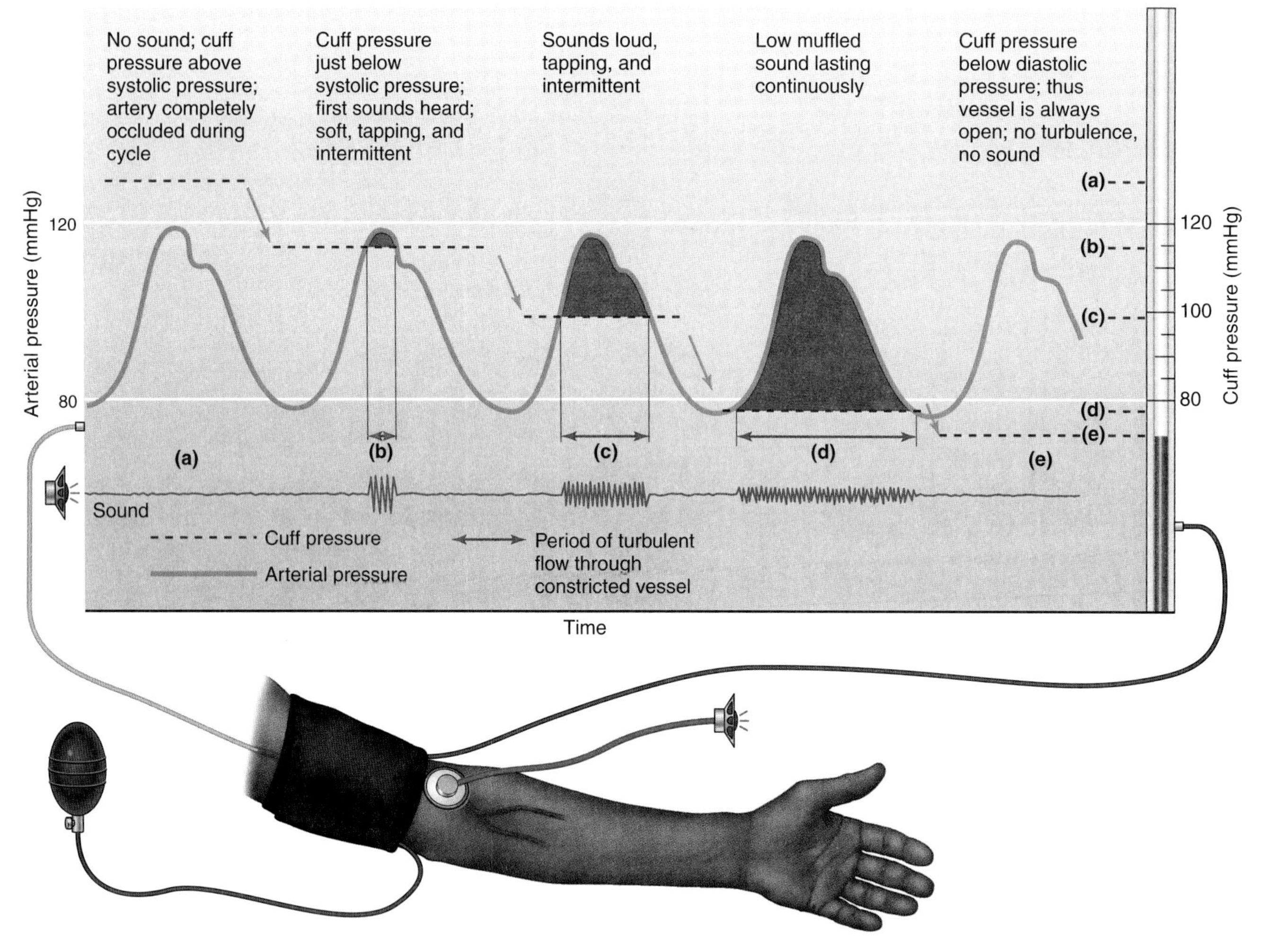

Figure 11-6 Sounds heard through a stethoscope as the cuff pressure of a sphygmomanometer is gradually lowered. Sounds are first heard at systolic pressure, and they disappear at diastolic pressure.

between systolic and diastolic pressure. Calculate the **pulse pressure** and record this result in Results Table 11.2.

Pulse pressure = Systolic pressure − Diastolic pressure

- Now have your lab partner slowly (to avoid fainting) stand up. Repeat the previous steps to obtain measures of systolic, diastolic, and pulse pressures in this upright, standing position. Record these data in Results Table 11.2 of your Laboratory Report.
- Repeat these steps to obtain data on the effects of body position on blood pressure for one or two more of your lab partners. Record these data in Results Table 11.2 of your Laboratory Report.
- Obtain data from the entire class for statistical analyses.

EXPERIMENT 11.3 Calculation of Cardiovascular Variables

In this experiment, you will use your initial measurements of heart rate and systolic and diastolic pressures from Experiments 11.1 and 11.2 to calculate pulse pressure, mean arterial pressure, cardiac output, and total peripheral resistance.

- Complete the following calculations and record your answers in the results section of your Laboratory Report.
- Use your initial recording (i.e., while reclining) of systolic and diastolic pressures to calculate pulse pressure.

  Pulse pressure (mmHg) = Systolic pressure − Diastolic pressure

- Use your reclining systolic and diastolic pressures to calculate **mean arterial**

pressure (MAP). MAP is an average of arterial pressure during the cardiac cycle. However, the period of diastole is approximately twice as long as that of systole during the cardiac cycle. Thus, MAP is best approximated using the following equation:

$$\text{MAP (mmHg)} = \frac{\text{Systolic pressure} + 2\,(\text{Diastolic pressure})}{3}$$

This value is important because the difference between MAP and the venous pressure is what forces blood though the capillary beds of organs.

- Use your initial heart rate recorded in Experiment 11.1 to calculate your **cardiac output.** Cardiac output is the volume of blood (in milliliters) that is pumped by the heart per unit time (minute). To calculate cardiac output we must first know the volume of blood that is pumped by the heart during a single cycle. This is called **stroke volume** and can be approximated by multiplying your body mass (in kg) by 0.92, a constant used for the calculation of human heart stroke volume.

  Cardiac output (ml/min) = Heart rate (beats/min) × Stroke volume (ml/beat)

- Now you can calculate your **total peripheral resistance** by using the following equation:

  Total peripheral resistance (mmHg/ml/min) = MAP/Cardiac output

  This resistance represents the pressure that your heart must overcome in a forced contraction of the ventricles to push blood through the arterioles and capillaries of the body.

COMPARATIVE NOTE *Laboratory 11*

Open Circulatory Systems: Spilling Blood

There is an overwhelming diversity of animals that do not have circulatory systems (e.g., Protozoans, Mesozoans, Cnidaria, Platyhelminthes). Most of these animals share one important characteristic: they are small. Animals one millimeter or less in diameter can obtain nutrients and respire simply by diffusion. Larger animals require a circulatory system to transport and deliver nutrients and respiratory gases because diffusion alone is inadequate to support the metabolic demand of the animals' tissues. Some jellyfish are an exception. These larger animals, without a circulatory system, cope with an inefficient transport system for nutrients and gases by having a very low metabolic rate.

You are quite familiar with the human circulatory system in which vessels (arteries, veins, and capillaries) are used to carry and direct the transport of respiratory gases, nutrients, metabolic wastes, and hormones through the body. This type of system is referred to as a closed circulatory system. However, many invertebrates have a different type of system that is called an open circulatory system. Blood flow in an open circulatory system empties into an open fluid cavity called the hemocoel, where tissues are directly bathed with blood. Depending on the taxonomic group, blood is then siphoned from the hemocoel and recirculated by a dorsal blood vessel that functions as a heart. Some of the physiological limitations of an open circulatory system are overcome with the use of other physiological systems. For example, insects overcome the problem of oxygen delivery with the use of spiracles and tracheoles for oxygen diffusion directly into the body.

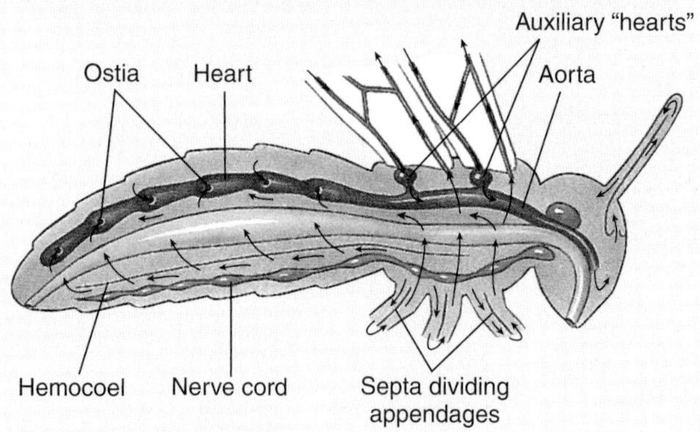

A picture of a common housefly and a general illustration of an insect's open circulatory system.

RESEARCH OF INTEREST

Abe, K. and J. Vannier. 1995. Functional morphology and significance of the circulatory system of ostracoda, exemplified by *Vargula hilgendorfii* (Myodocopida). Marine Biology 124:51–58.

McMahon, B. R. 2001. Control of cardiovascular function and its evolution in crustacea. Journal of Experimental Biology 204:923–932.

Paul, R.J., S. Bihlmayer, M. Colmorgen, and S. Zahler. 1994. The open circulatory system of spiders (*Eurypelma californicum, Pholcus phalangioides*): a survey of functional morphology and physiology. Physiological Zoology 67:1360–1382.

Ulrich, T., O. Schmidt, K. Soderhall, and M. S. Dushay. 2004. Coagulation in arthropods: defense, wound closure, and healing. Trends in Immunology 25: 289–294.

West, L. S. 1951. The housefly: its natural history, medical importance, and control. Cornell Univ. Press.

RESULTS AND DISCUSSION

LABORATORY REPORT 11

EXPERIMENT 11.1

The Effect of Caffeine on Heart Rate

Results Table 11.1 Class Data for Experiment 11.1: The Effect of Caffeine on Heart Rate

Student (Name)	Resting Heart Rates (Beats per Minute)			
	Control Beverage Group No Caffeine		Treatment Beverage Group Caffeine	
	Before Ingestion	15 Min After Ingestion	Before Ingestion	15 Min After Ingestion
1				
2				
3				
4				
5				
6				
7				
8				
9				
10				
11				
12				
13				
14				
15				
16				
17				
18				
19				
20				
21				
22				
23				
24				
25				
Mean (±SE)				

Problem Set 11.1: The Effect of Caffeine on Heart Rate

Using the data from Results Table 11.1, create a bar graph showing mean heart rate before and after the caffeine treatments. Be sure to include 95% CI for each of the experimental groups. Your graph should look similar to Figure 11-7.

Using these data, perform the appropriate statistical test to examine the effect of caffeine on mean heart rate. These data are most appropriately analyzed using a two factor repeated measures ANOVA. However, for simplicity you should use a paired t-test to investigate the two experiments separately (i.e., test for a difference in mean heart rate before and after consumption of no caffeine and then test for a difference in mean heart rate before and after consumption of caffeine).

a. State the null hypotheses.

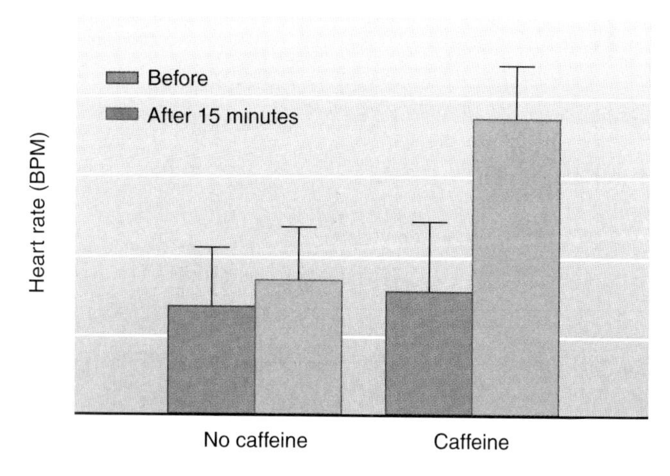

Figure 11-7 Example graph showing the effects of caffeine on heart rate.

b. Should you accept or reject the null hypotheses? Support your conclusions with the appropriate statistical results.

c. What is the mechanism by which caffeine affects heart rate?

EXPERIMENT 11.2
The Effect of Body Position on Blood Pressure

Results Table 11.2 Class Data for Experiment 11.2: The Effect of Body Position on Blood Pressure.

	Blood Pressure (mmHg)					
	Reclining			Standing		
Student (Name)	Systolic Pressure	Diastolic Pressure	Pulse Pressure	Systolic Pressure	Diastolic Pressure	Pulse Pressure
1						
2						
3						
4						
5						
6						
7						
8						
9						
10						
11						
12						
13						
14						
15						
16						
17						
18						
19						
20						
21						
22						
23						
24						
25						
Mean (±SE)						

Problem Set 11.2: The Effect of Body Position on Blood Pressure

Using the data from Results Table 11.2, create a simple bar graph showing the mean pulse pressures for both the lying down and standing body positions. Be sure to include the 95% CI for each of the body positions. Using these data, perform a paired *t*-test to examine possible differences in pulse pressure between the two body positions.

a. State the null hypothesis.

b. Should you accept or reject the null hypothesis? Support your conclusions with the appropriate statistical results.

c. When you stand up from a reclining position, there is a substantial shift of blood (500–700 ml) from the veins of the thoracic cavity to the veins of your lower extremities. This pooling of blood reduces venous return and hence cardiac output, resulting in a decrease in blood pressure. If you happened to accept the null hypothesis, can you offer an explanation for why you did not see a difference in pulse pressure between the two body positions? Give a complete and thorough answer.

d. Does your reasoning in question c explain why a healthy person does not faint upon quickly standing? Why?

Problem Set 11.3: Calculation of Cardiovascular Variables

a. Calculations: Show all work in determining the following cardiovascular variables.

What is your systolic pressure obtained from the sphygmomanometer? _______ mmHg

What is your diastolic pressure obtained from the sphygmomanometer? _______ mmHg

What is your pulse pressure? _______ mmHg

What is your mean arteriole pressure (MAP)? _______ mmHg

What is your cardiac output? _______ ml/min

What is your total peripheral resistance? _______ mmHg/ml/min

b. Do you think there is a difference between the cardiac output of males and females? Explain.

c. During the average life span of humans, females have similar or lower blood pressures than males. Thus, do females have higher or lower total peripheral resistances than males of the same age? Explain.

d. Some researchers calculate MAP with the following equation:

(1) MAP = 1/3(Pulse pressure) + Diastolic pressure

rather than the equation:

$$(2)\ \text{MAP} = \frac{\text{Systolic pressure} + 2\ (\text{Diastolic pressure})}{3}$$

As you learned in Experiment 11.2, pulse pressure is equal to the difference between systolic and diastolic pressure. Using this information, prove that equation (1) equals equation (2) and can be used to calculate MAP. Once you have derived equation (2) from equation (1), use your measures of blood pressure to see if you obtain the same values of MAP.

LABORATORY 12

Physiology of Blood

PURPOSE

This laboratory will introduce the physiological functions and the major components of blood. You will determine blood type for two blood samples using the ABO and Rh systems and you will learn about the interactions between antigens and antibodies in agglutination reactions. An experiment will demonstrate how hematocrit is determined and will introduce factors that may influence the hematocrit level. You will also examine blood smears for differential leukocyte counts.

Learning Objectives

- Identify blood as a type of connective tissue and list its functions in transportation, regulation, and protection.
- Identify the formed elements of blood and describe their functions.
- Describe the functional morphology of the red blood cell and the role of hemoglobin in gas exchange.
- Understand the physiological significance of the hematocrit level.
- Describe the ABO system of blood typing and understand the importance of the agglutination reaction in blood typing.
- Identify the physiological role of leukocytes in immune function and the influence of disease on the differential leukocyte count.

Laboratory Materials

Experiment 12.1: Blood Typing

1. Blood samples from unknown patients
2. Blood typing tray
3. Anti-A, anti-B, and anti-Rh antibodies
4. Plastic toothpicks for stirring samples

Experiment 12.2: Hematocrit Levels in Male and Female Patients

1. 12 male blood samples
2. 12 female blood samples
3. Microscope slides
4. Capillary tubes
5. Critoseal tray
6. Hematocrit centrifuge
7. Ruler

Experiment 12.3: Mr. Jones

1. Blood sample from Mr. Jones
2. Microscope slides
3. Capillary tubes
4. Critoseal tray
5. Hematocrit centrifuge
6. Ruler

Experiment 12.4: Differential Leukocyte Count

1. Microscope slide of a blood smear from a healthy adult male
2. Microscope slides of blood smears from ill adult males
3. Microscope

A NOTE REGARDING SAFETY

All exercises in this laboratory will be performed with simulated blood products to eliminate concerns about blood-borne pathogens and infectious diseases. However, we encourage you to use clinical safety practices as if you were handling real biological samples in a diagnostic laboratory.

Introduction and Pre-Lab Exercises

Blood is a connective tissue that serves many functions in the circulatory system. Although these functions are integrated like most physiological processes, we can divide them into three broad categories: transportation, regulation, and protection. Transportation is the primary function of blood; without transportation, regulation and protection would not be possible. Most substances in the body require transport from one area to another. These substances include glucose, hormones, respiratory gases, metabolic wastes, drugs, and antibodies, to name a few. Even heat, although not a substance but rather a state of kinetic energy, is transported to areas of the body for temperature regulation. The following outline summarizes the three major functions of the blood. Note that transportation is an integral part of regulation and protection.

I. **Transportation**—Blood is essential in moving substances throughout the body.
   - Respiration—Red blood cells (**erythrocytes**) transport O_2 from the lungs to the cells of the body for cellular respiration and CO_2 from the cells of the body to the lungs for exhalation.
   - Nutrition—During digestion, nutrients are absorbed from the small intestine and transported by the blood to the cells of the body.
   - Excretion—blood transports waste products (e.g., urea) to the kidneys where they are removed from the blood and excreted.

II. **Regulation**—Blood is essential to the negative feedback systems that regulate hormone production and body temperature.
   - Hormones—Hormones are secreted from cells and are transported to their target tissues by the blood.
   - Temperature—Body temperature is regulated by directing blood flow from peripheral to deep vessels to maintain temperature. Blood is directed to the periphery for cooling. This transport of warm or cool blood helps in regulating body temperature.

III. **Protection**—Blood is essential for protection against blood loss and against toxins and the invasion of microbes.
   - Blood clotting—Blood contains platelets (**thrombocytes**) and fibrinogen, which allow for platelet plug formation and clotting following an injury.
   - Immune response—White blood cells (**leukocytes**) are transported throughout the body to protect against disease-causing agents.

Blood contains formed and unformed elements suspended in the plasma. The **formed elements,** or the cellular portion of the blood, include the anucleated red blood cells (erythrocytes), nucleated white blood cells (leukocytes), and platelets (thrombocytes—actually cell fragments). The **unformed elements** generally include nutrients, metabolic products, hormones, cytokines, and proteins, including antibodies that are transported in the plasma portion of the blood.

The percentage of the total blood volume that is composed of red blood cells is known as the **hematocrit.** The hematocrit is determined by centrifuging a blood sample to separate the blood into three layers: the red blood cells, the white blood cells and platelets (buffy coat), and the plasma (Figure 12-1). By measuring the percentage of the blood sample represented by red blood cells (the hematocrit), you can clinically assess blood function.

Anemia is a symptom of many different erythrocyte disorders and is defined as a deficiency in the oxygen-carrying capacity of the blood. Although there are many different causes of anemia, it generally results from a reduction in overall blood hemoglobin, either by a decrease in the number of red blood cells or a reduction in functional hemoglobin.

Aplastic anemia, for example, results from the destruction of the bone marrow (as from radioactivity). **Sickle-cell anemia** is a hereditary recessive trait characterized by the change of a sin-

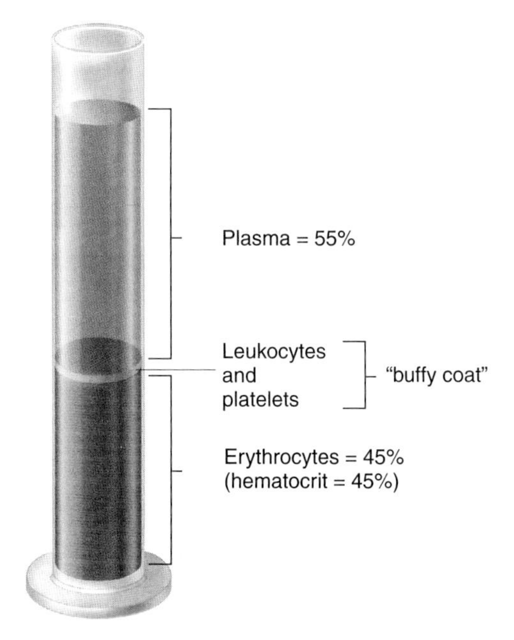

Figure 12-1 Measurement of the hematocrit by centrifugation.

gle hydrophilic amino acid (glutamate) to a hydrophobic amino acid (valine) in the hemoglobin molecule. This change in amino acid sequence changes the shape of hemoglobin and affects its oxygen-carrying capacity. This condition is known as sickle-cell anemia because it causes normally biconcave red blood cells to become sickle-shaped (Figure 12-2). The biconcave shape of normal red blood cells maximizes surface area, allowing respiratory gases to diffuse more rapidly. Sickle-shaped red blood cells have a low surface area-to-volume ratio and a significantly slower diffusion rate of respiratory gases. In addition, this sickle shape decreases the efficiency of blood flow through the circulatory system.

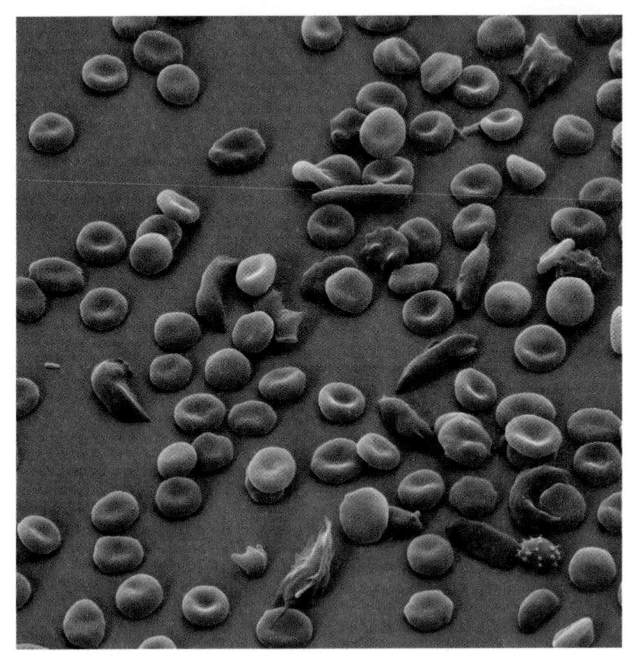

Figure 12-2 Electron micrograph of (a) normal erythrocytes and (b) erythrocytes from a patient with sickle-cell anemia.

12-1 Maturation anemia results from various vitamin deficiencies or problems in vitamin absorption. For example, **pernicious maturation anemia** results from insufficient secretion of intrinsic factor by the stomach. What is the role of intrinsic factor in red blood cell production?

Like all cells in the body, erythrocytes have glycoproteins (potential antigens) on the surface of their plasma membranes. Although there are at least 30 other common erythrocyte antigens, the A, B, and Rh factors are the most antigenic, and they can cause major problems during transfusions. **Agglutination** is the clumping of erythrocytes that results from antibody-antigen interactions (Figure 12-3). Antibodies against the A and B antigens, if not present on one's red blood cells, are automatically formed soon after birth, even without prior exposure to the antigens. Note that the O blood type denotes an absence of both A and B antigens on erythrocytes, and therefore the blood will contain both anti-A and anti-B antibodies (Table 12-1). Antibodies to other erythrocyte antigens not present on one's blood cells, such as the Rh factor (named for its discovery in rhesus monkeys), are formed only after exposure to the antigen.

12-2 Erythroblastosis fetalis, or **hemolytic disease of the newborn,** may occur when an Rh^- mother gives birth to an Rh^+ child (who inherited the dominant allele for the Rh factor from the father). At birth, the mother may be exposed to the blood of the fetus and develop antibodies to the Rh factor. If a mother is exposed to the fetus's Rh^+ blood during birth, what will happen if this mother conceives any subsequent Rh^+ children?

Rh^- mothers now routinely receive Rh-immune globulin (also called RhoGAM®) shortly after the birth of an Rh^+ fetus. RhoGAM is a preparation of anti-Rh antibodies that sequester the Rh epitopes (the antigenic portion of a molecule) on the surface of the fetal erythrocytes so that they do not induce an immune reaction in the mother.

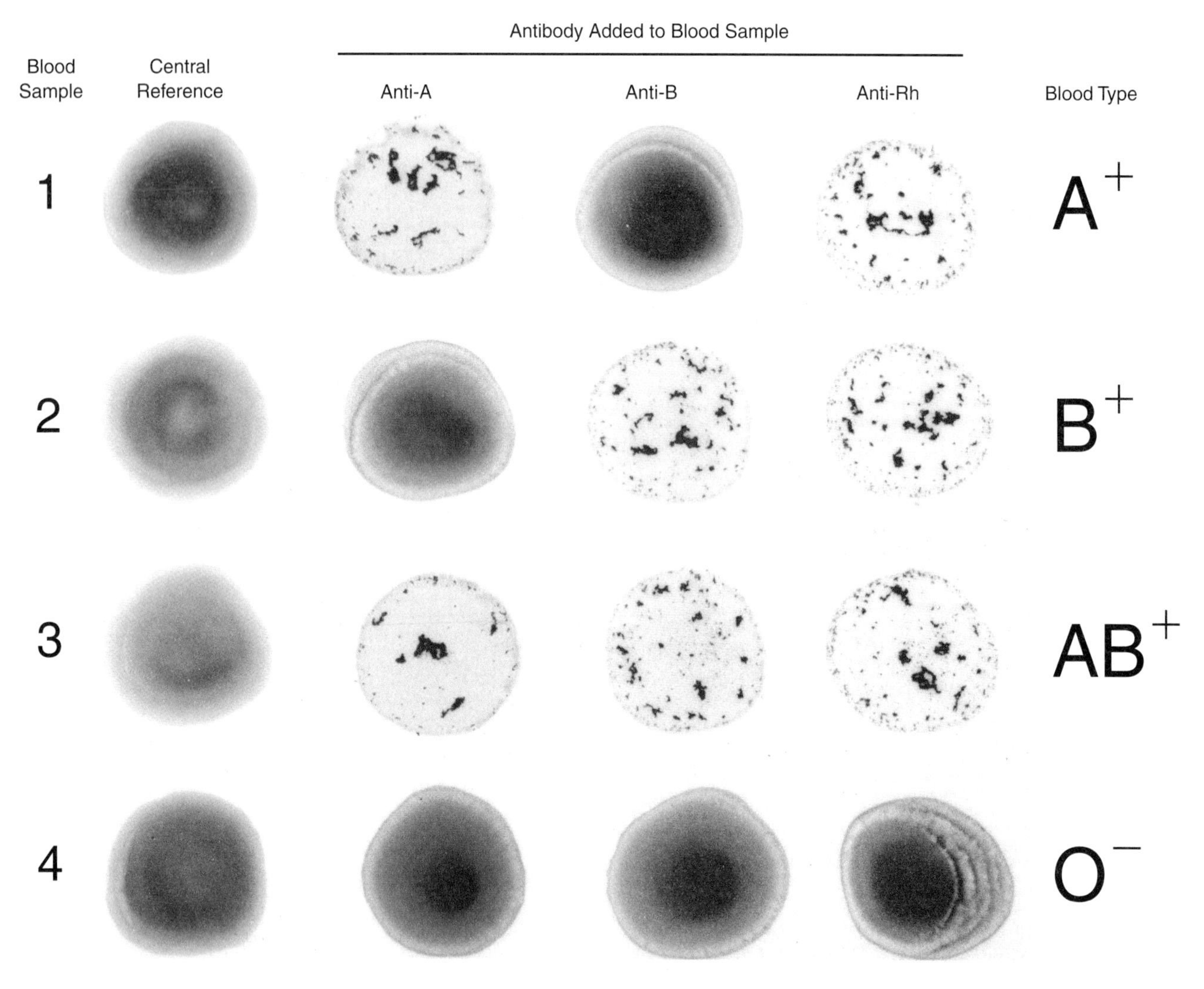

Figure 12-3 Sample agglutination reactions resulting from antibody-antigen interactions for the A, B, AB, and O blood types. When antibodies for an antigen are added to a blood sample containing that antigen, the erythrocytes clump together, or agglutinate.

Table 12-1 Summary of Factors for Blood Types A, B, AB, and O.

Blood Type	Genotype	Antigen on Erythrocyte	Antibodies in Plasma
A	$\mathbf{I^AI^A}$ or $\mathbf{I^Ai}$	A	Anti-B
B	$\mathbf{I^BI^B}$ or $\mathbf{I^Bi}$	B	Anti-A
AB	$\mathbf{I^AI^B}$	A and B	Neither Anti-A nor Anti-B
O	**ii**	Neither A nor B	Anti-A and anti-B

Another component of the formed elements of blood is white blood cells, or **leukocytes.** These are nucleated cells that function in immunity. There are two categories of leukocytes: granulocytes and agranulocytes, with several different types in each. **Granulocytes** contain numerous membrane-bound granules within the cytoplasm and appear grainy under a microscope. They are also called polymorphonuclear leukocytes because of the oddly shaped nuclei with multiple lobes. Neutrophils, eosinophils, and basophils are all types of granulocytes.

Neutrophils are the most abundant type of leukocyte, comprising 50 to 70% of the leukocytes in the blood. The nuclei of neutrophils usually consist of three to six lobes. The name neutrophil literally means "neutral-loving", and is derived from the fact that its two types of granules stain with both basic (blue) and acidic (red) dyes. These granules are very fine and, together, give the cytoplasm a light purple color. Neutrophil granules contain hydrolytic enzymes and antibiotic-like proteins called **defensins.** Neutrophils function as phagocytes and play an important role in combating bacterial infections.

Eosinophils contain large, coarse granules that stain deep red with acid (e.g., eosin) dyes. These lysosome-like granules contain digestive enzymes that help destroy invading parasitic worms such as tapeworms, flukes, and hookworms. The nucleus of an eosinophil has two lobes that are connected by an isthmus of nuclear material. Eosinophils normally comprise only 1 to 4% of the leukocytes in the blood.

Basophils are the rarest of the white blood cells and comprise only 0.1 to 0.3% of the total leukocytes. Their cytoplasm contains large, coarse granules that stain dark purplish-blue with basic dyes (basophil means base loving). The nucleus is usually S- or U-shaped, but it may be difficult to see among the many darkly-stained granules. Basophils release histamine, a vasodilator that promotes inflammation and attracts other leukocytes to the inflamed site. Basophils also contain heparin, an anticoagulant.

In contrast to granulocytes, **agranulocytes** lack visible granules within the cytoplasm, and their nuclei are typically round or kidney-shaped. The two types of agranulocytes are lymphocytes and monocytes. **Lymphocytes** are the second most common type of leukocyte, comprising 20 to 40% of the leukocytes in the blood. They are small cells (half the size of neutrophils) with large, round nuclei and little cytoplasm. Lymphocytes are responsible for specific immune defenses; they include the B lymphocytes (which initiate antibody-mediated immune responses), T lymphocytes (which initiate antigen-specific responses to virus-infected, cancer, and tissue-implant cells), and natural killer cells (which initiate nonspecific responses to virus-infected and cancer cells). **Monocytes,** the second type of agranulocyte, are the largest of the leukocytes. They are relatively rare in the bloodstream, normally comprising only 2 to 8% of the leukocytes, and they possess kidney- or horseshoe-shaped nuclei. Monocytes function in phagocytosis; upon leaving the bloodstream they are transformed into highly mobile macrophages. These macrophages are an important component of the body's defense against invading microbes and debris. They are also important in activating lymphocytes during specific immune responses.

In Experiment 12.4 of this laboratory, you will examine differential leukocyte counts in different patients. You should identify examples of each leukocyte type in the blood smear from the healthy male patient. Identifying which type(s) of leukocytes are affected in the ill patients, combined with your knowledge of leukocyte function, will help you determine the disease condition of each patient.

12-3 For each of the following conditions, identify which type of leukocyte plays the primary role in regulating the immune response.

________________ Release histamine to promote inflammation at the site of injury.

________________ Transform into plasma cells and secrete antibodies.

________________ Release antibiotic-like proteins to destroy bacterial invaders, such as in bacterial meningitis.

________________ Transform into tissue macrophages and activate lymphocyte-mediated immune responses.

________________ Initiate an immune response to (i.e., reject) a kidney transplant.

________________ Kill invading tapeworms. ■

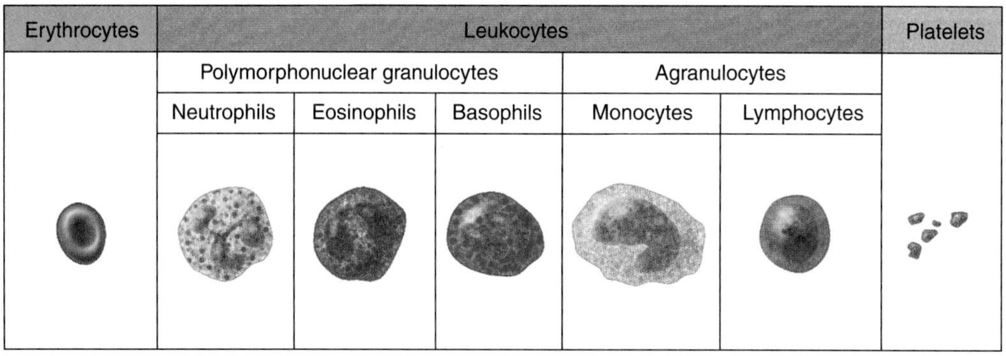

Erythrocytes	Leukocytes					Platelets
	Polymorphonuclear granulocytes			Agranulocytes		
	Neutrophils	Eosinophils	Basophils	Monocytes	Lymphocytes	

Figure 12-4 Classes of blood cells.

Figure 12-4 shows examples of the major classes of blood cells. Table 12-2 summarizes their numbers and distribution.

Table 12-2 Numbers and Distributions of Erythrocytes, Leukocytes, and Platelets in Normal Human Blood

Total erythrocytes = 5,000,000 per mm^3 of blood
Total leukocytes = 7000 per mm^3 of blood
Percent of total leukocytes
Polymorphonuclear granulocytes
Neutrophils 50–70%
Eosinophils 1–4%
Basophils 0.1–0.3%
Monocytes 2–8%
Lymphocytes 20–40%
Total platelets = 250,000 per mm^3 of blood

Methods and Materials

EXPERIMENT 12.1 Blood Typing

In this experiment, you will determine the blood types of two randomly-selected unknown patients.

- If the blood typing tray is transparent, place it on a white sheet of paper to help you to see the agglutination reactions.
- Place two drops of the anti-A antibody into the well labeled "A."
- Place two drops of the anti-B antibody into the well labeled "B."
- Place two drops of the anti-Rh antibody into the well labeled "Rh."
- Now place two drops of the patient's blood into each of the three labeled wells.
- Using a plastic toothpick, stir the blood into the antibody solution. Use a different toothpick for each reaction, being careful not to cross-contaminate the reactions.
- Examine each well for an agglutination reaction. If the sample appears cloudy or grainy, then the antibody has reacted with that specific antigen on the patient's red blood cells.
- Record your observations in Results Table 12.1 of your Laboratory Report.
- Repeat these procedures to determine the blood type of a second unknown patient.
- From your observations, determine the blood type of each of your two patients. Check with your instructor to determine if you have correctly identified the blood type of each patient.
- After you have correctly identified each patient's blood type, wash the blood typing trays and toothpicks and return the cleaned items to your laboratory table.

EXPERIMENT 12.2 Hematocrit Levels in Male and Female Patients

In this experiment, you will examine differences in the hematocrit levels of male and female patients.

- Each laboratory group should determine the hematocrit for two male and two female patients.

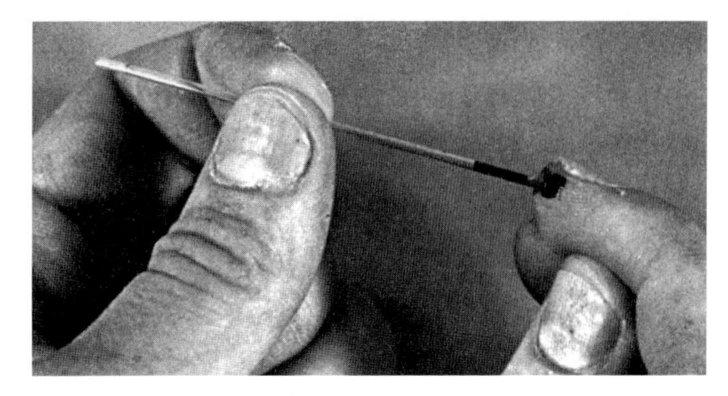

Figure 12-5 Method for filling a capillary tube with blood from the fingertip.

- For each patient, place three drops of blood on a microscope slide. Remember to keep track of which capillary tube belongs to each patient.
- To fill a capillary tube with the patient's blood, hold the capillary tube at a 45° angle to the microscope slide. Place the end of the capillary tube into the patient's blood. Allow the capillary tube to fill with blood (Figure 12-5).
- Seal one end of the capillary tube by sticking it into the Critoseal tray.
- When your laboratory group's capillary tubes have been prepared for two male and two female patients, place the capillary tubes into the hematocrit centrifuge. Be sure the sealed ends point toward the outer edge of the centrifuge, with the unsealed ends of the capillary tubes closest to the center of the centrifuge (otherwise, your blood sample will be splattered around the inside of the centrifuge when the centrifugal force pushes the heavier formed elements toward the outer edge of the centrifuge).
- Be sure to record the slot numbers where your group's capillary tubes are located as well as the slot numbers of your individual patients.
- When all capillary tubes are correctly loaded into the hematocrit centrifuge, your instructor will centrifuge the capillary tubes for approximately five minutes.
- When the centrifuge stops, retrieve your patients' hematocrit samples. You will find that each blood sample is now separated into three distinct regions: the erythrocytes, the buffy coat, and the plasma. The centrifugal force during centrifuging causes the heavier formed elements

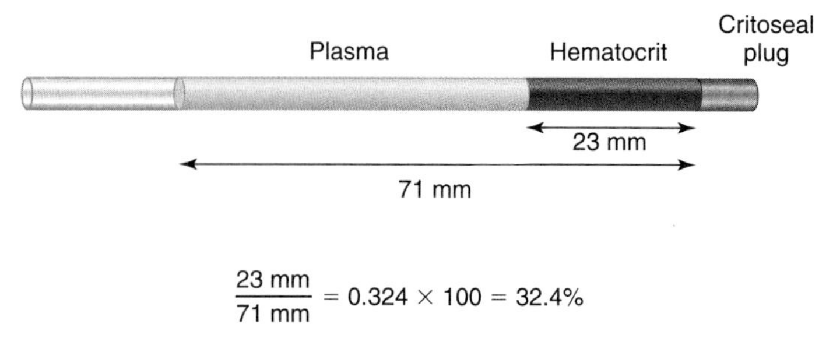

$$\frac{23\text{ mm}}{71\text{ mm}} = 0.324 \times 100 = 32.4\%$$

Figure 12-6 Sample calculation of hematocrit from blood in a capillary tube.

(namely the erythrocytes) to be packed into one end of the capillary tube. If you examine closely the border between the red blood cells and the plasma, you can sometimes see a distinct layer of leukocytes, called the buffy coat.

- Using a ruler, measure the length of the entire blood sample to the nearest millimeter.
- Now measure the length of the erythrocyte region to the nearest millimeter.
- Use the following formula to calculate the hematocrit, or the percentage of each blood sample occupied by red blood cells (Figure 12-6).

$$\text{Hematocrit} = \frac{\text{Length of the erythrocyte region}}{\text{Length of the blood sample}} \times 100$$

- Record the hematocrit for each of your patients in the results section of your Laboratory Report.

EXPERIMENT 12.3 Mr. Jones

Mr. Jones goes to see the doctor with complaints of being easily fatigued and always feeling lethargic and tired. The doctor orders a blood sample to be drawn and analyzed for hematocrit level. In the laboratory, you receive Mr. Jones's blood sample for analysis.

- Place three drops of Mr. Jones's blood on a microscope slide.
- Using the methods described in Experiment 12.2, fill a capillary tube with Mr. Jones's blood and seal one end of the capillary tube using the Critoseal tray.
- Place the capillary tube into the hematocrit centrifuge and record in which slot your laboratory group's capillary tube is located.
- When all capillary tubes are correctly loaded into the hematocrit centrifuge, your instructor will centrifuge the capillary tubes for approximately five minutes. When the centrifuge stops, retrieve your group's capillary tube and measure the hematocrit level for Mr. Jones's blood sample.
- Record your data in Problem Set 12.3 of your Laboratory Report.

EXPERIMENT 12.4 Differential Leukocyte Count

In this experiment you will examine the relative frequency of leukocytes that are present in blood smears from different patients.

- Using a microscope, examine the blood smear from the healthy male patient. Estimate the relative ratio of erythrocytes to leukocytes (for example, 500 to 1).
- Identify an example of each of the five primary leukocytes.
- Now examine the blood smears from the ill male patients. Estimate the relative ratio of erythrocytes to leukocytes. Which type(s) of leukocytes are affected?
- Record your observations in the results section of your Laboratory Report.

RESULTS AND DISCUSSION

LABORATORY REPORT 12

EXPERIMENT 12.1
Blood Typing

Results Table 12.1 Individual Group Data For Experiment 12.1: Blood Typing. Use the following table to record your data from the blood typing experiment for your two randomly selected patients. Enter "Yes" if the patient's blood showed an agglutination reaction with the antibody and "No" if there was no reaction. Use your observations to discern the blood type of each patient.

Patient Number	Agglutination Reaction with:			Blood Type
	Anti-A Antibody?	Anti-B Antibody?	Anti-Rh Antibody?	

Problem Set 12.1: Blood Typing

a. Which blood type is considered the universal donor? Why?

b. Which blood type is considered the universal recipient? Why?

EXPERIMENT 12.2
Hematocrit Levels in Male and Female Patients

Results Table 12.2 Class Data for Experiment 12.2: Hematocrit Levels in Male and Female Patients.

Laboratory Group	Patient	Hematocrit (% Erythrocytes)	
		Male Patients	Female Patients
A	**1**		
	2		
B	**3**		
	4		
C	**5**		
	6		
D	**7**		
	8		
E	**9**		
	10		
F	**11**		
	12		
	Mean (±SE)		

Problem Set 12.2: Hematocrit Levels in Male and Female Patients

Using the data from Results Table 12.2, create a bar graph showing mean hematocrit level for the male and female patients. Be sure to include a 95% CI. Using these data, perform the appropriate statistical test to determine whether there are differences in the hematocrit levels of males and females.

a. State the null hypothesis.

b. Should you accept or reject the null hypothesis? Support your conclusions with the appropriate statistical results.

c. What physiological significance does a higher hematocrit have? Explain.

d. One factor that contributes to the sex difference in hematocrit levels is the higher concentration of testosterone in males, which stimulates the production of erythropoietin. Explain the role of erythropoietin in determining hematocrit level.

Problem Set 12.3: Mr. Jones

a. What is Mr. Jones's hematocrit? Show your calculations.

b. Compare Mr. Jones's hematocrit level to the 95% confidence interval you calculated for normal male patients in Experiment 12.2. Is Mr. Jones's hematocrit level normal for an adult male?

c. What do you think is wrong with Mr. Jones? Describe why your lab results help explain his symptoms.

d. As you are writing up the summary of lab results for the doctor, you notice that the patient label actually reads "Ms. Jones." Does this change your diagnosis? Why or why not?

e. How is it possible that Ms. Jones is complaining of symptoms that are characteristic of anemia, yet she has a normal hematocrit level? Could Ms. Jones still be suffering from anemia? Explain your answer thoroughly and provide an example.

Problem Set 12.4: Differential Leukocyte Count

a. How does the relative ratio of erythrocytes to leukocytes compare for the different male patients?

b. What do you think may be wrong with the ill male patients? Explain.

COMPARATIVE NOTE *Chapter 12*

Blood or Bloods?

Not all animals have the same blood properties as humans. There are different types of blood, depending on the respiratory pigment and where this pigment is located. Some animals, such as earthworms and *Planorbis* (e.g., Escargot) have their respiratory pigment (oxygen-carrying protein) in the plasma rather than enclosed within red blood cells. In addition, there are four different respiratory pigments. One of these, hemocyanin, uses copper rather than iron as the oxygen-binding element. Iron is what gives human blood its red color when blood is oxygenated. What color do you think hemocyanin is when oxygenated? (*Hint:* Think about what color a penny is when it becomes oxidized.) Animals with hemocyanin as their respiratory pigment include squid, gastropods, and the arthropod pictured below, the horseshoe crab (*Limulus polyphemus*). Of the four respiratory pigments, hemoglobin is the most common. The following table shows the different respiratory pigments and some of their characteristics.

			LOCATION	
Respiratory Pigment	**Protein Properties**	**Molecular Weight ($\times 10^3$)**	In Cells?	In Plasma?
Hemoglobin	Iron-porphyrin protein	17–3000	yes	yes
Chlorocruorin	Iron-porphyrin protein	2750–3400	no	yes
Hemerythrin	Iron-containing protein	66–108	yes	no
Hemocyanin	Copper-containing protein	300–9000	no	yes

Information in this table was modified from Schmidt-Nielsen (1983).

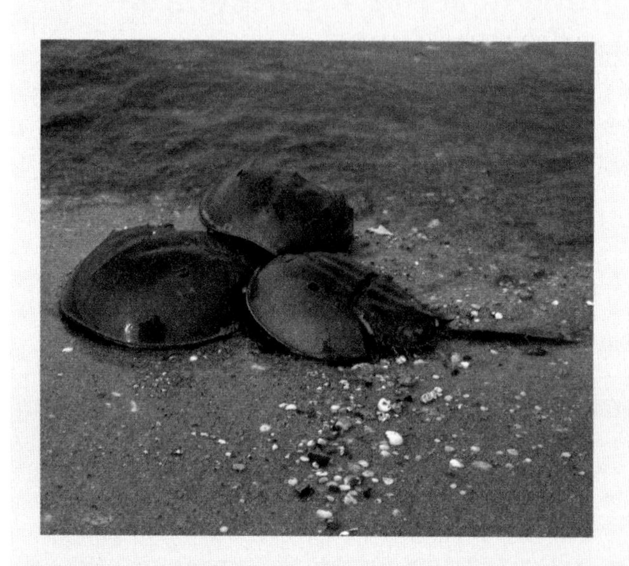

RESEARCH OF INTEREST

Ballarin, L., M. Dall'Oro, D. Bertotto, A. Libertini, A. Francescon, and A. Barbaro. 2004. Hematological parameters in *Umbrina cirrosa* (Teleostei, Sciaenidae): a comparison between diploid and triploid specimens. Comparative Biochemistry and Physiology Part A 138:45–51.

Rai, S. and C. Haldar. 2003. Pineal control of immune status and hematological changes in blood and bone marrow of male squirrels (*Funambulus pennanti*) during their reproductively active phase. Comparative Biochemistry and Physiology Part C 136:319–328.

Sergent, N., T. Rogers, and M. Cunningham. 2004. Influence of biological and ecological factors on hematological values in wild Little Penguins, *Eudyptula minor*. Comparative Biochemistry and Physiology Part A 138:333–339.

LABORATORY 13

Respiratory Physiology

PURPOSE

This laboratory will introduce the mechanics of breathing and the measures of pulmonary function. You will learn to use a spirometer to determine your lung volumes and capacities. The neurological control of respiration rate in response to changing carbon dioxide (CO_2) concentration and pH will be the major focus of this laboratory.

Learning Objectives

- Explain and understand the mechanics of breathing and identify the major muscles that are involved in inspiration and expiration.
- Define all of the lung volumes and capacities and identify these pulmonary measures on a spirogram recording.
- Demonstrate an ability to measure and calculate lung volumes and capacities using a spirometer and the appropriate mathematical relationships.
- Understand how blood pH is calculated and how this calculation differs from the pH calculations discussed in Laboratory 4 ("Enzyme Activity").
- Describe the mechanisms by which CO_2 concentration and pH serve as a neural stimulus for the control of ventilation rate.
- Define the terms *acidosis* and *alkalosis* and understand how these conditions relate to ventilation rates and metabolic rate.

Laboratory Materials

Experiment 13.1: Measures of Pulmonary Function

Spirometry equipment (e.g., Spirotest® or Spirocomp®)

Experiment 13.2: The Effect of CO_2 Concentration on Ventilation

Stopwatch

Introduction and Pre-Lab Exercises

Most cells need oxygen to perform the chemical reactions of ATP synthesis. Carbon dioxide, a by-product of these reactions, must also be eliminated from cells. Although unicellular organisms can exchange these gases directly with the environment, larger organisms need a system of specialized structures for gas transport to and from all cells. In mammals, the **respiratory system** consists of the oral and nasal cavities, the lungs, the airways connecting the oral and nasal cavities with the lungs, and the chest structures (e.g., skeletal muscles) involved in moving air into and out of the lungs during respiration.

The airways beyond the larynx can be divided into two zones: the conducting zone and the respiratory zone. The **conducting zone** extends from the superior trachea to the beginning of the respiratory bronchioles. With a volume of about 150 ml, the space within the conducting airways is also called the **anatomic dead space** because no gas exchange occurs here. The conducting zone helps to filter, warm, and humidify inhaled air. The **respiratory zone** consists of those structures that partake in gas exchange, namely the respiratory bronchioles, alveolar ducts, and alveolar sacs (Figure 13-1).

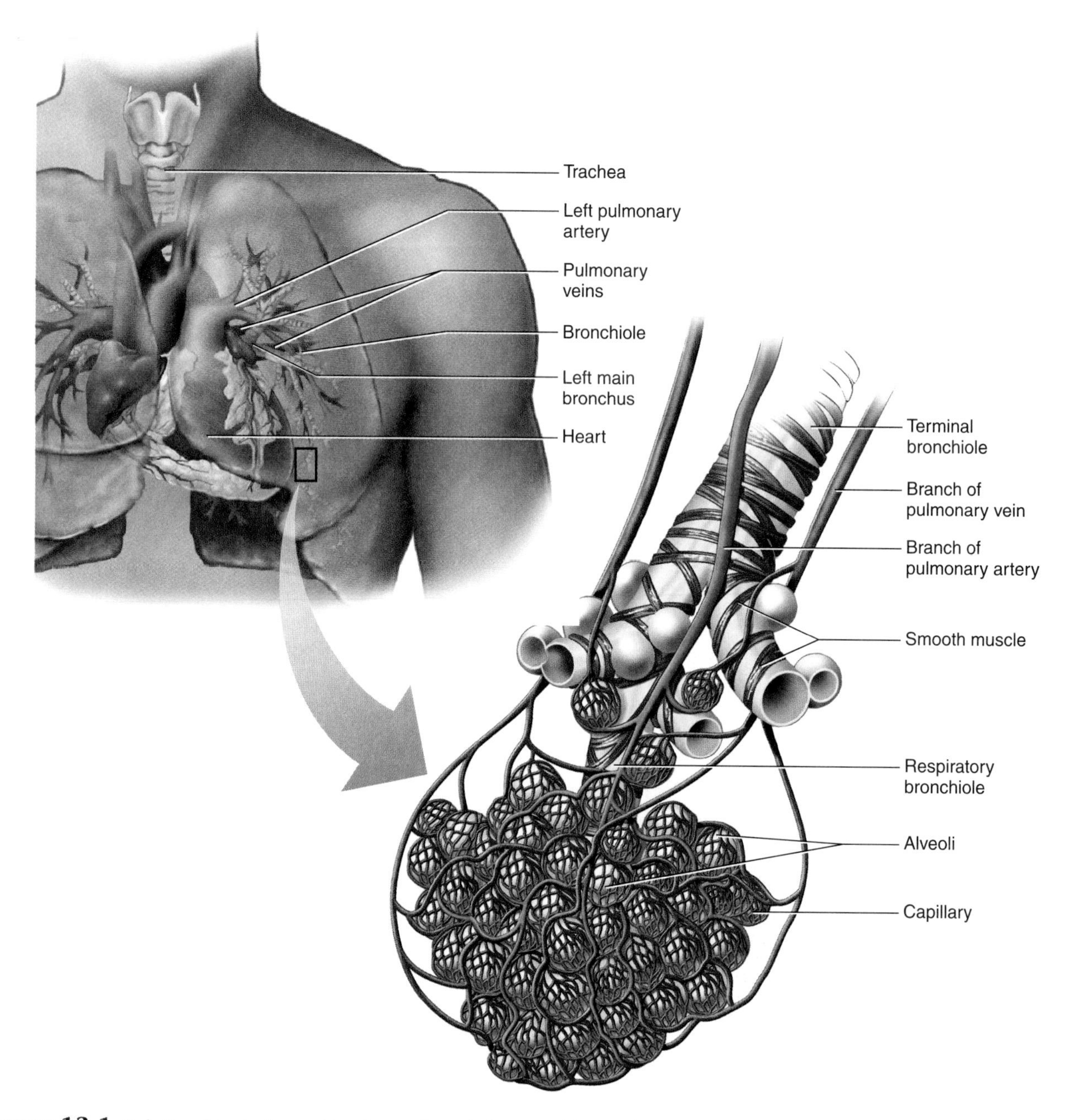

Figure 13-1 Relationships between blood vessels and airways with an enlargement that shows a cluster of alveoli and associated capillaries at the end of the airway.

During quiet or unforced ventilation, the movement of air into and out of the lungs results primarily from contraction and relaxation, respectively, of the diaphragm. The **diaphragm** is a dome-shaped skeletal muscle that separates the thorax (chest) from the abdomen. When the diaphragm contracts, it "flattens" and pulls the lungs toward the abdomen (Figure 13-2). This creates a larger pulmonary volume vertically, which in turn lowers intrapulmonary pressure (recall Boyle's law, which states that the pressure of a given quantity of gas is inversely proportional to its volume). Because intrapulmonary pressure is now lower than atmospheric pressure outside the lungs, air flows down its pressure gradient into the lungs until intrapulmonary and atmospheric pressures are equal. This is the process of **inspiration.** When the diaphragm relaxes, it regains its dome shape and "moves into" the thoracic cavity. This decreases the volume of the lungs and raises intrapulmonary pressure above atmospheric pressure. Thus, air will move out of the lungs until pressures equalize. This is the process of **expiration.**

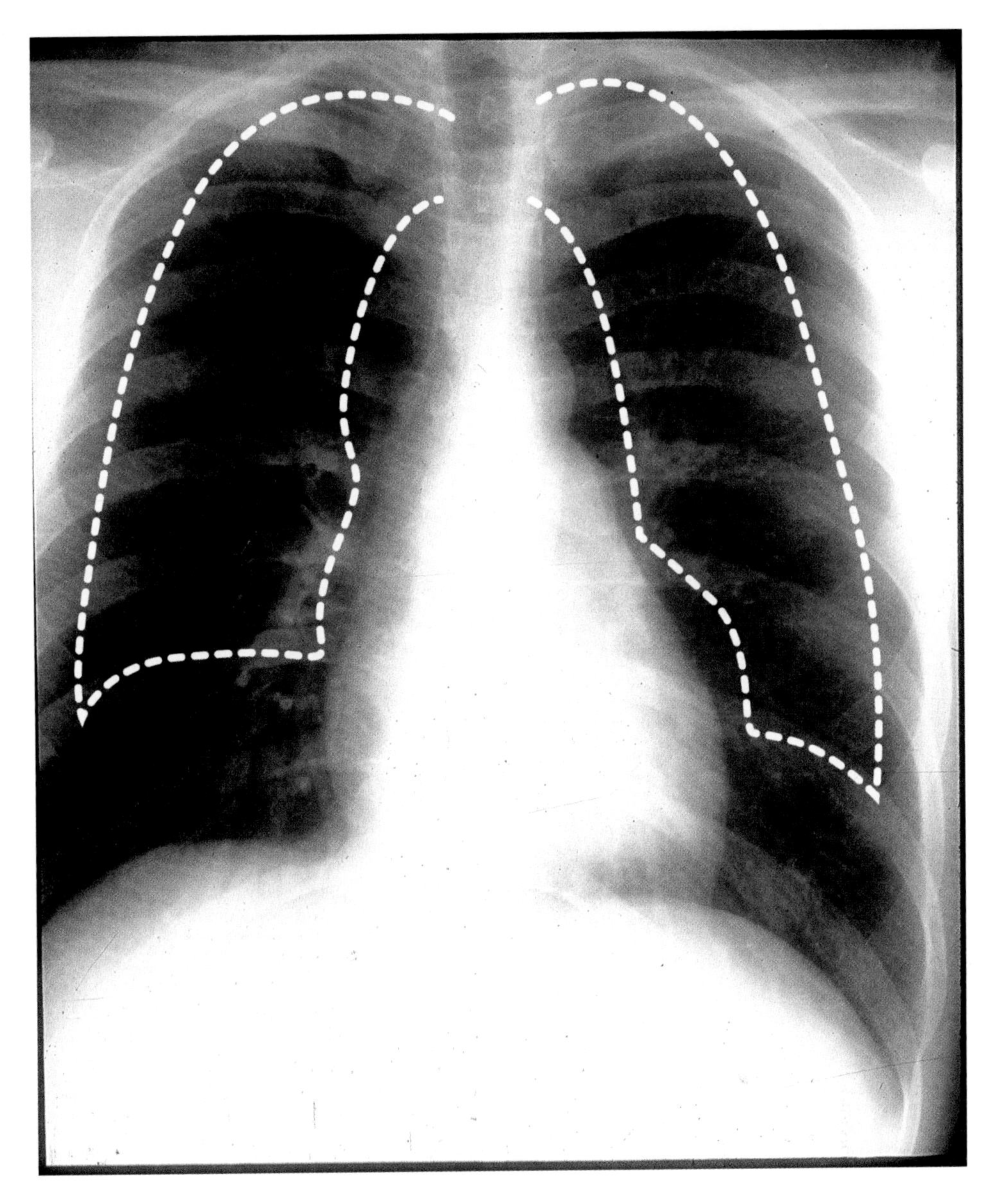

Figure 13-2 X-ray of chest at full inspiration. The dashed white line is an outline of the lungs when they are in full expiration.

Other skeletal muscles involved in the processes of ventilation include the internal and external intercostal muscles (Figure 13-3). The **external intercostal muscles** aid in inspiration when they contract and raise the ribs to expand the thoracic cavity laterally. Recall that the origin of an external intercostal muscle is the superior rib and the insertion (i.e., the movable point) is the inferior rib. The **internal intercostal muscles** aid in expiration by contracting to pull the ribs downward and decrease the volume of the thoracic cavity (internal intercostal muscles originate on the inferior rib and insert on the superior rib). In forceful (deep) breathing, other skeletal muscles, such as the **abdominal muscles, pectoralis minor, scalene muscles,** and the **sternocleidomastoid,** also become involved in the processes of ventilation. Although the muscles that aid in ventilation are skeletal muscles under voluntary control, ventilation is an involuntary process ultimately regulated by the nervous system. This neural regulation of the ventilation rate occurs in the **respiratory control center** in the **medulla oblongata,** a structure located in the brainstem. Thus, although you can modulate your breathing patterns and breathing rate voluntarily, you cannot completely override the neural control of ventilation.

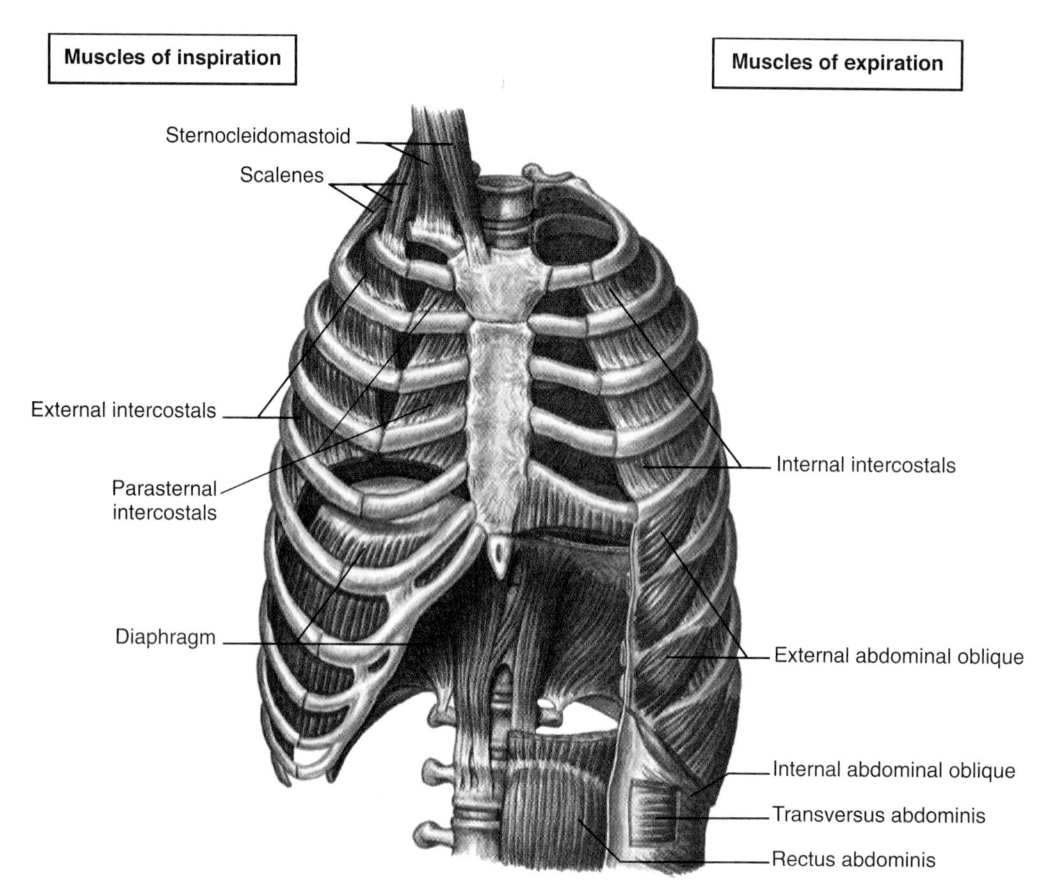

Figure 13-3 Muscles involved in inspiration and expiration.

13-1 The idle threat of a child to hold his breath until he turns blue is of little consequence to a concerned parent. Explain why. ■

The clinical assessment of pulmonary function is accomplished by measuring a variety of lung volumes. These volumes can be measured using a spirograph. For example, the **tidal volume (TV)** is the volume of air moved per breath (approximately 500 ml) during quiet breathing. This volume represents the volume of air (and hence the quantity of O_2) necessary to sustain resting metabolic activity. The tidal volume is far less than the total capacity of the lungs; this allows the respiratory volumes to be adjusted in response to changes in O_2 demand. For example, we have the capacity to inhale and exhale far more air than the tidal volume. The **inspiratory (IRV)** and **expiratory (ERV) reserve volumes** are equal to the additional volume of air that can be forcefully inhaled or exhaled, respectively, beyond the tidal volume. Thus, the capacity of the lungs to move air is described by the vital capacity. **Vital capacity (VC)** is the total volume of air that can be moved into and out of the lungs; it is equal to TV + IRV + ERV.

13-2 Label Figure 13-4 with the appropriate respiratory volumes by filling in the blank spaces provided. ■

13-3 What happens to an individual's tidal volume during physical exertion? What is the metabolic significance of this change in tidal volume? ■

13-4 What happens to an individual's inspiratory reserve volume during physical exertion? What happens to an individual's vital capacity during physical exertion? ■

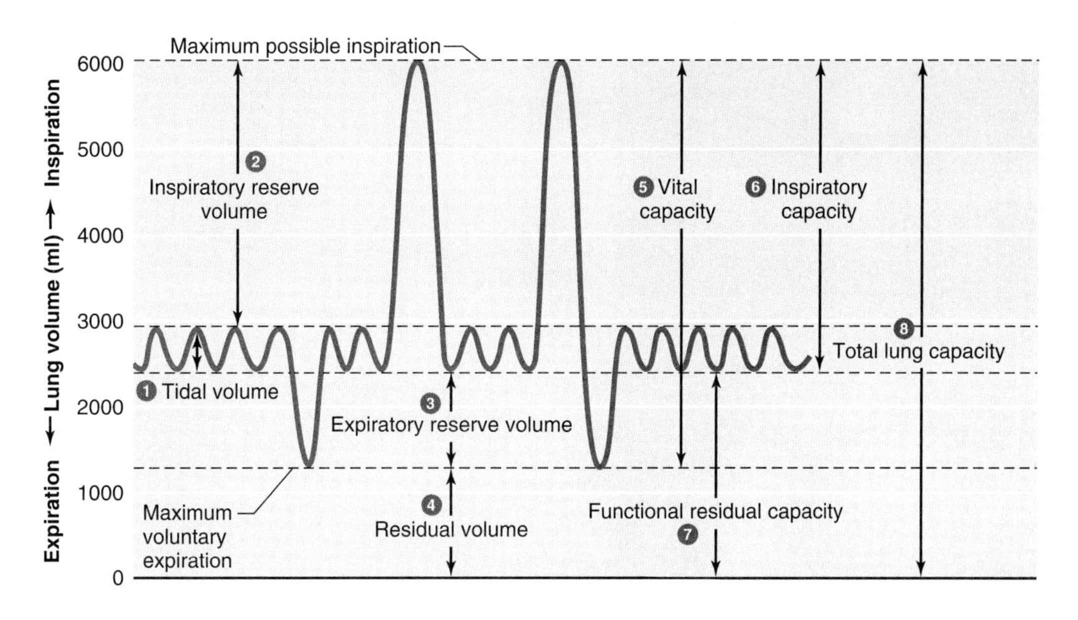

Respiratory Volumes and Capacities for an Average Young Adult Male			
	Measurement	Typical Value	Definition
			Respiratory Volumes
1	____________	500 ml	Amount of air inhaled or exhaled in one breath during relaxed, quiet breathing
2	____________	3000 ml	Amount of air in excess of tidal inspiration that can be inhaled with maximum effort
3	____________	1200 ml	Amount of air in excess of tidal expiration that can be exhaled with maximum effort
4	____________	1200 ml	Amount of air remaining in the lungs after maximum expiration; keeps alveoli inflated between breaths and mixes with fresh air on next inspiration
			Respiratory Capacities
5	____________	4700 ml	Amount of air that can be exhaled with maximum effort after maximum inspiration (ERV + TV + IRV); used to assess strength of thoracic muscles as well as pulmonary function
6	____________	3500 ml	Maximum amount of air that can be inhaled after a normal tidal expiration (TV + IRV)
7	____________	2400 ml	Amount of air remaining in the lungs after a normal tidal expiration (RV + ERV)
8	____________	5900 ml	Maximum amount of air the lungs can contain (RV + VC)

Figure 13-4 Lung volumes and capacities.

Note that the vital capacity is not the total amount of air contained in the lungs. The lungs retain a **residual volume (RV)** of approximately 1200 ml that remains in the lungs even after a maximal expiration (i.e., this volume cannot be removed from the lungs). This residual volume helps to keep the alveoli open and prevents lung collapse. **Total lung capacity (TLC)** is therefore the total volume of air contained in the lungs following a maximal inspiration. It is equal to VC + RV.

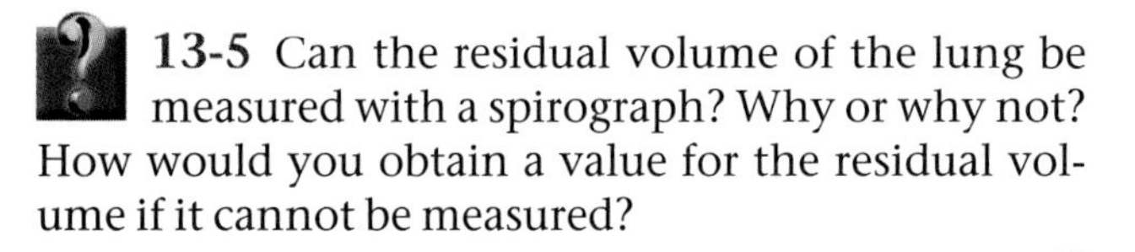

13-5 Can the residual volume of the lung be measured with a spirograph? Why or why not? How would you obtain a value for the residual volume if it cannot be measured?

■

13-6 What is the difference between a lung volume and a lung capacity?

■

13-7 Figure 13-5b is a spirometer recording from a patient. Examine the respiratory volumes

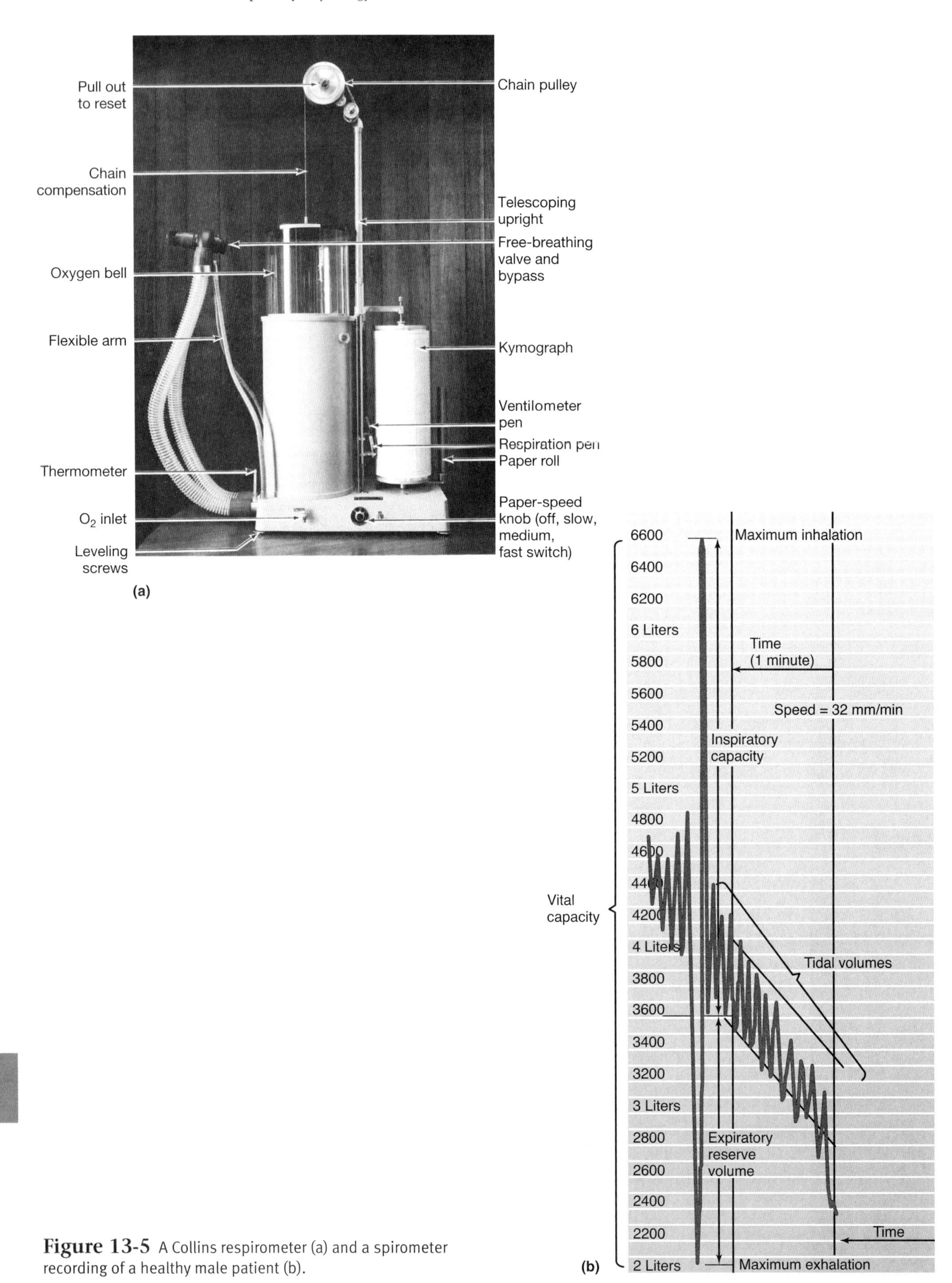

Figure 13-5 A Collins respirometer (a) and a spirometer recording of a healthy male patient (b).

Respiratory Volumes	Formula and Calculations	Volume (ml)
Tidal volume		
Alveolar volume		
Dead air space volume		140 ml
Inspiratory reserve volume		
Expiratory reserve volume		
Vital capacity		4300 ml
Total lung capacity		

shown on the tracing. Define and calculate the respiratory volumes listed in the table above. Use the information presented in both the tracing and the table to assist you in your calculations.

■

Measures of lung volumes and tests of pulmonary function aid in the diagnosis of lung disorders, including restrictive and obstructive disorders. In **restrictive disorders,** such as pulmonary fibrosis, airway resistance is normal but vital capacity is reduced because respiratory movements are impaired. Impaired respiratory movements may result from abnormalities or from damage to lung tissue or the chest wall, for example. In contrast, **obstructive disorders** are characterized by increased airway resistance and a significant reduction in the ability to inspire and expire air from the lungs. Because respiratory movements are normal and lung tissue is not damaged, patients with obstructive disorders typically have normal vital capacities. A common example of an obstructive disorder is the inflammation and bronchoconstriction that accompany asthma episodes or attacks.

13-8 Is snoring a sign of a restrictive or obstructive disorder? Explain.

■

As you learned in Laboratory 12 ("Physiology of Blood"), blood is a respiratory tissue that contains hemoglobin for the transport of O_2 and CO_2. In this laboratory, we will discuss respiratory physiology, focusing on the mechanisms by which CO_2 concentration and pH serve as a neural stimulus for the control of ventilation rate.

Carbon dioxide is a by-product of cellular respiration; it is produced when ATP is synthesized during the Krebs cycle. The CO_2 produced by respiring cells is highly soluble in the blood, where it reacts with water to form carbonic acid. One of the H^+ ions dissociates from carbonic acid to form bicarbonate and H^+.

$$CO_2 + H_2O \underset{\text{Carbonic anhydrase}}{\rightleftharpoons} \overset{\text{Carbonic acid}}{H_2CO_3} \underset{\text{Carbonic anhydrase}}{\rightleftharpoons} \overset{\text{Bicarbonate}}{HCO_3^-} + H^+$$

Recall from Laboratory 4 ("Enzyme Activity") that pH is calculated as the log of the inverse of the H^+ concentration. pH may also be calculated using the following formula:

$$\text{Blood pH} = \text{pK} + \log [HCO_3]/[CO_2]$$

For example, for a normal blood pH of 7.4, the ratio of bicarbonate to CO_2 concentration is 20:1. The pK is a constant that defines the isoelectric equivalence point of the log dissociation constant for a weak acid. For the calculation of blood pH, pK is 6.1. Thus,

$$\begin{aligned} \text{pH} &= 6.1 + \log 20/1 \\ &= 6.1 + 1.3 \\ &= 7.4 \end{aligned}$$

Most CO_2 is transported through the blood in the form of bicarbonate. Thus, if there is an excess of CO_2 in the system, the concentration of bicarbonate and H^+ in the blood increases, and blood pH will decrease. Likewise, if there is low partial pressure of CO_2, pH will increase. Hydrogen ions are transported, via hemoglobin, to the lungs. During the process of reoxygenation, hemoglobin releases the hydrogen ions, which then react with bicarbonate to form carbonic acid. Carbonic acid is then converted into CO_2 and H_2O by carbonic anhydrase, allowing CO_2 to be exhaled.

13-9 What effect does an exercising muscle have on the pH of the extracellular fluid? Explain how both aerobic and anaerobic metabolism in a muscle fiber affect pH.

■

Normal blood pH ranges from only 7.35 to 7.45. Deviations in blood pH are monitored by two sets of chemoreceptors: the central chemoreceptors and peripheral chemoreceptors. The **central chemoreceptors** are located in the medulla oblongata and monitor the pH of the brain's extracellular fluid. Central chemoreceptors are stimulated by an increase in the concentration of H^+; such changes in H^+ concentration result primarily from increases in blood CO_2 concentration. Thus, central chemoreceptors are usually described as being indirectly sensitive to increased CO_2 concentrations because they are directly sensitive to H^+.

The **peripheral chemoreceptors** include both the aortic bodies, located near the aortic arch, and the carotid bodies, located in the carotid artery. Both are contained within small nodules and monitor arterial blood pH. The location of the carotid bodies is particularly strategic for monitoring the O_2 concentration of the arterial blood supplying the brain. The peripheral chemoreceptors in the aortic and carotid bodies are sensitive to decreases in O_2 concentration (hypoxia) as well as increases in the H^+ concentration of the arterial blood. Thus, peripheral chemoreceptors can respond to an increase in H^+ concentration that results from either metabolic acidosis or respiratory acidosis (which again results mainly from an increase in CO_2 concentration).

Both central and peripheral chemoreceptors relay their sensory information to the respiratory control center in the medulla oblongata via excitatory synaptic input. In turn, the medulla oblongata, in conjunction with respiratory control centers located in the **pons,** modulates ventilation rate via motor neuron innervation of the respiratory muscles.

13-10 Identify the sensory receptors in the breathing reflex. What is (are) the effector(s)?

■

In the presence of a low pH, hemoglobin molecules experience a decreased affinity for O_2 (this is called the Bohr effect). Hemoglobin molecules release O_2 in exchange for H^+, and O_2 diffuses into the cells. When the chemoreceptors detect a decrease in pH, the medulla oblongata increases the breathing rate to reoxygenate the hemoglobin molecules and to rid the body of excess CO_2 more rapidly. The removal of the free H^+ from the area restores pH to normal levels, and the breathing rate returns to resting levels.

13-11 In response to the psychological stress of taking an exam, the person sitting next to you has begun to hyperventilate. The person's shallow, rapid breaths result in decreased blood O_2 as well as blood CO_2 levels because this person is expiring very rapidly but not inspiring normal quantities of air. Someone in the class gives the student a brown paper bag to breathe into. Will this help return the student's blood pH to normal? Explain the mechanism.

■

Methods and Materials

EXPERIMENT 13.1 Measures of Pulmonary Function

In this experiment you will measure and calculate lung volumes and capacities using spirometry.

Spirotest® Procedure (Non–Computer-Based Procedure)

- Connect a sterile mouthpiece to the nozzle of the Spirotest® spirometer (Figure 13-6). Remember to remove and dispose of your mouthpiece after obtaining your pulmonary measures.
- Check to see if the spirometer is zeroed. If not, rotate the scale in a clockwise or counter-clockwise direction until the pointer is on zero.
- Hold the spirometer in one hand so that the nozzle is facing you. Make sure the spirometer is held level to allow for proper rotation of the pointer. Also ensure that the vent holes are not covered by your hand, which would prevent normal air flow.
- **Measuring Vital Capacity:** From a sitting position, inhale as deeply as possible, fully filling your lungs with air. Hold your breath for a second or two and then place your lips on the mouthpiece and exhale continuously into the spirometer until you can no longer exhale. Record your vital capacity and repeat the procedure twice more to obtain a mean vital capacity.

  Trial 1: ________________ ml

  Trial 2: ________________ ml

  Trial 3: ________________ ml

  Mean vital capacity: ______________ ml

- **Measuring Expiratory Reserve Volume:** From a sitting position, inhale and exhale normally. At the end of a normal exhalation, place your lips on the mouthpiece and exhale the remaining air continuously into the spirometer until you can no longer exhale. Record your expiratory reserve volume and repeat the procedure twice more to obtain a mean expiratory reserve volume (ERV).

  Trial 1: ________________ ml

  Trial 2: ________________ ml

  Trial 3: ________________ ml

  Mean ERV: ________________ ml

Figure 13-6 Illustration for the use of Spirotest®.

- **Measuring and Calculating Tidal Volume:** Normal tidal volume is approximately 500 ml. Because the scale on the Spirotest® was designed to measure only vital capacity and not tidal volume, we must use the following method to obtain a measure of tidal volume:
  - Measure tidal and expiratory reserve volumes simultaneously by inhaling and exhaling normally. At the end of a normal inhalation, place your lips on the mouthpiece and exhale continuously into the spirometer until you can no longer exhale. This measure is both your tidal volume and your expiratory reserve volume. Subtract your mean ERV from this value to obtain tidal volume. Record this calculated tidal volume and repeat the procedure twice more to obtain a mean tidal volume (TV).

    Trial 1: ________________ ml

    Trial 2: ________________ ml

    Trial 3: ________________ ml

    Mean TV: ________________ ml

- Using your mean VC, ERV, and TV, calculate inspiratory capacity and inspiratory reserve volume and estimate your total lung capacity. Record all of your results in your Laboratory Report.

Spirocomp® Procedure (Computer-Based Procedure)

If you are using the Spirocomp® or a *Biopac* system, your laboratory instructor may have additional and more detailed procedures for spirometry. The following are the basic instructions for Spirocomp® spirometry.

- Raise and lower the bell of the spirometer several times to allow fresh air to be drawn under the bell.
- The Spirocomp® water respirometer can be used without the computer interface. However, it is sometimes difficult to read the scale during inhalation and exhalation. It may also be difficult to obtain pulmonary measures other than tidal volume.
- Spirocomp® Computer Procedure (modified from Fox 2002)
  - Press the "T" key on your computer keyboard and you will see the prompt ***Breathe Normal Cycles*** on your computer screen. After three or four normal tidal volume cycles of inhalation and exhalation, data will be calculated and presented on the computer screen.
  - Press the "E" key and you will see the prompt ***Breathe Normal Cycles*** on your computer screen. At the third ventilation cycle of inhalation and exhalation you will see a new prompt, ***Stop After Normal Exhale.***
  - Soon after this pause in breathing, a new prompt will appear, ***Exhale Forcefully,*** indicating that you should exhale as fully as possible.
  - Press the "V" key and you will be prompted to ***Inhale Maximally Then Press V*** and ***Exhale Forcefully.***
  - The computer program will display data that will allow you to calculate measures of pulmonary function and complete your Laboratory Report.

EXPERIMENT 13.2 The Effect of CO_2 Concentration on Ventilation

In this experiment you will investigate the effects of CO_2 concentration on ventilation using different breathing conditions.

- Determine which of your laboratory partners will participate in the experiment. At least two or three people from each laboratory group should participate to ensure that 10 to 15 subjects are acquired for the class data set.
- Each participant should determine randomly (e.g., by a coin toss) which respiratory condition will be measured first, either the normal breathing condition or the hyperventilation condition.
  - **Normal Breathing Condition:** Breathe normally for approximately 15 seconds. At the end of your last exhalation, hold your breath. Have your laboratory partner record the amount of time (in seconds) you can hold your breath. *To ensure that the data are not biased, make sure the subject is unaware of the time elapsed.*
  - **Hyperventilation Condition:** Hyperventilate (take shallow, quick breaths) for 15 seconds. At the end of the last exhalation, hold your breath. Have your

laboratory partner record the amount of time (in seconds) you can hold your breath. Again, make sure the subject is unaware of the time elapsed.

- Allow for a resting period of at least five minutes between measures of the two different breathing conditions.
- Record your data for breath-holding duration (in seconds) for each breathing condition in Results Table 13.2.

COMPARATIVE NOTE *Laboratory 13*

Ventilation Rate in Fishes: It's a Gas

You have learned that the partial pressure of carbon dioxide (P_{CO_2}) is extremely important in regulating the ventilation rate. Although there are slight differences in CO_2 sensitivity among air-breathing animals, this mechanism is highly conserved evolutionarily. This mechanism primarily detects changes in hydrogen-ion concentration that result from increases in CO_2 rather than detecting decreases in the partial pressure of O_2. This system is adaptive in that change is more easily detected with smaller partial pressures or concentrations of gas. The concentration ratio of CO_2 (0.03%) to O_2 (20.95%) in air is 1/698.3. Thus, you can see that detecting and monitoring changes in CO_2 would be easier than detecting changes in O_2. Consider the following scenario: There are two tables, one with 5 pennies and the other with 3492 (a 1 to 698.3 ratio). If the number of pennies on both tables is decreased by two, on which table would you easily detect a change in the number of pennies? That is correct, the table with the smaller "concentration" of pennies.

This same reasoning may explain why fishes monitor changes in O_2 rather than CO_2. Unlike air, the concentration ratio of CO_2 (1.9%) to O_2 (33.4%) in water is 1/17.6. Thus, aquatic organisms such as fishes generally respond to changes in decreasing concentrations of O_2 in the water.

Homeostasis of respiratory gases in the body is maintained by changing the ventilation rate. Thus, it seems reasonable that ventilation rate would respond to the particular respiratory gas for which small changes in concentration are easily detected. Thus O_2 is monitored for water breathers, whereas air breathers are adapted to monitor changes in CO_2.

A Comparison of Gases in the Respiratory Media Air and Fresh Water at 20°C

Gas	Air	Fresh Water
Oxygen	209.5 ml/L (20.95%)	6.98 ml/L (33.4%)
Carbon dioxide	0.3 ml/L (0.03%)	0.40 ml/L (1.9%)
Nitrogen	780.9 ml/L (78.09%)	13.5 ml/L (64.7%)
Argon	9.3 ml/L (0.93%)	

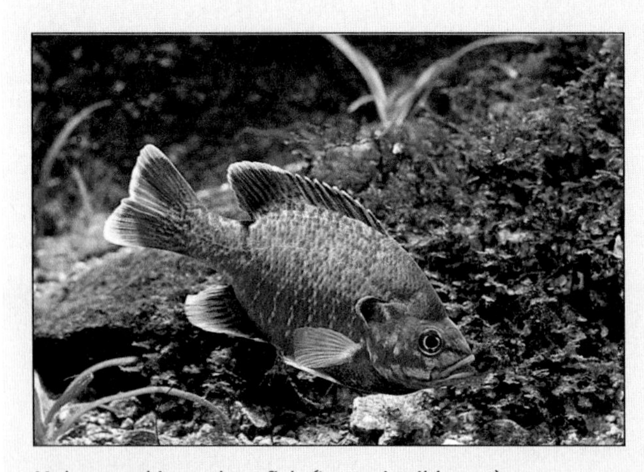

Male pumpkinseed sunfish *(Lepomis gibbosus)*

RESEARCH OF INTEREST

Cooper, A.R. and S. Morris. 2004. Hemoglobin function and respiratory status of the Port Jackson shark, *Heterodontus portusjacksoni,* in response to lowered salinity. Journal of Comparative Physiology B 174:223–236.

Florindo, L.H., S.G. Reid, A.L. Kalinin, W.K. Milsom, and F.T. Rantin. 2004. Cardiorespiratory reflexes and aquatic surface respiration in the neotropical fish tambaqui (*Colossoma macropomum*): acute responses to hypercarbia. Journal of Comparative Physiology B 174:319–328.

Gilmour, K.M. 2001. The CO_2/pH ventilatory drive in fish. Comparative Biochemistry and Physiology Part A 130:219–240.

RESULTS AND DISCUSSION

LABORATORY REPORT 13

Problem Set 13.1: Measures of Pulmonary Function.

a. What is your mean vital capacity?

b. What is your mean expiratory reserve volume?

c. What is your tidal volume? Show your calculations and demonstrate the logic of these calculations by sketching the spirograph for this experiment and labeling the lung volumes you used to calculate tidal volume.

d. What is your inspiratory reserve volume and inspiratory capacity? Show your calculations and demonstrate the logic of these calculations by sketching the spirograph for this experiment. Label the lung volumes you used to calculate inspiratory reserve volume and inspiratory capacity.

e. Estimate your total lung capacity. Show your calculations.

f. You are presented with spirograph tracings from two different patients. The first patient, a 30-year-old female with a height of 5 ft 11 in, has a vital capacity of 3390 ml. The second patient, a 50-year-old male with a height of 5 ft 6 in, has a vital capacity of 3420 ml.

Which patient has a normal vital capacity? Use the predicted vital capacities listed in Tables 13.1a and 13.1b to help you answer this question.

If the male patient was 64 years old, would the classification of his vital capacity be normal or abnormal?

Based upon their vital capacities, which patient would be better able to sustain moderate physical activity? Explain your reasoning thoroughly.

Which patient would experience a faster rate of CO_2 accumulation in the blood during physical activity? Explain.

g. The female patient is reporting shortness of breath, or dyspnea. You decide to conduct more tests. You measure the volume of air the patient can forcefully exhale, the **forced expiratory volume** (**FEV**) in 1 sec (**FEV_1**). The results of this test are shown in Figure 13-7. For reference, the results of an FEV_1 test for a healthy patient are shown in tracing (a) of figure 13-7. This reference patient has a normal vital capacity of 3385 ml and a normal FEV_1 of 80%.

Using the results shown in (b) of Figure 13-7, calculate the female patient's FEV_1. Recall from question (f) that her vital capacity is 3390 ml.

Based upon this person's vital capacity and an FEV_1 of 80% for a normal patient, what volume of air (in ml) should this female patient be able to forcefully exhale in 1 sec? What volume of air (in ml) is this patient currently able to forcefully exhale in 1 sec?

What type of lung disorder (restrictive or obstructive) do you suspect this person is suffering from? Explain.

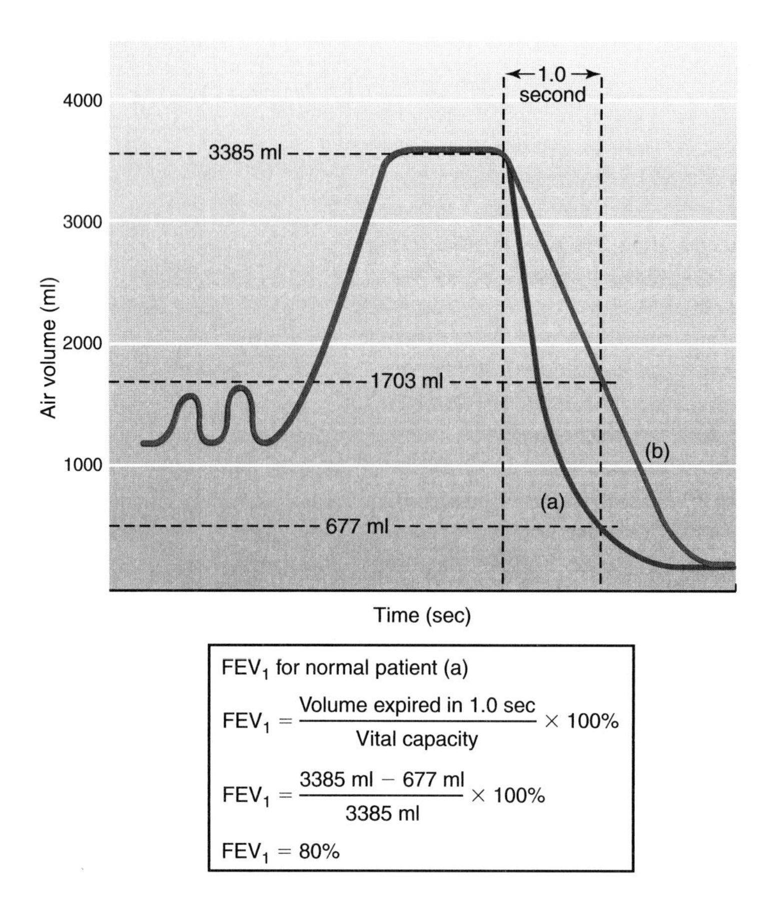

Figure 13-7 Computer tracing of air volumes inspired and expired by two patients during tests of the one-second forced expiratory volume (FEV_1). Tracing (a) demonstrates a normal FEV_1 volume of 80% of the patient's vital capacity. Tracing (b) may indicate a lung disorder.

EXPERIMENT 13.2
The Effect of CO_2 Concentration on Ventilation

Results Table 13.2 Class Data for Experiment 13.2: The Effect of CO_2 Concentration on Ventilation.

Student (Name)	Breath-Holding Duration (sec)	
	Normal Breathing Condition	Hyperventilation Condition
1		
2		
3		
4		
5		
6		
7		
8		
9		
10		
11		
12		
13		
14		
15		
16		
17		
18		
19		
20		
21		
22		
23		
24		
25		
Mean (±SE)		

Problem Set 13.2: The Effect of CO_2 Concentration on Ventilation

Using the data from Results Table 13.2, create a simple bar graph showing the mean breath-holding duration for both normal breathing and hyperventilation conditions. Be sure to include the 95% CI. Using these data, perform the appropriate statistical test to examine possible differences in breath-holding duration between the two breathing conditions. Remember that each subject is serving as his or her own control.

a. State the null hypothesis.

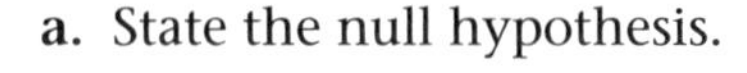

b. Should you accept or reject the null hypothesis? Support your conclusions with the appropriate statistical results.

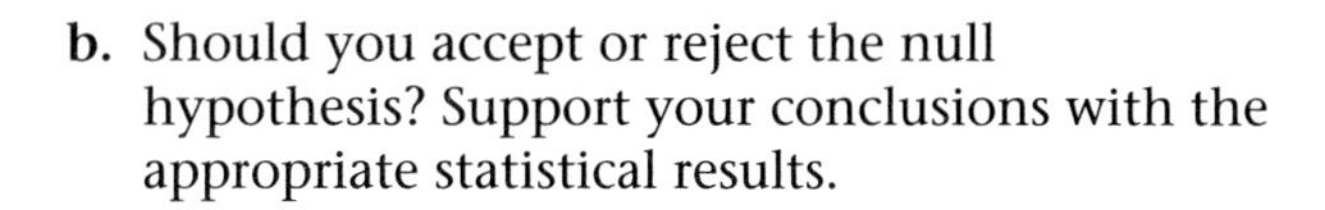

c. What is the difference between the normal breathing condition and the hyperventilation breathing condition? What is the physiological significance of this difference?

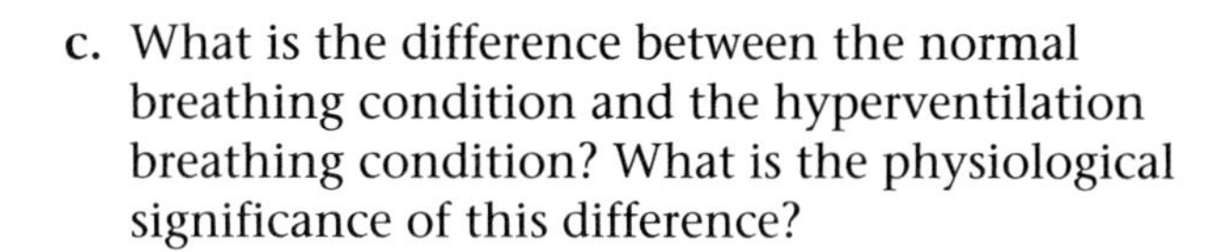

d. Why does hyperventilation affect the breath-holding duration? Explain the homeostatic mechanism thoroughly.

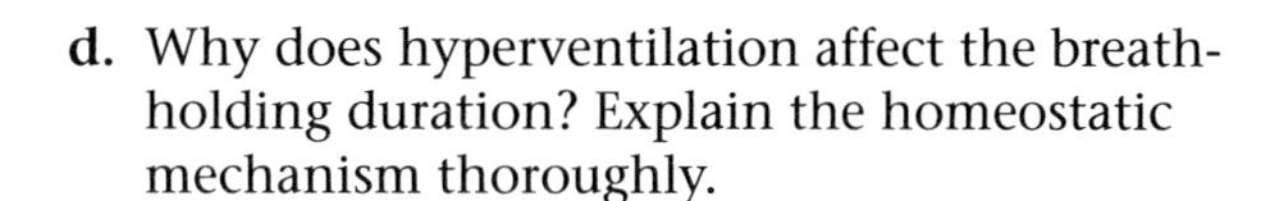

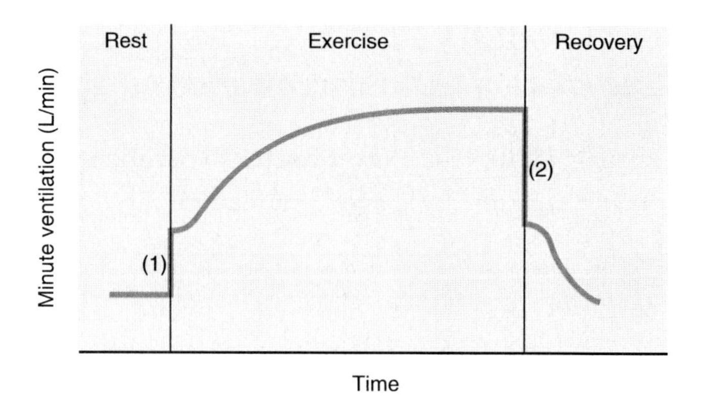

Figure 13-8 Ventilation changes during exercise. Note (1) the abrupt increase at the onset of exercise and (2) the equally abrupt but larger decrease at the end of exercise.

e. If a person continued to hyperventilate, would she begin to experience respiratory acidosis or alkalosis? Explain.

f. Based upon your experimental results, why do island pearl hunters hyperventilate prior to a dive?

g. As shown in Figure 13-8, ventilation changes dramatically during exercise. The volume of air respired per minute (the **minute ventilation,** measured in L/min) is calculated by multiplying the number of breaths per minute by the tidal volume (in ml). During exercise the minute ventilation can increase to as much as 100 to 200 L/min. Draw a flowchart summarizing the factors that stimulate changes in ventilation during exercise.

Table 13.1a Predicted Vital Capacities, Females (mL)

	Height in Centimeters																							
Age	146	148	150	152	154	156	158	160	162	164	166	168	170	172	174	176	178	180	182	184	186	188	190	192	194
16	2950	2990	3030	3070	3110	3150	3190	3230	3270	3310	3350	3390	3430	3470	3510	3550	3690	3630	3570	3715	3755	3800	3840	3880	3920
17	2935	2975	3015	3055	3095	3135	3175	3215	3255	3295	3335	3375	3415	3455	3495	3535	3575	3615	3655	3695	3740	3780	3820	3860	3900
18	2920	2960	3000	3040	3080	3120	3160	3200	3240	3280	3320	3360	3400	3440	3480	3520	3560	3600	3640	3680	3720	3760	3800	3840	3880
20	2890	2930	2970	3010	3050	3090	3130	3170	3210	3250	3290	3330	3370	3410	3450	3490	3525	3565	3605	3645	3695	3720	3760	3800	3840
22	2860	2900	2940	2980	3020	3060	3095	3135	3175	3215	3255	3290	3330	3370	3410	3450	3490	3530	3570	3610	3650	3685	3725	3765	3800
24	2830	2870	2910	2950	2985	3025	3065	3100	3140	3180	3220	3260	3300	3335	3375	3415	3455	3490	3530	3570	3610	3650	3685	3725	3765
26	2800	2840	2880	2920	2960	3000	3035	3070	3110	3150	3190	3230	3265	3300	3340	3380	3420	3455	3495	3530	3570	3610	3650	3685	3725
28	2775	2810	2850	2890	2930	2965	3000	3040	3070	3115	3155	3190	3230	3270	3305	3345	3380	3420	3460	3495	3535	3570	3610	3650	3685
30	2745	2780	2820	2860	2895	2935	2970	3010	3045	3085	3120	3160	3195	3235	3270	3310	3345	3385	3420	3460	3495	3535	3570	3610	3645
32	2715	2750	2790	2825	2865	2900	2940	2975	3015	3060	3090	3125	3160	3200	3235	3275	3310	3350	3385	3425	3460	3495	3535	3570	3610
34	2685	2725	2760	2795	2835	2870	2910	2945	2980	3020	3055	3090	3130	3165	3280	3240	3275	3310	3350	3385	3425	3460	3495	3535	3570
36	2655	2695	2730	2765	2805	2840	2875	2910	2950	2985	3020	3060	3095	3130	3165	3205	3240	3275	3310	3350	3385	3420	3460	3495	3530
38	2630	2665	2700	2735	2770	2810	2845	2880	2915	2950	2990	3025	3060	3095	3130	3170	3205	3240	3275	3310	3350	3385	3420	3455	3490
40	2600	2635	2670	2705	2740	2775	2810	2850	2885	2920	2955	2990	3025	3060	3095	3135	3170	3205	3240	3275	3310	3345	3380	3420	3455
42	2570	2605	2640	2675	2710	2745	2780	2815	2850	2885	2920	2955	2990	3125	3060	3100	3135	3170	3205	3240	3275	3310	3345	3380	3415
44	2540	2575	2610	2645	2680	2715	2750	2785	2820	2855	2890	2925	2960	2995	3030	3060	3095	3130	3165	3200	3235	3270	3305	3340	3375
46	2510	2545	2580	2615	2650	2685	2715	2750	2785	2820	2855	2890	2925	2960	2995	3030	3060	3095	3130	3165	3200	3235	3270	3305	3340
48	2480	2515	2550	2585	2620	2650	2685	2715	2750	2785	2820	2855	2890	2925	2960	2995	3030	3060	3095	3130	3160	3195	3230	3265	3300
50	2455	2485	2520	2555	2590	2625	2655	2690	2720	2755	2785	2820	2855	2890	2925	2955	2990	3125	3060	3090	3125	3155	3190	3225	3260
52	2425	2455	2490	2525	2555	2590	2625	2655	2690	2720	2755	2790	2820	2855	2890	2925	2955	2990	3020	3055	3090	3125	3155	3190	3220
54	2395	2425	2460	2495	2530	2560	2590	2625	2655	2690	2720	2755	2790	2820	2855	2885	2920	2950	2985	3020	3050	3085	3115	3150	3180
56	2365	2400	2430	2460	2495	2525	2560	2590	2625	2655	2690	2720	2755	2790	2820	2855	2885	2920	2950	2980	3105	3045	3080	3110	3145
58	2335	2370	2400	2430	2460	2495	2525	2560	2590	2625	2655	2690	2720	2750	2785	2815	2850	2880	2920	2945	2975	3010	3040	3075	3105
60	2305	2340	2370	2400	2430	2460	2495	2525	2560	2590	2625	2655	2685	2720	2750	2780	2810	2845	2875	2915	2940	2970	3000	3035	3065
62	2280	2310	2340	2370	2405	2435	2465	2495	2525	2560	2590	2620	2655	2685	2715	2745	2775	2801	2840	2870	2900	2935	2965	2995	3025
64	2250	2280	2310	2340	2370	2400	2430	2465	2495	2525	2555	2585	2620	2650	2680	2710	2740	2770	2805	2835	2865	2895	2925	2955	2990
66	2220	2250	2280	2310	2340	2370	2400	2430	2460	2495	2525	2555	2585	2615	2645	2675	2705	2735	2765	2800	2825	2860	2890	2920	2950
68	2190	2220	2250	2280	2310	2340	2370	2400	2430	2460	2490	2520	2550	2580	2610	2640	2670	2700	2730	2760	2795	2820	2850	2880	2910
70	2160	2190	2220	2250	2280	2310	2340	2370	2400	2425	2455	2485	2515	2545	2575	2605	2635	2665	2695	2725	2755	2780	2810	2840	2870
72	2130	2160	2190	2220	2250	2280	2310	2335	2365	2395	2425	2455	2480	2510	2540	2570	2600	2630	2660	2685	2715	2745	2775	2805	2830
74	2100	2130	2160	2190	2220	2245	2275	2305	2335	2360	2390	2420	2450	2475	2505	2535	2565	2590	2620	2650	2680	2710	2740	2765	2795

Courtesy of Warren E. Collins, Inc., Braintree, MA.

Table 13.1b Predicted Vital Capacities, Males (mL)

	Height in Centimeters																								
Age	146	148	150	152	154	156	158	160	162	164	166	168	170	172	174	176	178	180	182	184	186	188	190	192	194
16	3765	3820	3870	3920	3975	4025	4075	4130	4180	4230	4285	4335	4385	4440	4490	4540	4590	4645	4695	4745	4800	4850	4900	4955	5005
18	3740	3790	3840	3890	3940	3995	4045	4095	4145	4200	4250	4300	4350	4405	4455	4505	4555	4610	4660	4710	4760	4815	4865	4915	4965
20	3710	3760	3810	3860	3910	3960	4015	4065	4115	4165	4215	4265	4320	4370	4420	4470	4520	4570	4625	4675	4725	4775	4825	4875	4930
22	3680	3730	3780	3830	3880	3930	3980	4030	4080	4135	4185	4235	4285	4335	4385	4435	4485	4535	4585	4635	4685	4735	4790	4840	4890
24	3635	3685	3735	3785	3835	3885	3935	3985	4035	4085	4135	4185	4235	4285	4330	4380	4430	4480	4530	4580	4630	4680	4730	4780	4830
26	3605	3655	3705	3755	3805	3855	3905	3955	4000	4050	4100	4150	4200	4250	4300	4350	4395	4445	4495	4545	4595	4645	4695	4740	4790
28	3575	3625	3675	3725	3775	3820	3870	3920	3970	4020	4070	4115	4165	4215	4265	4310	4360	4410	4460	4510	4455	4605	4655	4705	4755
30	3550	3595	3645	3695	3740	3790	3840	3890	3935	3985	4035	4080	4130	4180	4230	4275	4325	4375	4425	4470	4520	4570	4615	4665	4715
32	3520	3565	3615	3665	3710	3760	3810	3855	3905	3950	4000	4050	4095	4145	4195	4240	4290	4340	4385	4435	4485	4530	4580	4625	4675
34	3475	3525	3570	3620	3665	3715	3760	3810	3855	3905	3950	4000	4045	4095	4140	4190	4225	4285	4330	4380	4425	4475	4520	4570	4615
36	3445	3495	3510	3555	3605	3650	3695	3745	3790	3840	3885	3930	3980	4025	4070	4120	4165	4210	4260	4305	4350	4400	4445	4495	4540
38	3415	3465	3510	3555	3605	3650	3695	3745	3790	3840	3885	3930	3980	4025	4070	4120	4165	4210	4260	4305	4350	4400	4445	4495	4540
40	3385	3435	3480	3525	3575	3620	3665	3710	3760	3805	3850	3900	3945	3990	4035	4085	4130	4175	4220	4270	4315	4360	4410	4455	4500
42	3360	3405	3450	3495	3540	3590	3635	3680	3725	3770	3820	3865	3910	3955	4000	4050	4095	4140	4185	4230	4280	4325	4370	4415	4460
44	3315	3360	3405	3450	3495	3540	3585	3630	3675	3725	3770	3815	3860	3905	3950	3995	4040	4085	4130	4175	4220	4270	4315	4360	4405
46	3285	3330	3375	3420	3465	3510	3555	3600	3645	3690	3735	3780	3825	3870	3915	3960	4005	4050	4095	4140	4185	4230	4275	4320	4365
48	3255	3300	3345	3390	3435	3480	3525	3570	3615	3655	3700	3745	3790	3835	3880	3925	3970	4015	4060	4105	4150	4190	4235	4280	4325
50	3210	3255	3300	3345	3390	3430	3475	3520	3565	3610	3650	3695	3740	3785	3830	3870	3915	3960	4005	4050	4090	4135	4180	4225	4270
52	3185	3225	3270	3315	3355	3400	3445	3490	3530	3575	3620	3660	3705	3750	3795	3835	3880	3925	3970	4010	4055	4100	4140	4185	4230
54	3155	3195	3240	3285	3325	3370	3415	3455	3500	3540	3585	3630	3670	3715	3760	3800	3845	3890	3930	3975	4020	4060	4105	4145	4190
56	3125	3165	3210	3255	3295	3340	3380	3425	3465	3510	3550	3595	3640	3680	3725	3765	3810	3850	3895	3940	3980	4025	4065	4110	4150
58	3080	3125	3165	3210	3250	3290	3335	3375	3420	3460	3500	3545	3585	3630	3670	3715	3755	3800	3840	3880	3925	3965	4010	4050	4095
60	3050	3095	3135	3175	3220	3260	3300	3345	3385	3430	3470	3500	3555	3595	3635	3680	3720	3760	3805	3845	3885	3930	3970	4015	4055
62	3020	3060	3110	3150	3190	3230	3270	3310	3350	3390	3440	3480	3520	3560	3600	3640	3680	3730	3770	3810	3850	3890	3930	3970	4020
64	2990	3030	3080	3120	3160	3200	3240	3280	3320	3360	3400	3440	3490	3530	3570	3610	3650	3690	3730	3770	3810	3850	3900	3940	3980
66	2950	2990	3030	3070	3110	3150	3190	3230	3270	3310	3350	3390	3430	3470	3510	3550	3600	3640	3680	3720	3760	3800	3840	3880	3920
68	2920	2960	3000	3040	3080	3120	3160	3200	3240	3280	3320	3360	3400	3440	3480	3520	3560	3600	3640	3680	3720	3760	3800	3840	3880
70	2890	2930	2970	3010	3050	3090	3130	3170	3210	3250	3290	3330	3370	3410	3450	3480	3520	3560	3600	3640	3680	3720	3760	3800	3840
72	2860	2900	2940	2980	3020	3060	3100	3140	3180	3210	3250	3290	3330	3370	3410	3450	3490	3530	3570	3610	3650	3680	3720	3760	3800
74	2820	2860	2900	2930	2970	3010	3050	3090	3130	3170	3200	3240	3280	3320	3360	3400	3440	3470	3510	3550	3590	3630	3670	3710	3740

Courtesy of Warren E. Collins, Inc., Braintree, MA.

LABORATORY 14

Renal Physiology

PURPOSE

This laboratory will introduce you to the functional anatomy of the kidney and will address the physiological mechanisms of osmotic regulation and ionic balance. An experiment will demonstrate how renal function maintains homeostasis through changes in the specific gravity and the production of urine.

Learning Objectives

- Describe the functional anatomy of the kidney.
- Describe how the kidney regulates water and ions in the body.
- Describe renal clearance and how it is clinically determined.
- Describe and illustrate the chemical structure of urea and explain urea's physiological importance.
- Understand how the ingestion of water and salt influences the specific gravity and production of urine.

Laboratory Materials

Experiment 14.1: Effect of Salt Loading on the Specific Gravity and Production of Urine

1. Dose of NaCl (salt pills) with ingestion of water.

   Hypotonic = 0.00 M = 0% NaCl solution = 0.00 g/250 ml

   0 salt tablets (452 mg/tablet) with 250 ml water

   Isotonic = 0.15 M = 0.9% NaCl solution = 2.25 g/250ml

   5 salt tablets (452 mg/tablet) with 250 ml water

   Hypertonic = 0.26 M = 1.5% NaCl solution = 3.86 g/250 ml

   8 salt tablets (452 mg/tablet) with 250 ml water

2. Volumetric urine collection cups
3. Urinometers or specific gravity meter

NOTE TO THE STUDENT:
To obtain the best results for Experiment 14.1, volunteers should not drink or eat for at least two hours prior to laboratory. All students are encouraged to participate unless health conditions (e.g., hypertension, diabetes) will be exacerbated by salt intake.

CAUTION: In this laboratory you will be working with bodily fluid. Although urine presents few health risks, caution should be used in this laboratory; handle only your own urine samples. Your instructor will provide additional instructions regarding disposal and cleanup of the laboratory materials.

Introduction and Pre-Lab Exercises

The kidneys perform several functions that are vital to life. They play a central role in regulating the volume and concentration of body fluids. The kidneys also remove metabolic waste products as well as foreign chemicals (e.g., drugs) from the blood by excreting them into the urine. Other functions of the kidneys include gluconeogenesis during times of prolonged fasting as well as the production of enzymes and hormones. Every day, your kidneys filter 180 liters of water! Fortunately, 99% of this water is reabsorbed, so on average only 1.8 liters of urine are produced per day.

The two kidneys are retroperitoneal in position. That is, each kidney lies posterior to the peritoneal membrane that lines the abdominal cavity.

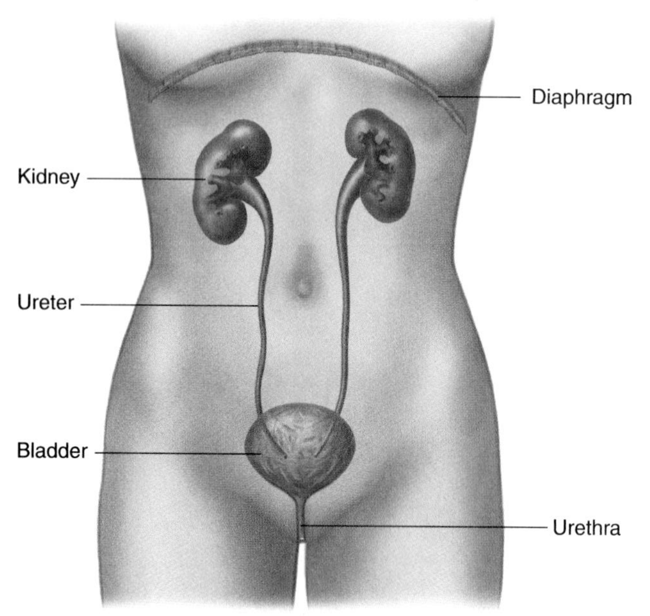

Figure 14-1
Urinary system in a female. In the male, the urethra passes through the penis.

The kidneys are therefore not contained within the abdominal cavity and are protected from injury by the inferior posterior rib cage (Figure 14-1).

The functional unit of the kidney is called a **nephron** (Figure 14-2). Each kidney contains approximately one million nephrons, each of which consists of a filtering component and a tubule. The filtering component, called the **renal corpuscle,** forms a filtrate from the blood that is free of cells and large proteins. Each renal corpuscle is composed of a series of interconnected capillary loops called the **glomerulus** and a fluid-filled capsule called the **glomerular** or **Bowman's capsule.** The filtrate is formed when blood enters the glomerulus, which protrudes into Bowman's capsule. Approximately 20% of the blood plasma is filtered into Bowman's capsule as blood flows through the glomerulus. This filtrate then enters the **renal tubule,** which is continuous with Bowman's capsule. As the filtrate moves through the renal tubule, water and ions are absorbed and secreted so that a small volume of concentrated filtrate is produced.

The renal tubule itself is divided into different anatomical sections, each of which is specialized for a particular function. The renal tubule is a narrow, hollow cylinder lined by simple (one-layer) epithelial tissue. Differences in the structure and function of this epithelial tissue contribute to the varying functions of each renal tubule segment.

The portion of the renal tubule that Bowman's capsule drains into is called the **proximal convoluted tubule.** The next portion of the renal tubule is the **loop of Henle.** This renal segment is a long, hairpin-like loop that consists of both the **descending limb** from the proximal tubule and the **ascending limb** that leads to the next tubular segment, the **distal convoluted tubule.**

Filtrate reaching the distal convoluted tubule drains into a **collecting duct.** The collecting ducts from multiple nephrons merge so that the renal filtrate eventually drains into the kidney's central cavity, or **renal pelvis,** via only a few hundred large collecting ducts (Figure 14-3). The renal pelvis is continuous with the **ureter,** which carries the renal filtrate into the **urinary bladder.** Urine then exits the body from the urinary bladder via the **urethra.**

Each renal tubule is surrounded by a capillary bed all along the tubule's length. These capillaries, called the **peritubular capillaries,** are important in the reabsorption of ions and water from the renal tubules into the bloodstream. The peritubular capillaries also play an important role in the secretion of substances from the blood into the renal tubule for excretion in the urine.

The primary mechanism by which the kidneys produce hyperosmotic urine is by the reabsorption of filtered Na^+ into the interstitial fluid. Sodium reabsorption is an active process that occurs in all segments of the renal tubule except the descending limb of the loop of Henle. Active transport of Na^+ from the renal tubule into the interstitial fluid creates a high osmolarity in the interstitial fluid. The high osmolarity of the interstitial fluid creates an osmotic gradient across the renal tubule, which in turn causes water to diffuse from the renal tubule into the interstitial fluid. The loops of Henle are **countercurrent multiplier systems,** in which the flow of fluid in the descending limb is opposite to that in the ascending limb, producing a countercurrent flow (Figure 14-4). Secondly, the moderate concentration differences between the descending and ascending limbs at any one point produce a multiplicative effect along the entire length of the loop, as shown in Figure 14-4.

Note that the reabsorption of water is always a passive process that depends upon the reabsorption of Na^+ into the interstitial fluid. Water reabsorption occurs in all segments of the renal tubule except the ascending limb of the loop of Henle. Recall from the "Diffusion, Osmosis, and Tonicity" laboratory (Laboratory 3) that net water movement occurs from an area of low solute concentration to an area of high solute concentration. Thus, water "follows" Na^+ as Na^+ is reabsorbed from the kidney tubules into the interstitial fluid.

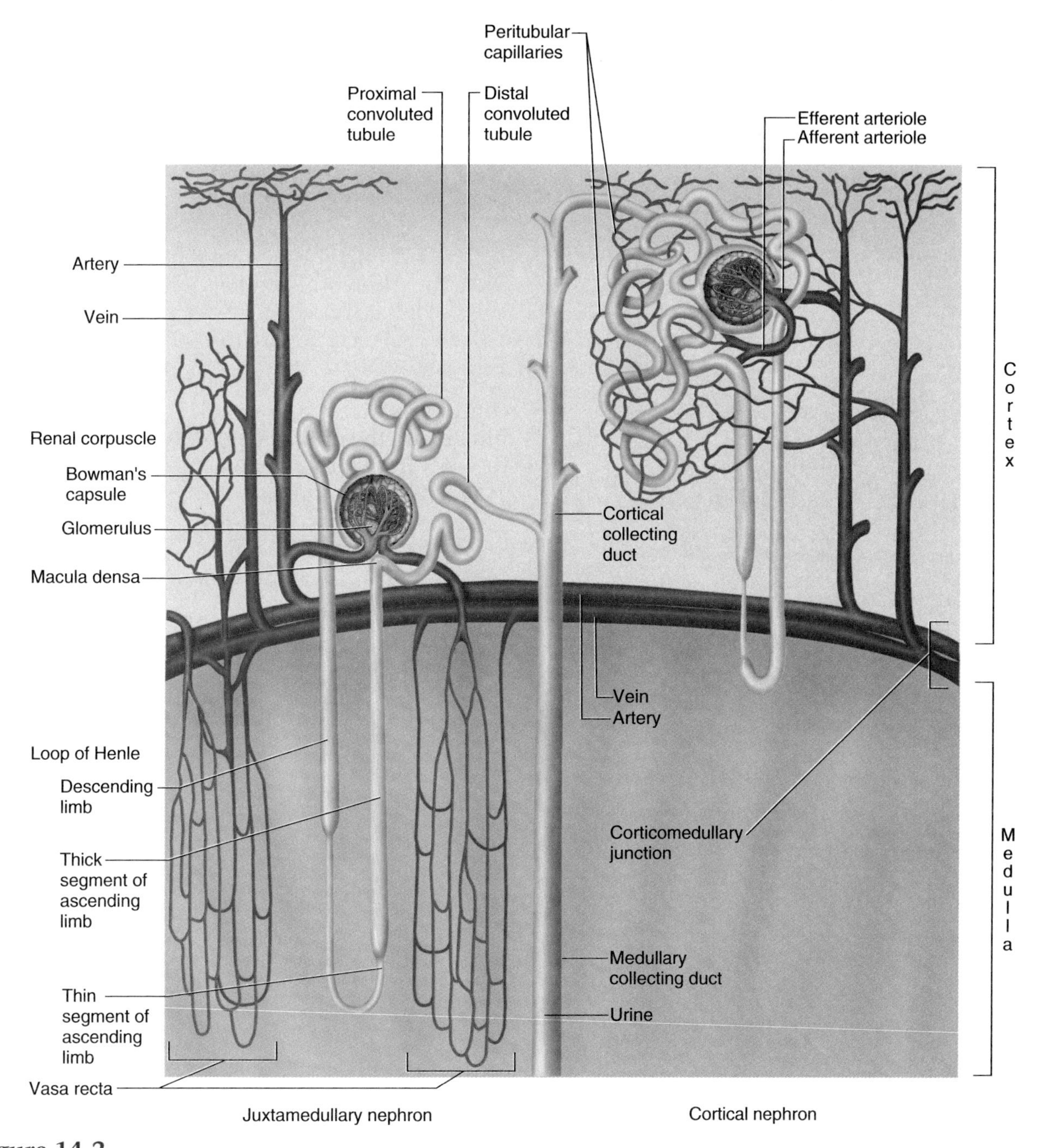

Figure 14-2
Functional anatomy and anatomical organization of a nephron. Two types of nephrons are shown. The juxtamedullary nephrons have long loops of Henle that penetrate deeply into the renal medulla, while the cortical nephrons have short or no loops of Henle.

14-1 You are stranded on a desert island with only seawater as a water source. You know that drinking seawater will make you even more dehydrated and thirsty. Why?

The permeability of each tubular segment to water varies and depends largely on the presence of water channels in the plasma membranes of the tubular epithelial cells. These water channels are comprised of proteins called **aquaporins.** The more aquaporins present in the plasma membranes of cells, the more water can be reabsorbed (and hence the less water excreted as urine). In the

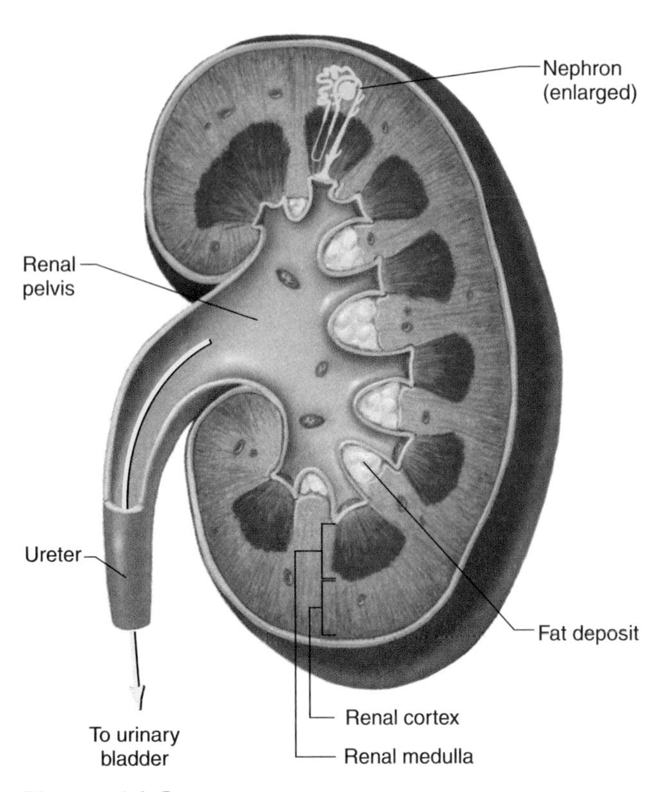

Figure 14-3
Cross section of a kidney.

collecting ducts, water permeability is largely regulated by the hormone **antidiuretic hormone** (ADH; also called vasopressin). ADH, secreted by the posterior pituitary gland, stimulates the insertion of aquaporin water channels into the collecting tubule epithelium of the kidney, thus leading to increased water reabsorption.

14-2 Ethanol inhibits the release of ADH by the posterior pituitary gland. Explain the physiological consequence of alcohol consumption to water balance. Be sure to explain the mechanism thoroughly.

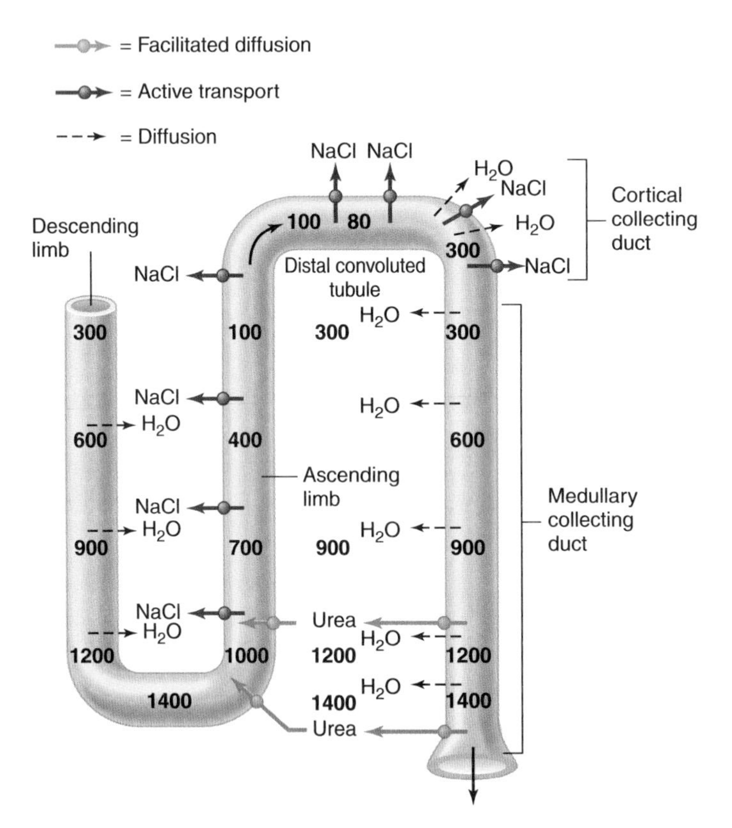

Figure 14-4
Simplified depiction of the renal countercurrent multiplier system and its role in the formation of hyperosmotic urine. Fluid osmolarity (mOsmol/L) is indicated by the bold numbers.

Of the three foodstuffs (i.e., carbohydrate, fat, and protein), only the catabolism or breakdown of amino acids from proteins produces the toxic waste product ammonia (NH_3). If ammonia were allowed to accumulate in the body, it would poison our system and cause death. Thus, our liver, with the use of CO_2, ATP, and enzymes such as arginase, converts ammonia molecules to **urea** in the **ornithine cycle.**

Urea

$$NH_2—\underset{\underset{O}{\|}}{C}—NH_2$$

Once ammonia is converted to urea, it may remain in the body until it can be excreted in the urine through the process of urination (micturition). Urea is filtered by glomeruli and enters the renal tubules for excretion.

The **glomerular filtration rate** (**GFR**) is influenced by many factors, including neural and hormonal stimuli, blood pressure, dilation and constriction of glomerular capillaries, the permeability of the glomerular capillaries, and the size of the kidneys. GFR determines, in part, the rate of urine production. It can be examined by knowing the concentration of a solute, such as urea, in the urine, the concentration of that solute in the plasma, and the volume of urine produced over time. The normal GFR for humans is 125 ml/min. Thus, 180 liters of fluid are filtered from the glomeruli into Bowman's capsules per day.

14-3 Although we normally filter 180 liters of fluid per day into the renal tubules, we only excrete about 1.8 liters of urine per day. What has happened to the other 178 liters of fluid? Why is this so important?

■

14-4 A doctor is interested in measuring your GFR to determine if you have normal kidney function. He tells you that renal clearance of a plasma-borne substance is equal to the amount of that substance filtered from the blood by the kidneys per unit time. A substance that is freely filtered into the nephron but is neither reabsorbed nor secreted by the kidney tubules permits the calculation of GFR. The doctor now injects a large polysaccharide called inulin into your bloodstream. Your body does not produce inulin, thus the concentration injected into your bloodstream is the only source of inulin. After 30 minutes the doctor asks for a urine and blood sample. The doctor informs you that you have produced 90 ml of urine containing 30 mg of inulin per ml. Your blood sample contains 0.5 mg of inulin per ml. Use the following equation to determine if you have a normal GFR.

$$GFR = \frac{U \times V}{P}$$

U = concentration of inulin in urine (mg/ml)

V = rate of urine production (ml/min)

P = concentration of inulin in the blood plasma (mg/ml)

■

COMPARATIVE NOTE Laboratory 14

Nitrogenous Waste Production: To Pee or Not to Pee?

All animals produce metabolic wastes from the digestion of food. One of these wastes is the amino group (+NH_2), formed during the breakdown of amino acids in proteins and nucleic acids. This amino group can easily react with hydrogen to form the toxic compound ammonia (NH_3). Therefore all animals that catabolize proteins metabolically produce a potentially poisonous and harmful compound. So how do animals rid their systems of this toxic metabolic waste?

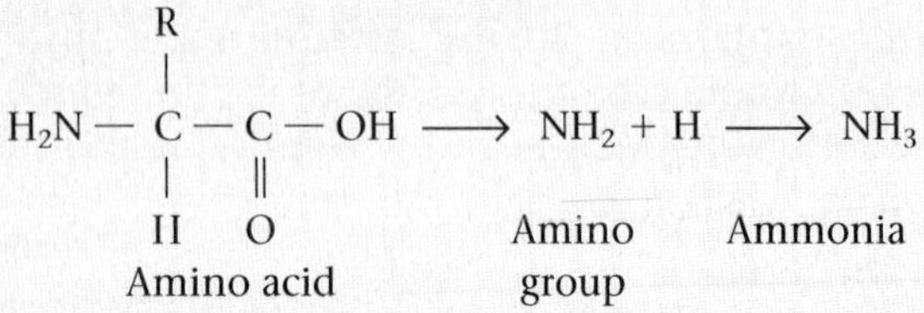

$$H_2N-\underset{H}{\overset{R}{C}}-\underset{O}{\overset{}{C}}-OH \longrightarrow NH_2 + H \longrightarrow NH_3$$

Amino acid — Amino group — Ammonia

The types of nitrogenous wastes that animals produce are strongly correlated with their habits and environments. Thus, we see interesting trends between behavior, habitat use, and nitrogenous waste production. Although ammonia is highly toxic, it is also extremely water soluble. Thus, many fishes produce ammonia and allow the blood to carry this toxic waste to the gills, where it is directly excreted into the water. Some teleost fishes excrete 10 times more nitrogen from their gills than from their kidneys. Fully aquatic amphibians also produce ammonia as a nitrogenous waste because of water availability.

Animals that are more terrestrial have special physiological processes to sequester NH_2 before it becomes NH_3. One process used by many animals, including mammals, is the urea or ornithine cycle. This is a complex metabolic and enzymatic process that aids in the binding of CO_2 to NH_2 to prevent the formation of NH_3.

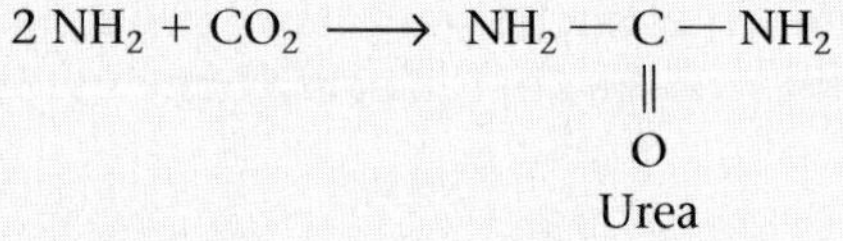

$$2\,NH_2 + CO_2 \longrightarrow NH_2-\underset{O}{\overset{}{C}}-NH_2$$

Urea

Urea is less toxic than ammonia and is collected in the bladder in urine. This allows an animal to carry this nitrogenous waste within the body until it can be excreted.

There is a third type of nitrogenous waste called uric acid. This nitrogenous waste is highly water insoluble and is excreted as a solid in the feces of birds and reptiles. Uric acid is so insoluble that it requires a liter of water to dissolve six milligrams. Uric acid production is common in animals that live in desert environments and must conserve water. Using large amounts of water to excrete nitrogenous waste such as urea would be highly nonadaptive in arid environments. It is also adaptive for birds to produce uric acid so that large amounts of water need not be carried during flight. Why are the solubility properties of uric acid important for the embryonic development of oviparous (egg-laying) reptiles and birds?

RESEARCH OF INTEREST

Lillywhite, H.B. and T.M. Ellis. 1994. Ecophysiological aspects of the coastal-estuarine distribution of acrochordid snakes. Estuaries 17:53–61.

Phillips, K. 2004. Dogmas and controversies in the handling of nitrogenous wastes. Journal of Experimental Biology 207:1–5.

Wright, P.A., P. Part, and C.M. Wood. 1995. Ammonia and urea excretion in the tidepool sculpin (*Oligocottus maculosus*): sites of excretion, effects of reduced salinity and mechanisms of urea transport. Fish Physiology and Biochemistry 14:111–123.

Methods and Materials

EXPERIMENT 14.1 Effect of Salt Loading on the Specific Gravity and Production of Urine

In this experiment you will investigate the effect of hypotonic, isotonic, and hypertonic solutions on the volume and ion concentration of urine produced. Ion concentration of each urine sample will be determined by measuring **specific gravity,** which is the ratio of the urine sample density to the density of the reference solution, water (1.0 g/ml). Because specific gravity is a ratio of solution densities, specific gravity has no units of measurement.

- Randomly determine which individuals in your lab group will participate in the experiment. It is best if each volunteer has not consumed any food or more than 12 ounces of fluid within two hours of beginning this laboratory session. Each laboratory group should have at least two volunteers, but all students are encouraged to participate.
- Before consuming any of the treatments, each volunteer should acquire an initial (time 0) measurement of urine volume and specific gravity using the following methods:
  - Take an empty 500 ml volumetric urine collection cup to the lavatory. Urinate into the collection cup to measure the volume of urine produced. If the urine volume exceeds the capacity of the collection cup, empty the cup and dispose of the excess urine properly. Be sure to make note of how much urine was disposed of and add it to the volume of urine still contained in the collection cup. Record urine volume in Results Table 14.1a of your Laboratory Report. *Make sure you completely void your bladder for accurate measures of urine formation.*
  - Bring the collection cup of urine to the lab to measure specific gravity. Place one drop of urine on the specific gravity meter and record the specific gravity of this urine sample in Results Table 14.1b of your Laboratory Report.
  - Rinse the specific gravity meter with distilled water after you are finished. Dispose of the excess urine properly.
- Now consume the selected treatment as quickly as possible:

  Hypotonic Treatment Group (0.00 M NaCl)

  (0% NaCl solution = 0.00 g/250 ml)

  0 salt pills (452 mg each) with 250 ml water

  Isotonic Treatment Group (0.15 M)

  (0.9% NaCl solution = 2.25 g/250ml)

  5 salt pills (452 mg each) with 250 ml water

  Hypertonic Treatment Group (0.26 M)

  (1.5% NaCl solution = 3.86 g/250 ml)

  8 salt pills (452 mg each) with 250 ml water

Note: Although the hypertonic treatment of 3.86 g or 8 salt pills may seem like a large salt dose, this dose is well within the realm of salt loads experienced during the intake of certain foods. For example, a 3.5 oz (100g) bag of beef jerky contains 4.3 g of sodium. A can of soup may contain as much as 2 g of sodium and the Burger King® meal shown in Problem Set 14.1 contains over 3.2 g of sodium.

- Twenty minutes following ingestion, measure urine volume and specific gravity using the methods described previously.
- Continue to measure urine volume and specific gravity every 20 minutes for 100 minutes.
- Record your data and the class data in Results Tables 14.1a and 14.1b of your Laboratory Report.

RESULTS AND DISCUSSION

LABORATORY REPORT 14

EXPERIMENT 14.1

Effect of Salt Loading on the Specific Gravity and Production of Urine

Results Table 14.1a Class Data for Experiment 14.1: Effect of Salt Loading on Total Urine Volume.

	Student (Name)	Volume of Urine Produced (ml)						Total Urine Volume between 20 and 100 min (ml)
		0 min	20 min	40 min	60 min	80 min	100 min	
Hypotonic	1							
	2							
	3							
	4							
	5							
	6							
	7							
	8							
	9							
	10							
	Mean of total urine volume (±SE) for hypotonic treatment:							

Isotonic	Student (Name)	Volume of Urine Produced (ml) 0 min	20 min	40 min	60 min	80 min	100 min	Total Urine Volume between 20 and 100 min (ml)
	1							
	2							
	3							
	4							
	5							
	6							
	7							
	8							
	9							
	10							
	Mean of total urine volume (±SE) for isotonic treatment:							

Hypertonic	Student (Name)	Volume of Urine Produced (ml) 0 min	20 min	40 min	60 min	80 min	100 min	Total Urine Volume between 20 and 100 min (ml)
	1							
	2							
	3							
	4							
	5							
	6							
	7							
	8							
	9							
	10							
	Mean of total urine volume (±SE) for hypertonic treatment:							

Results Table 14.1b Class Data for Experiment 14.1: Effect of Salt Loading on Specific Gravity of Urine.

		Specific Gravity of Urine					
	Student (Name)	0 min	20 min	40 min	60 min	80 min	100 min
Hypotonic	1						
	2						
	3						
	4						
	5						
	6						
	7						
	8						
	9						
	10						
	Mean specific gravity (±SE)						

Isotonic	Specific Gravity of Urine						
	Student (Name)	0 min	20 min	40 min	60 min	80 min	100 min
	1						
	2						
	3						
	4						
	5						
	6						
	7						
	8						
	9						
	10						
	Mean specific gravity (±SE)						

Hypertonic	Specific Gravity of Urine						
	Student (Name)	0 min	20 min	40 min	60 min	80 min	100 min
	1						
	2						
	3						
	4						
	5						
	6						
	7						
	8						
	9						
	10						
	Mean specific gravity (±SE)						

Problem Set 14.1: Effect of Salt Loading on the Specific Gravity and Production of Urine

Using the data from Results Table 14.1a, create a bar graph showing mean total urine volume for each of the three different treatments. Remember that this value excludes the 0 min measurement and is the total volume from 20 to 100 min. Be sure to include the 95% CI for each of the experimental groups. Using these data, perform the appropriate statistical test to examine the effect of salt loading on the total volume of urine produced.

a. State the null hypothesis.

b. Should you accept or reject the null hypothesis? Support your conclusions with the appropriate statistical results.

c. What is the mechanism by which salt loading affects the volume of urine produced? Explain in detail.

d. Using the data from Results Table 14.1b, create a line graph showing the change in mean specific gravity (±95% CI) of urine versus time for each of the three different treatments.

e. What is the mechanism by which salt loading affects the specific gravity of urine? Explain in detail.

f. Your graph of mean specific gravity versus time may show a point at which the specific gravity of urine in all three treatments approaches a similar value. What is indicated by this convergence of specific gravity?

g. The following table lists the sodium content for a value meal at Burger King®. How would the effects of consuming this meal compare to the effects of the salt loading treatments used in this experiment?

Food Item	Sodium Content (mg)
Whopper® with Cheese	1450
Large Fries	880
Dutch Apple Pie	270
Large Chocolate Shake	640
Total Sodium	**3240**

LABORATORY 15

Metabolic Rate

PURPOSE

This laboratory will demonstrate the relationship between body size and metabolic rate and will explain how the measures and calculations of metabolism are used to quantify the energetic costs of maintaining homeostasis.

Learning Objectives

- Define the terms *anabolism* and *catabolism*.
- Define and distinguish between *basal* and *standard metabolic rate*.
- Describe mass-specific metabolic rate.
- Describe and understand the relationship between body size and metabolic rate.
- Understand how to measure and calculate the metabolic rate of an organism.

Laboratory Materials

Experiment 15.1: Standard Metabolic Rate and Body Size

Physiology Interactive Lab Simulations (Ph.I.L.S.)

Introduction and Pre-Lab Exercises

Although the general term **metabolism** is used to refer to many different processes (e.g., glucose metabolism, protein metabolism), we will refer specifically to metabolism as **energy metabolism.** Thus, metabolism is the total energy expenditure from both anabolic and catabolic processes in the body. Many physiologists refer to this as the energy budget of an organism.

Those metabolic processes that require energy are anabolic processes, and those that liberate energy from the breakdown of complex macromolecules are catabolic processes. These general metabolic processes contribute to the generation of body heat. This heat production by the body can be quantified and used to determine the metabolic rate of an organism.

Although physiologists can use heat production to determine metabolic rate as shown in Figure 15-1, it is sometimes easier to use indirect calorimetry and measure the amount of O_2 consumed or CO_2 produced by an animal. Heat production from metabolism is directly related to O_2 consumption during respiration. During aerobic metabolism, oxidative phosphorylation requires O_2 as the final electron acceptor in the electron transport chain. The energy released by the transfer of electrons in the electron transport chain is ultimately used to produce adenosine triphosphate (ATP). When the rate of metabolic processes increases, the rate of ATP production, and hence the amount of O_2 consumed, also increases. Thus, the amount of O_2 consumed by an organism can be used to estimate its metabolic rate. All physiological processes that maintain the body's homeostatic environment require the expenditure of energy in the form of ATP. The measure of O_2 consumption attempts to quantify the total metabolic or energetic cost of all physiological processes within the body (Figure 15-2).

15-1 The metabolic rate of an organism can be determined by measuring either its heat production or its consumption of O_2. However, metabolic rate in the literature is commonly reported in joules, a unit of energy. How would you compare your results of metabolic rate determined with respirometry to results determined with calorimetry? Show the conversion of metabolic rate in ml/h to kcal/h. Show how you would convert ml/h and kcal/h to metabolic rate reported in joules.

An animal's metabolic rate changes constantly depending on physiological conditions. Your metabolic rate increases during activity or exercise and after eating. Likewise, metabolic rate decreases during rest or while sleeping. Thus, to estimate the energy expenditure of an animal, one must measure both active and basal metabolism and determine the amount of time the animal is active and at rest. When quantifying the metabolism of an animal, physiologists are often more concerned with determining the minimal energy required for the animal to maintain species-specific physiology. Thus, we measure **basal metabolic rate** or **BMR.**

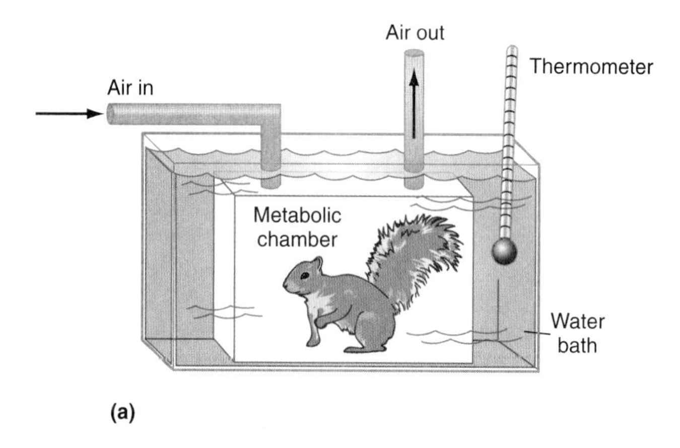

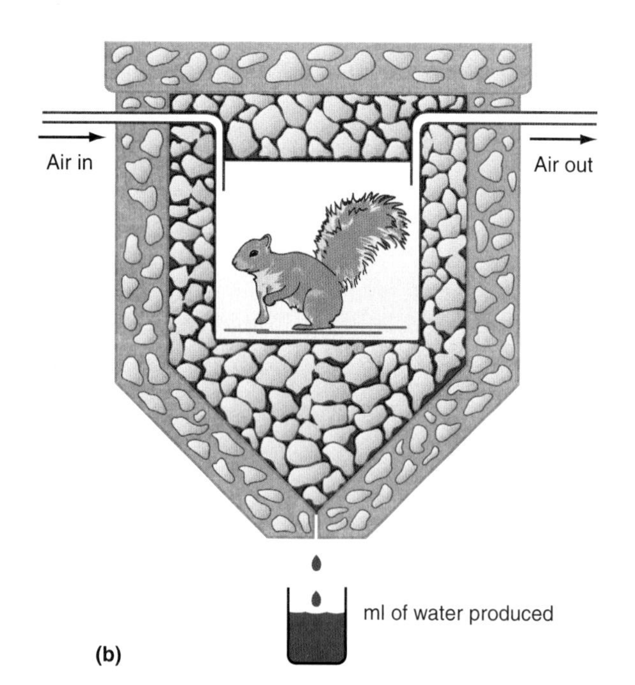

Figure 15-1
(a) A calorimeter is used to determine the amount of energy produced by an animal in calories. By placing an animal in the calorimeter, you can determine its metabolic rate by measuring the change in water temperature within the apparatus. If there is 1 liter of water in the apparatus, and the temperature is raised 1°C due to the animal's heat production, the animal has produced 1000 calories or 1 kilocalorie (kcal) of energy. The time required for the animal to produce this energy is the animal's metabolic rate in kcal/h. (b) Lavoisier's ice-jacket calorimeter works under the same principle whereby metabolic heat production is determined from the amount of water produced from the melting of ice (80 kcal of heat melts 1 kg of ice).

15-2 There are three important assumptions for measuring and reporting the BMR of an animal. List and describe these assumptions.

1.

2.

3.

Although these requirements for measuring BMR are controlled easily in humans, you might imagine the difficulty in meeting these requirements for wild or captive animals. For this reason, when referring to the baseline metabolic rate of animals other than humans, we refer to **standard metabolic rate** or **SMR.** SMR is the minimal metabolic rate of an animal at rest, but it does not require the same rigorous assumptions as BMR.

Every species has a unique SMR. This is based upon the species-specific requirements of its physiology. Although we can control the conditions for measuring BMR or SMR, metabolism can be influenced by many other factors related to activity, physiology, and morphology. Table 15-1 illustrates

Table 15.1 Energy Expenditure During Different Types of Activity for a 70 Kg (154 lb.) Person.

Form of Activity	Energy (kcal/h)
Lying still, awake	77
Sitting at rest	100
Typewriting rapidly	140
Dressing or undressing	150
Walking on level ground at 4.3 km/h (2.6 mi/h)	200
Bicycling on level ground at 9 km/h (5.3 mi/h)	304
Walking on 3 percent grade at 4.3 km/h (2.6 mi/h)	357
Sawing wood or shoveling snow	480
Jogging at 9 km/h (5.3 mi/h)	570
Rowing at 20 strokes/min	828

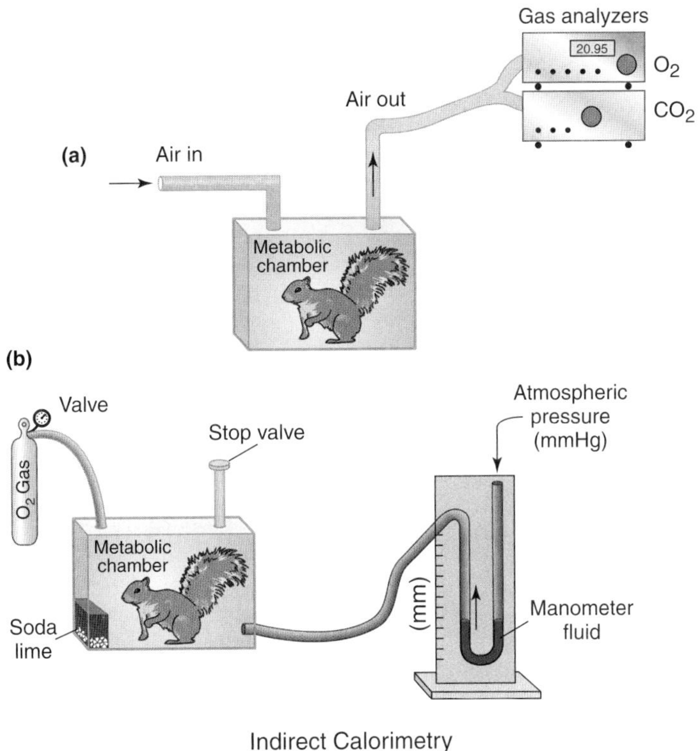

Figure 15-2
A respirometer is used to determine the amount of O_2 (ml) consumed (and/or the amount of CO_2 produced) by an animal. By placing an animal in the respirometer, you can determine its metabolic rate directly from the amount of O_2 it uses over time. (a) A respirometer shown with the use of gas analyzers. However, O_2 consumption can also be measured without the expense of gas analyzers by using a manometer. (b) With a manometer, a standardized concentration of O_2 is used to displace the air in the metabolic chamber and soda lime absorbs the CO_2 produced by the animal. The upward movement of the manometer fluid toward the chamber is due to the depletion of O_2 or a decrease in the partial pressure of O_2 in the chamber. Using this methodology, we report the animal's metabolic rate in ml O_2/h.

the energy expenditure during different activities in humans.

15-3 Using the energy expenditures for a 70 kg person listed in Table 15-1, calculate the difference in energy that would be expended while sitting at rest or jogging for an hour. How many more calories would be burned in a day if one decided to make jogging a part of their daily routine?

■

In regard to the influence of physiology on metabolic rate, many hormones influence specific metabolic processes. Cortisol, a hormone produced by the adrenal glands of humans (especially in response to stress), increases glucose metabolism. Prolactin, a hormone produced by the anterior pituitary, stimulates breast development and milk synthesis in women. Many nursing women will attest to the dramatic increase in metabolic rate stimulated by lactation. Leptin, a hormone produced by adipose cells, also regulates organic metabolism and metabolic rate.

15-4 In addition to the preceding examples, the thyroid hormones have widespread effects on metabolic processes throughout the body. Thyroid hormones increase the rate of ATP consumption by directly increasing the activity and number of Na^+/K^+-ATPase transporters in cell membranes. Thyroid hormones are therefore the most important hormones determining baseline metabolic rate. What are some clinical symptoms of hyperthyroidism? Relate these symptoms to the effects of hyperthyroidism on metabolic rate.

■

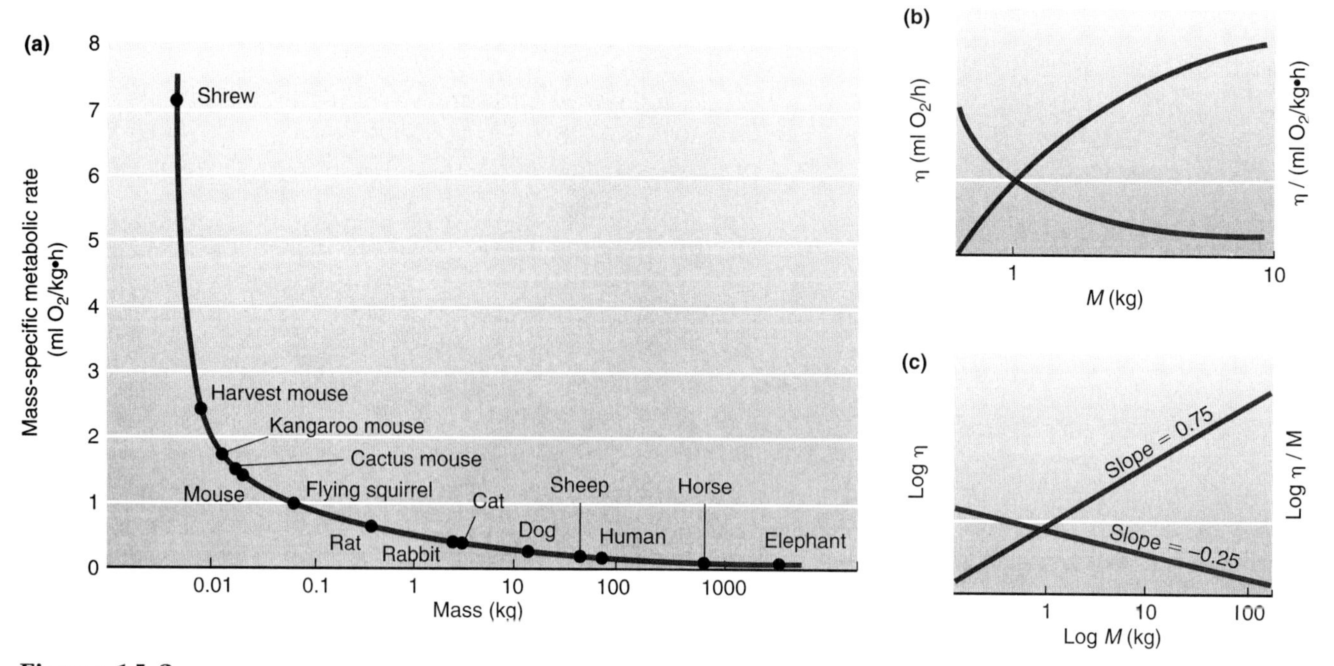

Figure 15-3
Relation between body mass (M_b) and O_2 consumption for different-sized animals at rest. (a) The classic "mouse-elephant" curve showing the decrease of mass-specific metabolic rate as body size increases. (b) The relationship between metabolic rate (blue line; ml/h) and mass-specific metabolic rate (orange line; ml/kg·h) versus M_b. (c) Log-log relationship between M_b and O_2 consumption showing the slopes of 0.75 and −0.25 for metabolic rate (blue line; ml/h) and mass-specific metabolic rate (orange line; ml/kg·h), respectively.

15-5 In Laboratory 10 ("Endocrine Physiology"), you investigated the influence of thyroxine on ventilation rate in fishes. How is the ventilation rate of goldfish similar to metabolic rate?

■

Metabolic rate also varies with morphology: both body size and body shape have a dramatic influence on metabolic rate. At the level of the whole organism, larger animals consume more O_2 per unit time (because they have a greater volume of tissues consuming O_2). If, however, metabolic rate is controlled for body mass, smaller animals consume more O_2 per unit time. This control for body mass in metabolism is called **mass-specific metabolic rate** because the amount of O_2 consumed is divided by body mass. Figure 15-3 demonstrates the difference between metabolic rate and mass-specific metabolic rate. Observe from this figure that although elephants consume more O_2 per unit time, elephants have a lower mass-specific metabolic rate than mice.

The relationship depicted in the "mouse-elephant" curve occurs because smaller animals have a much larger surface area-to-volume ratio than larger animals. Figure 15-4 demonstrates how the ratio of surface area to volume becomes smaller as body size increases. Thus, smaller animals tend to lose body heat much more quickly than larger animals. **Rubner's surface rule** states that cells in smaller animals will have a higher metabolic rate to compensate for the greater amount of heat loss due to their larger surface area-to-volume ratio. If an animal's metabolic rate is solely dependent upon surface area, then the slope for the relationship between metabolic rate and body mass should be equal to 0.67. Rubner's surface rule is therefore commonly referred to as the two-thirds rule. However, because an animal's metabolic rate is not solely governed by surface area, the relationship between metabolic rate and body mass demonstrates a slope of 0.75 as shown by the "mouse-elephant" curve.

We know that body size influences metabolic rate among different species as demonstrated by the "mouse-elephant" curve. However, how does body size influence metabolic rate within a species? If two bodies have different sizes, but are similar in geometric shape, then the relationship between metabolic rate and body mass may be described by the two-thirds rule. Similar to Max Rubner's research

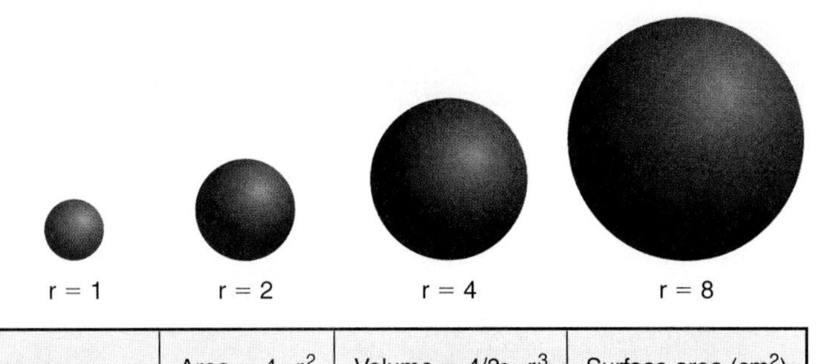

Radius (cm)	Area = $4\pi r^2$ (cm^2)	Volume = $4/3 \cdot \pi r^3$ (cm^3)	Surface area (cm^2) / Volume (cm^3)
1	12.6	4.2	3.00
2	50.3	33.5	1.50
4	201.1	268.1	0.75
8	804.2	2144.7	0.37

Figure 15-4
Demonstration of Rubner's surface rule.

examining the metabolic rates of dogs of various sizes, you will investigate the relationship between metabolic rate and body mass within mice. In this laboratory simulation, you will investigate the standard metabolic rates of four mice (*Mus musculus*) that differ in body mass. Your data will allow you to measure O_2 consumption and to complete statistical and graphical analyses to describe the relationship between standard metabolic rate and body mass within a species.

COMPARATIVE NOTE *Laboratory 15*

Hibernate and Save Energy

As you will observe in this laboratory, small animals tend to have much higher metabolic rates than larger animals. A higher metabolic rate consequently means that more food must be consumed to provide fuel for everyday physiological processes. The California pocket mouse (*Perognathus californicus*), for example, weighs only about 20 g, but it consumes more than 3 g of seeds per day (15% of its body mass). This would be the equivalent of an average adult human (70 kg) consuming 10.5 kg of food, or 36 Burger King Whoppers®, per day. During the winter months, food becomes unavailable and normal physiological processes cannot be maintained. Furthermore, maintaining homeostatic body temperatures in the face of decreasing environmental temperatures requires the expenditure of even more metabolic energy.

Many small mammals and some birds have found a solution to the constraints of low food availability during winter months in hibernation (the Latin word *hiberna* = winter). A few hibernating species include hamsters, pocket mice, dormice, bats, hedgehogs, hummingbirds, and insect-eating swifts. Hibernation is characterized by a greatly reduced metabolic rate, heart rate, and respiration rate. Because metabolic rate is dramatically reduced, body temperature decreases to the temperature of its surroundings, sometimes just a few degrees above 0°C. Without the burden of keeping warm during the winter, fat reserves accumulated during the summer and fall sustain the hibernating animal through the unfavorable winter months. Interestingly, most physiologists do not consider bears to be true hibernators. Bears show only a moderate decrease in metabolic rate during their "winter sleep." In addition, the body temperature of over-wintering bears decreases only a few degrees below that of their active state. Female bears also give birth to cubs during the winter. True hibernators are in a torpid state, showing little response to any external stimuli. Conversely, crawling into a bear's winter den expecting a safe encounter is a really bad idea.

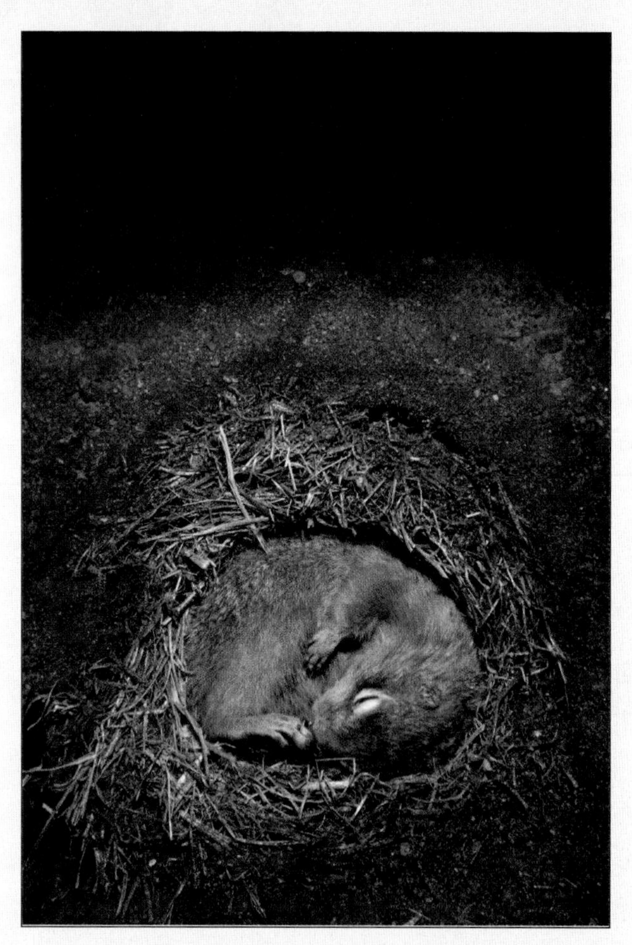

A hibernating artic ground squirrel *(Citellus parryi)* is unresponsive to external stimuli in its winter hibernaculum.

RESEARCH OF INTEREST

Harlow, H.J., T. Lohuis, R.C. Anderson-Sprecher, and T.D.I. Beck. 2004. Body surface temperature of hibernating black bears may be related to periodic muscle activity. Journal of Mammalogy 85:414–419.

Prunescu, C., N. Serban-Parau, J.H. Brock, D.M. Vaughan, and P. Prunescu. 2003. Liver and kidney structure and iron content in Romanian brown bears (*Ursus arctos*) before and after hibernation. Comparative Biochemistry and Physiology 134A:21–26.

Tucker, V. A. 1966. Diurnal torpor and its relation to food consumption and weight changes in the California pocket mouse *Perognathus californicus*. Ecology 47:245–252.

Methods and Materials

EXPERIMENT 15.1 Standard Metabolic Rate and Body Size

In this exercise you will conduct experiments on four different mice to investigate the effect of body size on Standard Metabolic Rate (SMR).

- Launch the Physiology Interactive Lab Simulations (Ph.i.L.S.). In the section entitled ***Metabolism***, select **2. *Size and Basal Metabolic Rate.***
- Read the *Objectives* of this simulation.
- Click the **Continue** button in this window to continue reviewing the introductory material. Be sure to click on and view all informational links presented within the *Introduction* window of this simulation. These links present valuable information and animations for increased understanding of the introductory material.
- Once you have completed your review of the introductory material, the **Continue** button will open the next tab entitled *Pre-Lab Quiz.* You may also click the *Pre-Lab Quiz* tab directly to enter this section of the simulation. Answer all questions in the Pre-Lab Quiz to test you comprehension and understanding of the introductory material. *Note:* The printed Lab Report will indicate your correct and incorrect responses on the *Pre-Lab Quiz* and report the number of correct responses out of the total number of questions possible. The simulation will only allow one attempt at the quiz, so answer each question carefully.
- Once you have completed the *Pre-Lab Quiz,* the **Continue** button within this window will open the next tab entitled *Wet Lab.* You may also click the *Wet Lab* tab directly to enter this section of the simulation. Read the material presented and be sure to click on and view all informational links presented within the *Wet Lab* window of this simulation. These links provide valuable information and video to introduce the apparatus and procedures necessary to conduct this experiment. *Note:* The printed Lab Report will indicate the number of video clips you viewed in the *Wet Lab* section.
- Once you have completed the *Objectives & Introduction, Pre-Lab Quiz,* and *Wet Lab* sections of the simulation, you are now ready to conduct the experiment. Click the **Continue** button to open the next tab entitled *Laboratory Exercise,* or click on the *Laboratory Exercise* tab directly to enter this section of the simulation. All of the experimental procedures are organized into 12 steps presented at the bottom of the simulation window. The simulation will indicate which step you are currently conducting and provide links for information about the apparatus. You may also click on **View All** to view the entire experimental procedure.

Setting Up the Apparatus

- Click on the **Power Switch** of the scale.
- Read the instructions at the bottom of the simulation screen and click on all informational links including the **animal rack** link to view further instructions on these items.
- Tare the scale by clicking on the **Tare** button. Select the first (top) mouse. Drag and place the mouse on the scale and enter its body mass (grams) in the Journal. The journal can be open and viewed by clicking the **Journal** icon. Close the Journal once you have enter the body mass of the first mouse.
- Drag the mouse from the scale to the respirometry chamber. Follow the instructions at the bottom of the simulation screen for placing bubble solution in the right-end opening of the tube.
- You are now ready to collect data on the O_2 consumption of this mouse. Notice the **Data Table** in the upper left-hand corner of the simulation window. This table has data cells for entering your readings every 15 seconds.
- Click on the **Location** link at the bottom of the simulation window to read about accurately measuring O_2 consumption.
- Now click on the **Data Table** link to read about proper data entry in the data table. *Note:* Although only 2 data points are required to calculate the amount of O_2 used per unit time, you will recall that the accuracy of a regression analysis is dependent upon sample size. Thus, you will want to collect as many data points to ensure your calculated relationship between O_2 consumption and time is most accurate.

- Enter the initial position of the bubbles within the tube for the first time entry (Time 0.00).
- Click the **Start** button and begin recording data every 15 seconds.
- Once you have collected the final data point for the 2 minute entry, click on the **CALC** button to calculate the relationship between O_2 consumption (ml) and time (min). Notice the simulation has reported the regression equation and the r^2 value for this relationship. Also notice that the slope of this regression equation is the rate O_2 is consumed by the mouse. This value or slope has been automatically entered into the *Trial 1* section of your journal. Click on the **Journal** icon and view the newly entered data.

$$y\,(\text{ml}) = \text{slope} \bullet x\,(\text{min}) + \text{b where slope} = \frac{\Delta y\,(\text{ml})}{\Delta x\,(\text{min})}$$

- Close your calculation and journal windows.
- Click on the **ERASE** button to clear the data from the data table for Trial 1. Notice the timer has been reset to zero.
- Click the **Start** button to repeat the experiment and collect data for both Trials 2 through 5 following the same procedures.
- Once you have collected data for all five trials, repeat these procedures to obtain measures of O_2 consumption for the other three mice in order from largest to smallest.
- Copy the data from your completed journal in the computer simulation to Results Table 15.1a in your Laboratory Report.
- Click on the **Graph** button within the Journal window. Notice that the average O_2 consumption (ml/min) for each mouse has already been calculated from your five data trials. Also note that mass-specific O_2 consumption (ml/g•h) has also been calculated.
- Within this Graph window, click the **Plot** button to plot your data and review carefully the results shown in both bar graphs. Study the format of these graphs so that you can reproduce them in a graphing program to be turned in with your laboratory report.
- Click on the *Post-Lab Quiz and Lab Report* tab. Answer the Post-Lab questions.
- Once you have completed the *Post-Lab Quiz*, a *Conclusion* will be presented for this experiment. Read the conclusion carefully and click the **Finish Lab** button.
- The *Print Your Laboratory Report* window will allow you to enter your name and course details. Click the **Print Lab** button at the bottom of the window. Once you have successfully printed your laboratory report, you have completed the laboratory simulation.

RESULTS AND DISCUSSION

LABORATORY REPORT 15

EXPERIMENT 15.1

Standard Metabolic Rate and Body Size

Results Table 15.1a Individual Group Data for Experiment 15.1: Standard Metabolic Rate and Body Size.

Mouse	Body Mass (g)	O_2 Consumption (ml/min)					
		Trial 1	Trial 2	Trial 3	Trial 4	Trial 5	**Mean (±SE)**
1							
2							
3							
4							

Problem Set 15.1a: Standard Metabolic Rate and Body Size

Using the data from Results Table 15.1a, create a scatter plot by graphing mean O_2 consumption (ml/min) on the y-axis and body mass on the x-axis. Now perform the appropriate statistical test to examine the relationship between the mean standard metabolic rate and body mass of mice.

a. State the null hypothesis.

b. Should you accept or reject the null hypothesis? Support your conclusions with the appropriate statistical results.

c. Why does body size influence O_2 consumption? Explain.

d. Using the data of mean O_2 consumption (ml/min) from Results Table 15.1a, create a log-log regression plot to investigate the relationship between O_2 consumption and body mass. Make sure you take the log of O_2 consumption and body mass prior to graphing. What is the slope of this regression?

e. How does this slope compare to the slope of the log-log relationship between O_2 consumption and body mass shown by the "mouse-elephant" curve (Figure 15-3c)? Is the slope of your data closer to that of Rubner's surface rule or closer to the 0.75 slope as shown in Figure 15-3c?

f. Can you explain why your slope may or may not agree with Rubner's surface rule? *Hint:* comparative differences in morphology between and among species.

g. Using the data from Results Table 15.1a, now calculate a mass-specific metabolic rate for each of the four mice. Record these calculations in Results Table 15.1b.

Results Table 15.1b Individual Group Data for Experiment 15.1: Mass-Specific Standard Metabolic Rate and Body Size.

Mouse	Body Mass (g)	Mass-Specific O_2 Consumption (ml/h/g)					
		Trial 1	Trial 2	Trial 3	Trial 4	Trial 5	**Mean (±SE)**
1							
2							
3							
4							

Problem Set 15.1b: Mass-Specific Standard Metabolic Rate and Body Size

Using the data from Results Table 15.1b, create a scatter plot by graphing mean mass-specific O_2 consumption (ml/h/g) on the y-axis and body mass on the x-axis. Now perform the appropriate statistical test to examine the relationship between the mean mass-specific standard metabolic rate and body mass of mice.

a. State the null hypothesis.

b. Should you accept or reject the null hypothesis? Support your conclusions with the appropriate statistical results.

c. What is the relationship between mass-specific O_2 consumption and body mass for the mice examined? Explain this relationship.

d. Is the relationship between mass-specific metabolic rate and body mass within a species similar to the relationship among species (illustrated in Figure 15-3a)? Explain.

e. All of the subjects used in today's experiment were female mice. How might the mean O_2 consumption (ml/min) of male mice compare to female mice? Explain two factors that help support your answer.

APPENDIX 1 *Scientific Writing and Preparing a Scientific Research Paper for Peer Review*

PURPOSE

This appendix provides guidelines and advice for preparing a well-written and organized scientific research paper. Formats are introduced as if you were preparing a manuscript for submission to a scientific journal and the peer review process. You will be introduced to the peer review and publication process in science as you properly prepare a laboratory research paper. An optional peer review exercise is outlined for use by your instructor.

Learning Objectives

- Understand the difference between popular and scientific writing.
- Prepare a formal manuscript formatted according to the "Guidelines to Authors" for a peer-reviewed journal in physiology.
- Participate in the peer review process for evaluating a scientific manuscript.

Introduction

A favorite movie quote is from *The Last Emperor,* where actor Peter O'Toole emphasizes the importance of conciseness and clarity by saying, "**If you cannot say what you mean, your majesty, you will never mean what you say and a gentleman should always mean what he says.**" Your ability to communicate concisely and clearly in both speech and writing is a key to success in any career. This skill is especially important to a scientist. Scientists must take very difficult and technical material and discuss it with other scientists and the general public. This is accomplished using a writing style known as technical or scientific writing. Many people find scientific writing to be dry and less poetic than popular writing. However, as scientists, your job is not to be a poet, but to communicate information concisely and clearly. The entertainment value of poetry is that it allows for individual interpretation. Scientists must write so that the reader is not forced to make his or her own interpretations. If you are not perfectly clear, then others may misinterpret your meaning and you ". . . will never mean what you say . . ."

How do you become a good scientific writer? The answer is simple: practice, practice, and more practice. If you wish to become a famous musician or an Olympic finalist, your goals can be accomplished only by dedication and practice. Scientific writing is no exception.

Keys to Good Scientific Writing

- **Review your words.** After you put your ideas and thoughts on paper, go back and think about word choice and sentence structure. Are your ideas clearly stated? Will your word usage or vocabulary be understood by your audience? Can you rewrite the sentence more clearly with fewer words?

- **Words cost money.** Literally, words cost money. As scientists, we not only have to be clear, we also have to be concise. Excessive word usage increases the length of a manuscript and in turn increases publication costs. Even abstract submissions for scientific meetings have word count limits to reduce the potential costs of lengthy meeting programs. Thus, you must be economical with word use and avoid unnecessary verbiage.
- **Minimize subject and verb separation.** Sentences are much clearer when the subject and verb are next to each other in a sentence.
- **Split infinitives.** In common spoken language, this error is regularly committed. Did you catch it? The last sentence is incorrect because the verb, *is committed,* is separated by the adverb *regularly.* Thus, this sentence may be unclear because it contains a split infinitive. It should be written: *In common spoken language, this error is committed regularly.*
- **First person.** Always write in the first person, as in version (b). Scientific writing is much clearer when the subject is placed at the beginning of the sentence.
  (a) *Each frog was selected randomly for measures of swimming endurance and lactate production.*
  (b) *I selected randomly each frog for measures of swimming endurance and lactate production.*
- **Order and organization.** Always present information chronologically and in a logical sequence. For example: smallest to largest, largest to smallest, simplest to most complex.
- **Consistency.** When you report numbers, use significant digits in the same manner throughout. Use similar units when reporting data, and always define technical terms and use them consistently.
- **Reference.** Always reference or cite the work of other scientists. We advance science by acknowledging the work of scientists who came before us. This encourages further discovery and the building of scientific knowledge.
- **Peer review.** Gain a new perspective on your writing and work by allowing a peer to critically review and evaluate your manuscript. This important peer review process in science ensures the quality of scientific work prior to its publication and presentation in scientific journals.

Catastrophes and Unprofessionalism in Scientific Writing

- **Plagiarizing, falsifying, and cheating.** The advancement of science and the discovery of truth are only possible through the practice of honesty. Never use the work of others and report this information as your own. Always make sure information gathered for your writing is properly cited, giving credit where credit is due. Never falsify data or cheat by making up data. This not only damages the trust that the general public and other scientists have in your work, but also disrupts the advancement of science. If scientists read and use your phantom study as a basis for future work, they may be led down the wrong path by your falsified data and conclusions, thus disrupting future discovery and the advancement of science. Science is based upon the pursuit of knowledge and the absolute truth; cheating has no place in this important endeavor.
- **Be complete and thorough.** Do not leave out details and important information. Limited details may lead a scientist to draw the wrong conclusions from your work. Describe your study completely and explain fully your results and conclusions. This will ensure that your ideas are expressed clearly and understood by your peers. If you don't express your ideas clearly, your results and conclusions may be incorrectly paraphrased and cited by your peers in future publications. This is one reason why it is always important to read and reference the original works you are citing, rather than relying on the interpretations and summaries of others who are citing these works.

Writing and Preparing a Scientific Research Paper for Peer Review

Manuscript Title—*What and Why?*

A scientific research paper should have an informative but concise title. This title should state: (1) the topic, problem, hypothesis, or question; (2) the type of data collected; and (3) the organism used in the study, indicated by both its common and scientific names.

Examples:

1. Time course analyses of the thermoregulatory responses to melatonin and chlorpromazine in bullsnakes (*Pituophis melanoleucus*).
2. The effect of surgically implanted transmitters upon the locomotory performance of the checkered garter snake, *Thamnophis m. marcianus*.
3. Physiological performance and stream microhabitat used by two Centrarchids (*Lepomis megalotis* and *Lepomis macrochirus*).
4. Melatonin and chlorpromazine: thermal selection and metabolic rate in the bullsnake, *Pituophis melanoleucus*.

Author Line and Byline—*Who did what and who is where?*

These lines indicate the author(s) of the manuscript and their affiliated academic/research institutions. If the manuscript is multiauthored, the order of authors is usually determined by who conducted the greatest amount of work on the project. The first author is usually the scientist who developed the questions and conducted the experiments described in the study. The byline gives credit to these authors' affiliated institutions for their direct and indirect support of the project.

Abstract—*What I did, what happened, and why you should care.*

Journal articles generally begin with an abstract. This is a summary or brief synopsis of the manuscript highlighting the purpose and results of the study. There are often philosophical differences among journals and authors regarding what should appear in an abstract. Thus, a journal's "instructions to authors" should be followed closely. Some journals require specific details and reports of statistical results, while others suggest an overview of your results and a brief report regarding the importance or significance of your findings. Regardless of these specifics, all abstracts summarize an entire study in about 200 to 300 words. You may now begin to appreciate the importance of learning how to write concisely in science. Within this limited space, you must provide the essential background, results, and importance of your experiments. Fellow scientists will often read only the abstract of a manuscript to evaluate whether the study is of interest to them. Thus, your abstract will often determine whether the rest of your presentation will be read.

Introduction—*What has been done in the past, what will I do, and why is it important?*

A manuscript's introduction introduces hypotheses and questions. However, the introduction must also provide the necessary background to enable the reader to understand these hypotheses and questions. This section will most likely contain more citations than any other part of your manuscript. To prepare a well-organized and informative introduction, first decide what your questions are and why these questions are important or unique. Use these questions to develop your background information and to direct your audience to the relevant literature. You may wish to consider the following steps in preparing an organized introduction:

a. Present the topic and its associated problems, concerns, or political issues and state the reasons why your audience should be interested.
b. Thoroughly review the related literature. Specifically, address what has been done previously in an attempt to answer related questions and/or explain how past data have led you to new questions. Provide your audience with enough background material to justify your investigation and to illustrate its importance.
c. Formally state the questions and hypotheses you will be addressing at the end of the introduction to outline what your study will address. Make sure you emphasize why your questions are important and unique.

Materials and Methods—*What I did and how I did it.*

The Materials and Methods section of a manuscript is a detailed account of the materials and methods used to gather the data that are presented in the results. To ensure that you are writing an accurate and complete Materials and Methods section, ask yourself one simple question: From what I have described, can someone repeat my experiment, exactly? A Materials and Methods section should include details about the supplier of chemicals, where experimental animals were caught or commercially obtained, and the exact conditions, treatments, and design of the experiments. These details may even include the light intensity in the laboratory where you conducted your experiments. If you cannot answer yes when asked whether someone could repeat your experiment based on the information you have provided, then you have not provided the detailed information required for a publishable Materials and Methods section. Remember, science is based upon repeatability. If you do not provide enough details for future experimentation, your findings may be disregarded as error.

Results—*What I found.*

A Results section is used to clearly state the findings of your experiments. Recall from Laboratory 1 ("Scientific Investigation") that there were several examples of results statements followed by the supporting statistical results.

Example: Heart rate was significantly different among individuals with low, normal, and high blood pressures (F = 33.14, df = 2,12; P < 0.05).

Be sure to state your findings simply and to reference any available data tables or figures to illustrate your results. Data tables and figures should be numbered sequentially in the order in which they are referenced within the Results section. Appropriate ways to refer to a table or figure within the text include:

The heart rates of patients with low and normal blood pressures are not significantly different from one another (Table 1).

Figure 1 demonstrates a significant relationship between testosterone treatment duration and bone density in men over 65 years of age (F = 137.4; df = 1,17; P < 0.05).

Each data table and figure should be placed on a separate page in the manuscript following the Literature Cited section; do not embed tables and figures within the manuscript text. In addition, each table and figure must be accompanied by a complete title or caption. A complete explanation of a figure, including a description of the variables measured and those that were manipulated, the subjects of study, abbreviations used within the figure, and any other relevant information typically requires several sentences. Figures should be able to stand alone—that is, a person should be able to interpret the findings presented in the figure without having to read the Results section for more information.

Many students make the mistake of explaining or interpreting their results within the Results section. A Results section should simply state the results of the experiments. All interpretations of these results and their biological significance are addressed in the discussion.

Discussion—*What do my findings mean?*

The Discussion section of the manuscript is where you offer an explanation for and interpret your results and findings. The Discussion is also where you should provide a detailed explanation of the significance of your findings and how they relate to previous work. You can also suggest future research to further address the results discussed in your study. However, your conclusions must be supported by and must follow logically from your results. Be careful not to extrapolate your results into "storytelling." It is sometimes tempting to make more out of your results than the data support. When drawing conclusions from your data, remember the principle of Occam's Razor: *With all things considered and equal, the simplest explanation is the explanation that is most likely correct.*

Literature Cited—*Who helped me introduce and discuss?*

Many students are more familiar with a Reference section than a Literature Cited section. Although journals seem to use these headings interchangeably, they do have slightly different meanings. **References** are articles or papers that you wish your reader to know about so that the reader can reference these for additional information if needed. In contrast, **literature cited** are articles or papers that have been specifically cited or referred to in a manuscript. Thus, when you refer to the findings of other scientists in the introduction or discussion of your scientific paper, it is more appropriate to list their works as Literature Cited. However, if you just wish to suggest articles to help someone better understand the topic or background, you may suggest a list of references. For example, these references provide additional reading material about scientific writing:

McMillan, V.E. 1988. Writing papers in the biological sciences. St. Martin's Press. New York. Pp. 142.

Pechenik, J.A. 2001. A short guide to writing about biology. Addison-Wesley Educational Publishers, Inc. New York. Pp. 318.

However, because we have not specifically cited these authors in writing this appendix, referring to their books as *Literature Cited* would be incorrect. Use the following questions to help you make the distinction between Literature Cited and References. Literature Cited—Who helped me introduce and discuss? References—Who can help the reader obtain additional information to better understand what I am presenting?

Example Format for Citing Literature

All journals have a guide or Instructions to Authors for formatting manuscripts prior to submission. Your instructor may ask you to follow the format of a particular journal such as *Comparative Physiology, American Journal of Physiology, The Journal of General Physiology,* or *Comparative Biochemistry and Physiology.* Here we present some of the standard formats for properly citing different types of literature. We also present an image of a published manuscript as an aid for properly preparing your scientific research paper and using citations within the text of your paper (pages 206 and 207). Note that the journal *Hormones and Behavior* requires that all the authors of a particular work be listed the first time the work is cited. This uncommon format is specific to this journal. When articles have more than two authors, most journals require that only the first author followed by *et al.* and the year be listed in the citation within the text.

Journal Article

Lutterschmidt, W.I. 1994. The effect of surgically implanted transmitters upon the locomotory performance of the checkered garter snake, *Thamnophis m. marcianus.* Herpetological Journal 4(1):11–14.

Lutterschmidt, D.I. and W.I. Lutterschmidt. 2002. Modifications for the successful use of thermocouples in studies of thermoregulation. Herpetological Review 33(2):110–112.

Lutterschmidt, D.I., W.I. Lutterschmidt, N.B. Ford, and V.H. Hutchison. 2002. Behavioral thermoregulation and the role of melatonin in a nocturnal snake. Hormones and Behavior 41(1):41–50.

Book

Lutterschmidt, W.I. and D.I. Lutterschmidt. 2007. Laboratory Exercises in Human Physiology: A Clinical and Experimental Approach. McGraw-Hill Companies, Inc., New York. Pp. 224.

Widmaier, E.P., H. Raff, and K.T. Strang. 2006. Human Physiology: The Mechanisms of Body Function. McGraw-Hill Companies, Inc., New York. Pp. 827.

Edited Book Chapter

Duvall, D., L.J. Guillete, Jr., and R.E. Jones. 1982. Environmental control of reptilian reproductive cycles. *In* Biology of the Reptilia, volume 13. C. Gans and F.H. Pough (Eds.), pp. 201–231. Academic Press, London.

Hutchison, V.H. and R.K. Dupre. 1992. Thermoregulation. *In* Environmental Physiology of the Amphibians. M.E. Feder and W.W. Burggren (Eds.), pp. 206–249. University of Chicago Press, Chicago, Illinois.

Mason, R.T. 1992. Reptilian pheromones. *In* Hormones, Brain, and Behavior: Biology of the Reptilia, volume 18. C. Gans and D. Crews (Eds.), pp. 114–228. University of Chicago Press, Chicago, Illinois.

Optional Peer Review Exercise

When research scientists seek publication of their work, they send a formatted manuscript to the editor of a scientific journal. This editor then makes a decision on the merits of the scientists' work and whether it is worthy of communication to the scientific community through publication. To accomplish this task, the editor must seek the opinions of scientists who conduct research related to the questions addressed in the manuscript. Editors will usually seek the reviews and professional advice of two or three researchers to judge the merits of the submitted manuscript. Some editors may remove the names and affiliations of authors so as not to bias the review process. Based upon the recommendations provided by peer reviewers, the editor makes an educated decision regarding the possible publication of the manuscript. This peer review process is essential to ensuring that only quality science is communicated to other scientists and the public.

Your instructor may wish for you to participate and gain practice in the peer review process. Although your scientific paper will be graded by your instructor (essentially the editor), he or she may wish to gain additional evaluations from your peers (your classmates). In this exercise, your instructor may ask you to peer-review a scientific research paper submitted by one of your classmates; this may be a class assignment or part of a participation grade. Your instructor may remove the student's name from the manuscript you receive in order to avoid bias. Your professional duty and responsibility as a reviewer is to evaluate the writing clarity, organization, and overall presentation of the research paper. It is important that you remain anonymous (as do reviewers) in the peer review process. Your instructor may have additional and specific instructions to enable you to gain the most experience from this exercise.

Three review sheets are provided on pages 208–210 to guide you through the peer review process of a submitted student research paper.

Hormones and Behavior **41,** 41–50 (2002)
doi:10.1006/hbeh.2001.1721, available online at http://www.idealibrary.com on 

Behavioral Thermoregulation and the Role of Melatonin in a Nocturnal Snake

Deborah I. Lutterschmidt,*[,1] William I. Lutterschmidt,† Neil B. Ford,‡ and Victor H. Hutchison*

*Department of Zoology, University of Oklahoma, Norman, Oklahoma, 73019; †Department of Biological Sciences, Sam Houston State University, Huntsville, Texas, 77341; and ‡Department of Biology, University of Texas, Tyler, Texas, 75799

Received February 8, 2001; accepted July 16, 2001

Daily and seasonal variations in hormone levels influence the complex interactions between behavior and physiology. Ectothermic animals possess the unique ability behaviorally to adjust body temperature (T_b) to control physiological rate processes. Thus, a hormone may indirectly influence a physiological rate by directly influencing the behaviors that adjust or control that rate process. Although many hormonal influences on behavioral regulation of T_b remain uninvestigated, melatonin (MEL) generally is considered a hormone that decreases mean preferred T_b. Many ectotherms demonstrate the selection of lower T_b's in response to increased MEL concentrations. Here, we examined the influence of MEL on the behavioral regulation of T_b in the nocturnal African house snake *Lamprophis fuliginosus.* A series of experiments with two injection regimes of MEL had no significant effect on the mean preferred T_b of *L. fuliginosus.* In addition, mean preferred T_b's during the photophase did not differ significantly from those during scotophase. Our findings suggest that *L. fuliginosus* does not respond to elevated concentrations of either endogenous or exogenous MEL. To verify that the African house snake is nocturnal, we investigated activity patterns of *L. fuliginosus* throughout the photoperiod. The activity period of *L. fuliginosus* occurs in the scotophase of the photoperiod, a pattern consistent with that of nocturnal species. This suggests that nocturnal organisms such as *L. fuliginosus* may not respond to MEL in the same manner as many diurnal species. Our results support the hypothesis that some animals, particularly nocturnal species, may have developed alternative responses to increased plasma concentrations of MEL.

[1] To whom correspondence and reprint requests should be addressed at (current address) Department of Zoology, Oregon State University, Corvallis, Oregon, 97331. E-mail: luttersd@bcc.orst.edu. Fax: (541) 737-0501.

Key Words: **snakes; African house snakes; *Lamprophis fuliginosus;* nocturnal; pineal; melatonin; activity; body temperature; thermal selection.**

The presence and success of organisms in both time and space depend upon a complexity of environmental factors (Odum, 1959). Of these factors, temperature is the most pervasive because it directly affects physiological rate processes. While endothermic homeotherms maintain a fairly constant body temperature (T_b), ectothermic poikilotherms are subject to very wide fluctuations in daily and seasonal environmental temperatures. Consequent fluctuation in ectothermic T_b causes variability not only in physiological processes but also in the behavioral capacities that are dependent upon them (Bennett, 1990). For example, temperature-induced changes in metabolic rate may result in changes in activity. Thus, if particular physiological functions are temperature sensitive, then behaviors associated with those processes will be modified as T_b changes (e.g., Bennett, 1990).

Precise behavioral thermoregulation by many ectothermic vertebrates is well documented. However, hormonal influences on thermoregulatory behavior and T_b have received less attention. Melatonin (5-methyl-*N*-acetyltryptamine) is one hormone involved in regulating T_b. Melatonin (MEL) is produced by the pineal gland and is thought to play a role in neuroendocrine transduction by converting light information into endocrine signals (Axelrod, 1974). Elevated levels of MEL are observed in response to darkness and significantly influence behavioral thermoregulation via a decrease in the set point of mean preferred T_b (Lutterschmidt, Lutterschmidt, Tracy, and Hutchison, 1998; Moyer, Firth, and Kennaway, 1995; Underwood,

0018-506X/02 $35.00

1985a; Vivien-Roels, Adrendt, and Bradtke, 1979; Vivien-Roels, Pévet, and Claustrat, 1988). These effects have been observed in the salamander *Necturus maculosus,* the turtle *Terrapene carolina triunguis,* and the lizard *Crotaphytus collaris* (Cothran and Hutchison, 1979; Erskine and Hutchison, 1981; Hutchison, Black, and Erskine, 1979). In addition to its influence on T_b and behavioral thermoregulation, MEL also affects circadian and circannual activity patterns, reproductive cycles, and thermal tolerance (Lynch, White, Grundel, and Berger, 1978; Ralph, 1978; Ralph, Firth, Gern, and Owens, 1979a; Ralph, Firth, and Turner, 1979b; Refinetti and Menaker, 1992; Underwood and Menaker, 1976; Vivien-Roels and Pévet, 1983). These effects of MEL, in combination with its seasonal rhythm of secretion in response to the longer duration of daylight in spring, function in synchronizing an animal's behavior and physiology with environmental cues (Crews, Hingorani, and Nelson, 1988; Mendonça, Tousignant, and Crews, 1995, 1996; Rismiller and Heldmaier, 1987; Underwood, 1981, 1985b; Underwood and Hyde, 1989).

Thus, there exists much complexity in which environmental stimuli and circadian and circannual cycles in hormone concentrations interact to regulate both physiological and behavioral thermoregulation. Understanding behavioral thermoregulation in ectotherms first requires an understanding of the complex interactions between behavioral adjustment of T_b and the hormonal regulation of T_b. We examined the role of MEL in the control of preferred T_b in African house snakes, *Lamprophis fuliginosus.* Interest in examining this species emanated from questions regarding the effects of MEL on the behavioral thermoregulation of nocturnal ectotherms. To investigate the hypothesis that nocturnal ectotherms exhibit different responses to MEL from diurnal species, we used the following experiments to address three independent questions: (1) Does the mean preferred T_b of *L. fuliginosus* differ between photophase and scotophase; (2) does MEL affect mean preferred T_b of *L. fuliginosus;* and (3) how does the response of *L. fuliginosus* to MEL differ from that of a diurnal snake species?

MATERIALS AND METHODS

Behavioral Thermoregulation Experiments

Animals, captive care, and acclimatization. A captive-bred population of African house snakes, *L. fuliginosus,* was obtained from the Ophidian Research Colony (University of Texas at Tyler). Snakes were housed in individually marked Rubbermaid containers (8-liter volume, 37.5 × 26.0 × 12.7 cm) to monitor closely body mass and treatment schedule. Snout–vent length (SVL) and body mass (M_b) of snakes prior to, during, and upon completion of all experimental trials were measured so that injection volumes could be adjusted. Snakes were fed mice weekly; water was provided *ad libitum.*

We used a 3-week acclimatization period with a constant temperature (25 ± 1°C) and photoperiod (L12:D12). Photophase was centered on 1200 h CST and began at 0600 h; scotophase began at 1800 h. Snakes also were fed weekly during the acclimatization period. We allowed for a minimum 1-week acclimatization between the last feeding and observations of T_b to ensure that all snakes were postabsorptive (Lysenko and Gillis, 1980; Regal, 1966).

We also allowed for a 1-week acclimatization between treatments to ensure that all snakes were free of residual MEL prior to new experimental treatments. In endotherms, the half-life of MEL after injection can be 1 h or less (Rollag and Stetson, 1982). Although the half-life in whole-animal ectotherms has not been measured, it is likely to be much longer because metabolic rate is as much as 10 times lower, depending on T_b (Filadelfi and Castrucci, 1996). Thus, we concluded 1 week would provide sufficient time for complete metabolism of previous MEL treatments. All experimental protocols were approved by the University of Oklahoma Animal Care and Use Committee (Assurance No. 73-R-100).

Experimental design. We examined the thermal selection of 16 untreated *L. fuliginosus* (i.e., snakes receiving no treatments) to determine if snakes select significantly different temperatures during the scotophase than during the photophase. All trials were conducted during the fall of 1998.

In the spring of 1999, eight snakes were randomly selected for additional experiments. We used these eight snakes in a repeated measures design [i.e., each snake was used in all treatments to eliminate problems associated with between-subject variation (Klugh, 1970)] to test the hypothesis that exogenous MEL would significantly influence thermal selection. To control for possible temporal shifts or responses in behavior due to treatment sequence and learning effects, we randomized all treatment sequences as required by a repeated measures design (Norusis, 1985).

Each snake received two MEL and two injected control (IC) treatments. For the $IC_{6\text{-d prior}}$ (5% ethanol in reptile Ringer's solution) and $MEL_{6\text{-d prior}}$ (5.0 mg kg^{-1}

Peer Evaluation of Student Research Paper

Evaluate each section of the manuscript using the following scale of 1 (poor) through 5 (excellent). Place your ratings for each section on the lines provided. Use the criteria outlined below each section to aid in your evaluation of each section.

Rating Scale	1	2	3	4	5
	Poor		Average		Excellent

Reviewer #_____

Title of Reviewed Manuscript: ____

____

____

Manuscript Section: **Peer Evaluation Total: _____**

Title: _____

Is the title informative? Does the title indicate the topic of study and the organism used as an experimental model?

Abstract: _____

Is this a concise summary (300 words or less) that clearly indicates what was done, what happened, and why the results are significant to the audience?

Introduction: _____

Is there an adequate and logical presentation of background information for you to fully understand the hypotheses and questions? Has the author adequately used the literature to develop this background? Are the hypotheses and/or questions clearly stated and their importance described?

Materials and Methods: _____

Are the materials and methods presented in sufficient detail to enable repeatability? Is the presentation of the methods organized, and are there illustrations or diagrams to help with explaining the experimental apparatus and protocols?

Results: _____

Are the findings clearly presented with statistical support? Are tables and/or figures used to illustrate data? Are these tables and figures referenced appropriately within the text [e.g., heart rate did not differ among groups (Figure 1)]. Did the author remember not to discuss or explain the results in this section?

Discussion: _____

Are the results explained and conclusions drawn? Are the stated hypotheses addressed? Are explanations for the results presented clearly and with reference to previous work in the area? Is there a clear discussion of why this research is important? Are there recommendations for additional studies and indications of how the information could be used by future scientists?

Literature Cited: _____

Does the literature used provide adequate support for the introduction and discussion of this manuscript? Is the literature cited correctly in the body of the manuscript? Is the Literature Cited section formatted correctly? Are there citations in the body of the manuscript that are not listed in the Literature Cited section? Is there literature listed in the Literature Cited section that is not cited in the body of the manuscript?

Overall Presentation: _____

Is the manuscript written clearly and concisely? Did you enjoy reading the manuscript or did you feel it was a difficult read? Was the manuscript well organized with a neat and clean presentation?

Additional Comments for Improving the Manuscript:

Peer Evaluation of Student Research Paper

Evaluate each section of the manuscript using the following scale of 1 (poor) through 5 (excellent). Place your ratings for each section on the lines provided. Use the criteria outlined below each section to aid in your evaluation of each section.

Rating Scale	1	2	3	4	5
	Poor		Average		Excellent

Reviewer #_____

Title of Reviewed Manuscript: ____

____

____

Manuscript Section: **Peer Evaluation Total:** _____

Title: _____

Is the title informative? Does the title indicate the topic of study and the organism used as an experimental model?

Abstract: _____

Is this a concise summary (300 words or less) that clearly indicates what was done, what happened, and why the results are significant to the audience?

Introduction: _____

Is there an adequate and logical presentation of background information for you to fully understand the hypotheses and questions? Has the author adequately used the literature to develop this background? Are the hypotheses and/or questions clearly stated and their importance described?

Materials and Methods: _____

Are the materials and methods presented in sufficient detail to enable repeatability? Is the presentation of the methods organized, and are there illustrations or diagrams to help with explaining the experimental apparatus and protocols?

Results: _____

Are the findings clearly presented with statistical support? Are tables and/or figures used to illustrate data? Are these tables and figures referenced appropriately within the text [e.g., heart rate did not differ among groups (Figure 1)]. Did the author remember not to discuss or explain the results in this section?

Discussion: _____

Are the results explained and conclusions drawn? Are the stated hypotheses addressed? Are explanations for the results presented clearly and with reference to previous work in the area? Is there a clear discussion of why this research is important? Are there recommendations for additional studies and indications of how the information could be used by future scientists?

Literature Cited: _____

Does the literature used provide adequate support for the introduction and discussion of this manuscript? Is the literature cited correctly in the body of the manuscript? Is the Literature Cited section formatted correctly? Are there citations in the body of the manuscript that are not listed in the Literature Cited section? Is there literature listed in the Literature Cited section that is not cited in the body of the manuscript?

Overall Presentation: _____

Is the manuscript written clearly and concisely? Did you enjoy reading the manuscript or did you feel it was a difficult read? Was the manuscript well organized with a neat and clean presentation?

Additional Comments for Improving the Manuscript:

Peer Evaluation of Student Research Paper

Evaluate each section of the manuscript using the following scale of 1 (poor) through 5 (excellent). Place your ratings for each section on the lines provided. Use the criteria outlined below each section to aid in your evaluation of each section.

Rating Scale	1	2	3	4	5
	Poor		Average		Excellent

Reviewer #____

Title of Reviewed Manuscript: ____

____

____

Manuscript Section: **Peer Evaluation Total:** ____

Title: ____

Is the title informative? Does the title indicate the topic of study and the organism used as an experimental model?

Abstract: ____

Is this a concise summary (300 words or less) that clearly indicates what was done, what happened, and why the results are significant to the audience?

Introduction: ____

Is there an adequate and logical presentation of background information for you to fully understand the hypotheses and questions? Has the author adequately used the literature to develop this background? Are the hypotheses and/or questions clearly stated and their importance described?

Materials and Methods: ____

Are the materials and methods presented in sufficient detail to enable repeatability? Is the presentation of the methods organized, and are there illustrations or diagrams to help with explaining the experimental apparatus and protocols?

Results: ____

Are the findings clearly presented with statistical support? Are tables and/or figures used to illustrate data? Are these tables and figures referenced appropriately within the text [e.g., heart rate did not differ among groups (Figure 1)]. Did the author remember not to discuss or explain the results in this section?

Discussion: ____

Are the results explained and conclusions drawn? Are the stated hypotheses addressed? Are explanations for the results presented clearly and with reference to previous work in the area? Is there a clear discussion of why this research is important? Are there recommendations for additional studies and indications of how the information could be used by future scientists?

Literature Cited: ____

Does the literature used provide adequate support for the introduction and discussion of this manuscript? Is the literature cited correctly in the body of the manuscript? Is the Literature Cited section formatted correctly? Are there citations in the body of the manuscript that are not listed in the Literature Cited section? Is there literature listed in the Literature Cited section that is not cited in the body of the manuscript?

Overall Presentation: ____

Is the manuscript written clearly and concisely? Did you enjoy reading the manuscript or did you feel it was a difficult read? Was the manuscript well organized with a neat and clean presentation?

Additional Comments for Improving the Manuscript:

APPENDIX 2 Reference Tables

Table 1 Normal Values for Pulmonary Function Tests

Measurement	Value
Vital capacity	4–5 L (men); 3–4 L (women)
Inspiratory capacity	2–4 L
Expiratory reserve volume	1–2 L
Residual volume	1–2 L
Functional residual capacity	2–3 L
Total lung capacity	6–7 L (men); 5–6 L (women)
Forced expiratory volume, 1 second ($FEV_{1.0}$)	over 3 L (men); over 2 L (women)
$FEV_{1.0}$ as a percent of vital capacity	over 60% (men); over 70% (women)
Arterial oxygen tension (Pa_{O_2})	95 ± 5 mm Hg
Arterial carbon dioxide tension (Pa_{CO_2})	40 ± 2 mm Hg

Table 2 Normal Values for Cardiac Function and Blood Gas Measurements

Measurement	Value
Ejection fraction (SV/EDV)*	0.55–0.78
End-diastolic volume	75 mL/m^2 of body surface area
Cardiac output	2500–3600 mL/min./m^2 of body surface area
Percent oxygen saturation	97% (artery); 60–85% (vein)
Arterial pH	7.38–7.44
Oxygen tension (P_{O_2})	80–100 mm Hg
Carbon dioxide tension (P_{CO_2})	35–45 mm Hg
Bicarbonate concentration	21–30 mEq/L

**SV = stroke volume; EDV = end-diastolic volume*

Table 3 Normal Values for Renal Function Tests and Urine Constituents

Measurement	Value
Renal Function Tests	
Inulin clearance (GFR), males	124 ± 25.8 mL/min.
Inulin clearance (GFR), females	119 ± 12.8 mL/min.
Creatinine clearance	91–130 mL/min.
Urea clearance	60–100 mL/min.
Urine Constituents	
Specific gravity	1.002–1.028
Protein	under 150 mg/L
Potassium	25–100 mEq/L (varies)
Sodium	100–260 mEq/L (varies)
pH	5–7.5

Table 4 Normal Values for Constituents of Blood Plasma

Measurement	Value
Cholesterol, bound to LDL	under 130 mg/dL
Cholesterol, total	under 200 mg/dL
Creatinine	under 1.5 mg/dL
Enzymes	
Amylase, serum	60–180 units/L
Creatine phosphokinase, serum	10–70 units/L (females) 25–90 units/L (males)
Lactate dehydrogenase, serum	25–100 units/L
Glucose, fasting	75–115 mg/dL
Hormones	
Aldosterone	under 8 ng/dL
Cortisol (8 a.m.)	5–25 μg/dL
Epinephrine	under 50 μg/dL
Estradiol, in women	20–60 pg/mL
Testosterone, in men	3–10 ng/mL
Insulin, fasting	6–26 μU/mL
Thyroxine	5–12 μg/dL
Osmolality, plasma	285–295 mOsm
Protein, total serum	5.5–8 g/dL
Triglycerides	under 160 mg/dL
Urea nitrogen	10–20 mg/dL

Table 5 Normal Values for Erythrocyte and Leukocyte Measurements

Measurement	Value
Hemoglobin	13–18 g/dL (males); 12–16 g/dL (females)
Hematocrit	42–52% (males); 37–48% (females)
Erythrocyte count	4.5–6.0 $\times$ $10^6/mm^3$ (males); 4.0–5.5 $\times$ $10^6/mm^3$ (females)
Leukocyte count	$5 \times 10^3 - 10 \times 10^3/mm^3$
Differential Leukocyte Count	
Neutrophils	55–75%
Eosinophils	2–4%
Basophils	0.5–1%
Lymphocytes	20–40%
Monocytes	3–8%

APPENDIX 3 *SI Unit Prefixes and Symbols*

SI Unit Prefixes and Symbols

Prefix	Symbol	Factor
Yotta	Y	10^{24}
Zetta	Z	10^{21}
Exa	E	10^{18}
Peta	P	10^{15}
Tera	T	10^{12}
Giga	G	10^{9}
Mega	M	10^{6}
Kilo	k	10^{3}
Hecto	h	10^{2}
Deka	da	10^{1}
Deci	d	10^{-1}
Centi	c	10^{-2}
Milli	m	10^{-3}
Micro	μ	10^{-6}
Nano	n	10^{-9}
Pico	p	10^{-12}
Femto	f	10^{-15}
Atto	a	10^{-18}
Zepto	z	10^{-21}
Yocto	y	10^{-24}

APPENDIX 4 *Schematic for an Osmosis Chamber*

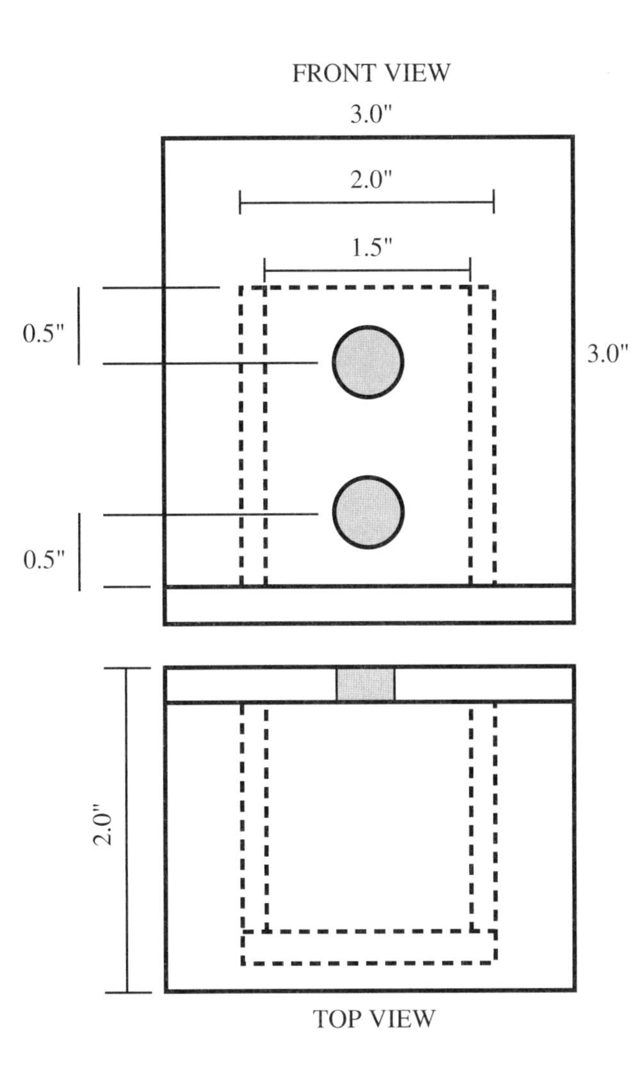

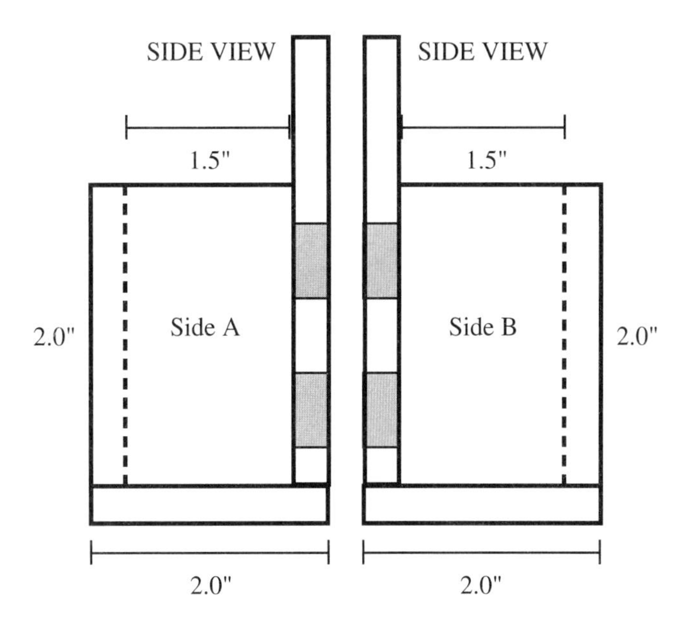

Plexiglass Pieces: The following pieces are required to build both sides (Side A and B) of the osmosis chamber. The chamber is constructed with 1/4" plexiglass.

Two 3" x 3" Pieces with 1/2" holes precisely drilled as shown in the FRONT VIEW.

Two 2" x 3" Pieces for chamber bottoms.

Two 2" x 2" Pieces for chamber backs.

Four 1.5" x 2" Pieces for chamber sides.

PHOTO CREDITS

Fig. 3.2: © David M. Phillips/Photo Researchers, Inc.; **Fig. 3.3a:** Courtesy Bill Lutterschmidt; **P. 40:** © J. Schmidt/National Park Service; **Fig. 7.8:** Courtesy Bill Lutterschmidt; **P. 86:** Courtesy Bill Lutterschmidt; **Fig. 8.1a:** © Ed Reschke, Photography; **Fig. 8.1b:** © Ed Reschke, Photography; **Fig. 8.1c:** © McGraw-Hill Higher Education, Inc./Dennis Strete, photographer; **Fig. 8.3:** Photograph used by permission of Dr. Marion L. Greaser, University of Wisconsin-Madison; **P. 115:** John C. Stroud, Copyright 2005, All rights reserved; **Fig. 10.3c:** © CNR/Phototake; **P. 131 top:** © Andrew J. Martinez/Photo Researchers, Inc.; **P. 131 bottom:** © Frederick R. McConnaughey/Photo Researchers, Inc.; **Fig. 11.1:** © Ed Reschke, Photography; **P. 142:** © Eye of Science/Photo Researchers, Inc.; **Fig. 12.2a:** © Bruce Iverson; **Fig. 12.2b:** © Eye of Science/Photo Researchers, Inc.; **12.3: ©** Phototake; **Fig. 12.5:** © Satum Stills/Photo Researchers, Inc.; **P. 158:** Andrew J. Martinez/Photo Researchers, Inc.; **Fig. 13.2:** © SIU; **Fig. 13.5a:** Courtesy Warren E. Collins, Inc.; **Fig. 13.6:** Courtesy Riester; **P. 170:** © Breck P. Kent/ Animals Animals; **P. 196:** © Charles George/ Visuals Unlimited

INDEX

Page numbers for boxes are indicated by b, figures by f, and tables by t.
95% confidence intervals (95% CI), 5–6, 6f

A

A band (sarcomere), 93, 93f
A factor (blood), 149, 150t
Abdominal muscles, 161, 162f
Absolute refractory period, 50
Abstract writing, 203
Accommodation, ocular, 74, 75f
Accuracy, definition of, 3
Acetylcholine
  in heart rate regulation, 137
  in motor neuron activation, 96, 107
  receptor, 107
Acidosis, metabolic, 37
ACTH (adrenocorticotropic hormone)
  manipulation in rats (experiment), 123–125, 129t
  secretion of, 122, 122f
Actin, 95
Action potentials, 45–58
  in cardiac muscle contraction, 134–135, 136
  compound, 50, 50f
  conduction velocity and axon myelination, 54
  determining velocity of, 62–63
  electrical events in, 47–48, 136
  experiments, 51–53, 55t, 57t, 58t
  membrane potential and permeability changes, 47–48, 49f
  in muscle contraction, 96
  Na^+/K^+ pump in, 45
  positive feedback in, 49
  refractory period in, 49
  resting membrane potential, 46–47, 46f
  threshold voltage in, 49, 108
  voltage-gated ion channels in, 47–48
Activation energy, 37–38, 38f
Active transport, 24
Adenosine triphosphate (ATP)
  metabolic rate and production of, 191
  in muscle contraction, 97, 109
ADH (antidiuretic hormone), 131, 180
Adrenal glands, 122, 122f
Adrenocorticotropic hormone (ACTH)
  manipulation in rats (experiment), 123–125, 129t
  secretion of, 122, 122f
Afferent (sensory) neurons, 59
Affinity (enzymes), 37
After-hyperpolarization, 48
Afterimages
  problem set, 91
  testing of, 81, 82f–84f
Agglutination, 149, 150f
Agonists (drug), 137
Agranulocytes, 151
Airway, anatomy of, 159
Airway resistance, 165
Alcohol consumption, 180
Alcohol dehydrogenase, 38
Alkalosis, metabolic, 37
Alveolar sacs (alveoli), 159, 160f
Amino acids, 182
Ammonia (NH_3), 180, 182
Amphibians, 182
Anabolic enzymes, 35
Anabolism, 191
Analysis of variance (ANOVA), 7, 7b
Anatomic dead space, 159
Anemia, 148–149
Antagonists (drugs), 137
Antidiuretic hormone (ADH), 131, 180
Aortic bodies, 166
Aplastic anemia, 148
Aquaporins, 179
Aquatic mammals, 64
Arginine vasopressin (AVP), 131
Arginine vasotocin (AVT), 131
Ascending distance, 79
Ascending limb (kidney tubule), 178, 179f
Asthma, 165
ATP (adenosine triphosphate)
  metabolic rate and production of, 191
  in muscle contraction, 97, 109
Atrioventricular (AV) node, 134
Atrophy, 123
Auditory canal, external, 73, 73f
Auditory system, 72–73
  anatomy of, 73, 73f
  clinical investigation of, 79–80
  tuning fork for testing, 72, 72f
Autocrine stimulation, 118
Autonomic nervous system, 59, 60f, 61
AV node, 134
AVP (arginine vasopressin), 131
AVT (arginine vasotocin), 131
Axons, 54

B

B factor (blood), 149, 150t
Basal metabolic rate (BMR), 192
Basophils, 151
Bears, hibernation in, 196
Bell curve, 5–6, 5f
Benedict's reagent, 38, 39f
Bernard, Claude, 13
Beta-adrenergic receptors, 137
Bicarbonate, 165
Bladder, urinary, 178, 178f
Blind spot
  measurement of, 80–81, 82f
  problem set, 91
  in visual field, 74
Blood
  clotting, 148
  as connective tissue, 147
  experiments, 152–154, 155t, 156t
  formed vs. unformed elements of, 148
  major functions of, 148
  normal values, 210t
  physiology of, 147–158
  problem sets, 155, 156–157
  respiration and, 165–166

Blood cells
 classes of, 151, 152f
 normal values, 210t
 numbers and distribution of, 151, 152t
Blood factors and types, 149, 150t
Blood gases, normal values, 209t
Blood plasma, normal values, 210t
Blood pressure
 body position and, 45t, 138–140
 measurement of, 138, 138f, 140f
Blood typing, 153, 155, 155t
Blue-headed wrasse *(Thalassoma bifasciatum),* 131
BMR (basal metabolic rate), 192
Body position, blood pressure and, 139–140, 145t
Body temperature
 blood and regulation of, 148
 comparative physiology and, 16
 enzyme activity and, 37, 41, 44
 homeostasis in, 13
 set point for, 13
Body water, 23
Bohr effect, 166
Bowman's capsule, 178
Boyle's law, 160
Bronchioles, 159, 160f
Bronchus, 160f
Bundle of His, 134

C

Caffeine, heart rate and, 137, 143t, 144
Calcitonin, 118
Calcium ions (Ca^{2+}), 96–97, 109
Calorimetry, 191, 192f, 193f
Capillary tube, filling of, 153, 153f
Carbon dioxide (CO_2)
 concentration and ventilation, 168–169, 173t, 174, 174f, 175t–176t
 human vs. fish ventilation rates, 170
 respiration and, 165–166
Carbonic acid, 165–166
Carbonic anhydrase, 166
Cardiac cycle, 134–137
Cardiac muscle
 anatomy, 93, 94f, 133, 134f
 contraction of, 134–135, 135f
 refractory period, 133–134
Cardiovascular physiology, 133–146. *See also* heart
 blood pressure, 138, 138f
 cardiac cycle, 133–137
 cardiac function, normal values, 209t
 cardiac output, 141
 drugs and endogenous factors affecting, 137
 experiments, 139–141, 143t, 145t
 heart sounds, 138
 open circulatory systems in invertebrates, 142
 problem sets, 144, 146
 variables, calculation of, 140–141, 146
Carotid bodies, 166
Catabolic enzymes, 35
Catabolism, 191
Cause-and-effect relationships, 8
Cell membrane permeability, 24, 47–48, 49f
Central chemoreceptors, 166
Chemoreceptors, 166
Chlorocruorin, 158
Cochlea, 73, 73f
Collecting duct, 178, 179f
Comparative phylogenetic method, 2
Comparative physiology
 blood, 158
 circulatiory systems, 142
 comparative phylogenetic method in, 2
 conduction velocity and axon myelination, 54
 enzyme activity, 49
 hibernation, 196
 homeostasis, 16
 hormone activity, 131
 muscles, 102, 115
 nitrogenous waste production, 182
 osmoregulatory mechanisms in fish, 33
 reflexes, 64
 sensory, 86
 sex change in fish, 131
 ventilation rate in fishes, 170
Compound action potentials, 50, 50f, 52, 57, 57t
Compression (sound waves), 72, 72f
Concentration gradients, 45–46
Conducting zone, airway, 159
Conduction deafness, 79–80
Conduction velocity, 54
Cones (eye), 74
Confidence intervals, 95%, 5–6, 6f
Contralateral reflex, 61
Cornea, 73
Correlation (statistics), 8–9, 9b, 9f
Corticotropin-releasing hormone (CRH), 122, 122f
Cortisol
 glucose metabolism and, 193
 manipulation in rats (experiment), 123–125, 129t
 secretion of, 120, 122, 122f
Countercurrent multiplier systems, 178, 180f
Cross-bridges
 anatomy of, 95, 95f
 in cardiac cycle, 136
 in muscle contraction, 96–98, 107
 power stroke of cycling, 96–97
Crossed extensor reflex, 62, 63f
Curare, 107–108
Cutaneous receptors, 79

D

Deafness, testing for, 79–80
Defensins, 151
Depolarization, 48, 136
Descending distance, 79
Descending limb (kidney tubule), 178, 179f
Descriptive statistics, exercises, 4–5
Diaphragm
 anatomy of, 162f, 178f
 respiration and, 160, 161f
Diastole, 134, 136
Diastolic blood pressure, 138, 138f
Diffusion, 23–33. *See also* osmosis
 cell membrane permeability in, 24
 experiment, 27, 29t–30t
 facilitated, 24
 problem set, 30
 rate of, 24
Diffusion chamber
 schematics and suppliers, 213–214
Dihydropyridine (DHP), 96
Diopters, 76
Distal convoluted tubule, 178, 179f
Diving reflex, 64

E

Ear, 73, 73f
Eardrum, 73, 73f
Ectotherms, 16
Effectors, 14, 59
Efferent (motor) neurons, 59, 108, 108f
Electrical gradients, 45–46
Electrocardiogram (ECG; EKG), 136, 136f
End diastolic volume, 135
End-plate potential, 96, 107
Endocrine gland, 117
Endocrine physiology, 117–131. *See also* hormones
 experiments, 123, 127t–128t
 homeostasis and, 117
 problem sets, 120, 128–129
 relationship with nervous system, 117
Endotherms, 16
Energy expenditure, 192, 192t
Energy metabolism, 191. *See also* metabolism
Enzymes, 35–44
 activation energy and catalysis, 37–38, 38f
 anabolic vs. catabolic, 35
 comparative physiology, 49

experiments, 41, 43–44
naming of, 38
pH effects on, 37, 41, 43–44
problem sets, 43–44
secondary structure, 36, 36f
specificity of, 35
substrate interactions, 35, 36f
temperature effects on, 41, 44
tertiary structure, 35, 36, 36f
Eosinophils, 151
Equilibrium potential, 47
ERV (expiratory reserve volume), 162, 167
Erythoblastosis fetalis, 149
Erythrocytes
antigens, 149, 150t
normal values, 210t
respiration and, 148
Excretion
blood transport and, 148
nitrogenous wastes, 182
urinary, 180–181
Exocrine system, 117–118
Expiration, 160
Expiratory reserve volume (ERV), 162, 167
Eye, anatomy of, 73–74, 73f

F

Facilitated diffusion, 24
Feedback loops
components of, 14b
definition, 14
mechanisms of, 13–14
negative, 14, 120–122, 121f, 122f
positive, 15, 49
Felsenstein, Joseph, 2
Filtrate, urinary, 178
Fishes
ammonia excretion in, 182
osmoregulatory mechanisms, 33
sex change in, 131
ventilation rate in, 170
Fluid compartments, 24f
Foot proteins (junctional feet), 96
Fovea centralis, 74
Frank-Starling law, 135
Frequency, 3
Fused tetanus, 109

G

Gap junctions, 133, 134f
Glomerular filtration rate (GFR), 181
Glomerulus, 178, 179f
Gluconeogenesis, 177
Glucose, 24, 193
Glycoproteins, 149
Goiter, 121, 121f
Goldfish *(Carassius auratus)*
hormone effects on ventilation, 123, 127t–128t
metabolism and ventilation, 122, 122f
Goldman-Hodgkin-Katz equation, 46
Gonadotropin-releasing hormone, 120
Granulocytes, 151
Growth hormone, 120

H

Hair cells, 73
Hearing. *See* auditory system
Heart, 160f. *See also* cardiac muscle; cardiovascular physiology
atrium, 135
cardiac output, 141
function, normal values, 209t
physiology, 133–135, 135f
ventricles, 135
Heart rate
athletes vs. non-athletes, 17, 19t
caffeine and, 137, 143t, 144
drugs and endogenous factors affecting, 137
Heart sounds, 138
Hematocrit
calculation of, 154f
measurement of, 148, 148f
sex differences in, 153–154, 153f, 156–157, 156t
Hemerythrin, 158
Hemocoel, 142
Hemocyanin, 158
Hemoglobin, 158, 165–166
Hemolytic disease of newborn, 149
Hermaphrodites (fish), 131
Hibernation, 196
His, bundle of, 134
Homeostasis, 13–21
comparative physiology, 16
definition, 1
endocrine system and, 117
experiments, 17, 19t
feedback loops, 13–15, 120–122, 121f, 122f
problem sets, 19–21
Hormones. *See also* endocrine physiology
autocrine vs. paracrine stimulation, 118
blood and regulation of, 148
definition, 118
experiments, 123–125, 129t
metabolic rate and, 193
problem set, 130
secretion rate, 120
target sites and actions, 118, 119t
Hydrogen ion concentration. *See* pH (hydrogen ion concentration)
Hyperopia (farsightedness), 75, 76f
Hyperpolarization, 48
Hypertonic solution, 26, 177, 183
Hypertrophy, 121, 121f, 123
Hyperventilation, 166
Hypothalamus
anatomy of, 120f
hormone secretion, 120, 122, 122f
Hypothyroidism, 121
Hypotonic solution, 26, 177, 183

I

I band (sarcomere), 93, 93f
Inactivation gate, 47–48, 48f
Incus, 73, 73f
Independent contrasts, 2
Indirect calorimetry, 191, 193f
Induced-fit model, 35, 36f
Infinite distance (accommodation), 74
Insects, open circulatory system in, 142
Inspiration, respiratory, 160
Inspiratory reserve volume (IRV), 162
Integrating center (feedback), 13
Intercalated discs, 133, 134f
Intercostal muscles, 161, 162f
Interstitial fluid, 178
Invertebrates, open circulatory system in, 142
Iodine deficiency, 121
Ion channels
acetylcholine receptor, 107
voltage-gated, 47–48, 48f
Ion pump, sodium/potassium, 45
Ipsilateral reflex, 61
IRV (inspiratory reserve volume), 162
Isotonic solution, 26, 177, 183

J

Joules, 191
Junctional feet (foot proteins), 96

K

Key-and-lock model (enzymes), 35, 36f
Kidney. *See also* renal physiology
anatomy of, 177–178, 178f
cross section of, 180f
function of, 177–181, 209t
Kinetic energy, 37

L

Lactate dehydrogenase, 38
Lactation, 193
Learned response time
experiments, 66, 66f, 68t–69t
problem set, 70
unlearned response vs., 62, 65–66

Lens, crystalline, 74
Leptin, 193
Leukocytes
 differential count, 157
 in immune response, 148
 normal values, 210t
 types of, 151
Loop of Henle, 178, 179f
Lung diseases, retrictive vs. obstructive, 165
Luteinizing hormone (LH), 123–125, 129t
Lymphocytes, 151

M

Malleus, 73, 73f
Mass-specific metabolic rate, 194–195, 194f
Maturation anemia, 149
Maximal stimulus voltage, 108–109
Mean arterial pressure (MAP), 140–141, 146
Mean (statistics), 3, 3b
Measures of central tendency, 4
Measures of dispersion, 4
Median (statistics), 3, 3b
Medulla oblongata, 166
Melatonin, 120
Membrane permeability, 24, 47–48, 49f
Membrane potentials
 action potentials and changes in, 47–48, 49f
 resting, 46, 46f
Metabolic acidosis, 37
Metabolic alkalosis, 37
Metabolic rate, 191–200
 calorimetry vs. respirometry, 191, 192f, 193f
 energy expenditure and activity, 192, 192t
 experiments, 197–198, 199t
 hibernation and, 196
 hormones and, 193
 mass-specific, 194–195, 194f
 physiologic conditions and, 191–192
 problem sets, 199–200, 200t
Metabolism, definition, 191
Micturition, 181
Middle ear, 73, 73f
Minute ventilation, 174, 174f
Mode, 3
Mode (statistics), 3b
Molarity, 25, 25b, 25t
Molecular transport, 24
Monocytes, 151
Monosynaptic reflex, 61
Morphology, metabolic rate and, 194–195, 194f
Motor (efferent) neurons, 59, 108, 108f
Motor unit recruitment, 108, 108f
 experiments, 111–112, 113t
 problem set, 114
"Mouse-elephant curve," 194, 194f
Mullins, Kary, 49
Muscarinic acetylcholine receptors, 137
Muscle contraction, 107–115
 after-loaded, 100–101, 100f, 105t
 cardiac muscle, 134–135, 135f
 direct-loaded, 99–100, 100f, 105t
 experiments, 99–101, 99f–100f, 103t–105t, 111, 114, 114f
 in flying insects, 115
 motor unit recruitment, 108, 108f, 111–112, 113t, 114
 muscle length-tension relationship, 99, 99f, 103t–104t, 111, 114f
 muscle work, calculation of, 101
 problem sets, 105–106, 113, 114
 reciprocal inhibition in, 61–62
 reflexes and, 61–62
 sliding-filament mechanism of, 96–98
 stimulus frequency, 109, 109f
 tetanus, 109, 112, 114
 wave summation, 109, 112, 114, 114f
Muscle fatigue, 109–110, 110f
Muscle fibers, 93, 95f, 102
Muscle spindles, 61, 62f
Muscle stretch reflexes
 experiments, 65, 65f
 physiology, 61–63, 62f, 63f
Muscle twitch, 109, 110f
Muscles
 comparative anatomy of, 102, 115
 in flying insects, 115
 functional anatomy of, 93–96
 respiratory, 161, 162f
 types of, 93, 94f, 102
Myocardium. *See* cardiac muscle
Myofibers. *See* muscle fibers
Myofibrils
 anatomy of, 93, 95f
 thick and thin filaments, 93, 95–96, 95f, 96f
Myopia, 75, 76f
Myosin, 95, 96–98

N

Na^+/K^+ pump in, 45
Nasal cavity, 159
Nearsightedness, 75, 76f
Negative feedback. See feedback loops, negative
Nephron, 178, 179f
Nernst equation, 47
Nervous system
 autonomic, reflexes and, 59, 60f, 61
 functional organization of, 60f
 hormones and, 117
 parasympathetic, heart rate and, 137
Net diffusion, 23–24
Neural pathways, reflex arc, 59
Neurons
 afferent (sensory), 59
 efferent (motor), 59, 108, 108f
Neutrophils, 151
Norepinephrine, 137
Normal probability distribution, 5–6, 5f
Null hypothesis (H_0), 3
Nutrients, blood transport of, 148

O

Obstructive lung diseases, 165
Open circulatory system, 142
Optic disc, 74, 81, 81f
Oral cavity, 159
Ornithine cycle, 180–181, 182
Osmolarity, 25, 25b, 25t
Osmoreceptors, 13
Osmoregulatory mechanisms, 33
Osmosis. *See also* diffusion
 definition, 24–25
 problem set, 31–32
 tonicity and, 27–28, 31t
Osmosis chamber, 27f
Osmotic pressure, 25
Osteoblasts, 118
Osteoclasts, 118
Oxidative phosphorylation, 191
Oxygen, blood transport of, 165–166
Oxygen consumption
 body mass and, 194–195, 194f
 human vs. fish ventilation rates, 170
 metabolic rate and, 191, 193f

P

P-value (probability), 7
P-wave (ECG), 136, 136f
Pacemaker cells, cardiac, 134
Paired *t*-test, 144, 146
Pancreas, 118
Para-nitrophenyl phosphate, 39f
Paracrine stimulation, 118
Parameter (statistics), 3
Parasympathetic nervous system, 137
Patellar tendon reflex arc, 61–62, 62f, 65
PCR (polymerase chain reaction), 49
Pectoralis minor muscle, 161, 162f
Peer review, 205–208
Peripheral chemoreceptors, 166
Peritubular capillaries, 178
Pernicious maturation anemia, 149
PH (hydrogen ion concentration)
 calculation of, 36
 chemoreceptors and monitoring of, 166
 enzyme activity and, 36, 41, 43–44, 43t
 normal values, 166
 respiration and, 165–166

Phosphatase, 39f
Photoreceptors, 74
Phylogeny, 2
Physical activity, 192, 192t
Physiology. *See also specific system* physiology
  definition of, 1
  phylogeny effects on, 2
Pituitary gland
  anatomy of, 120f
  negative feedback and hormone secretion, 120, 121f, 122f
Plagerism and falsified data, 202
Plasma, normal values, 210t
Platelets, 148
Polymerase chain reaction (PCR), 49
Polymorphonuclear leukocytes, 151
Population, 3
Positive feedback, 15, 49
Potassium channels, voltage-gated, 47–48, 48f
Potassium (K^+) ions, 51–52, 55t, 57
Precision, 3
Presbyopia, 75
Probability, 7
Prolactin, 193
Propylthiouracil, 123, 127t–128t, 128–129
Protein transporters, 24
Protoandrous hermaphrodite, 131
Protogynous hermaphrodite, 131
Proximal convoluted tubule, 178, 179f
Pulmonary arteries, 160f
Pulmonary fibrosis, 165
Pulmonary function
  measurement of, 162–165, 167–168, 171, 172f
  normal values, 209t
Pulmonary veins, 160f
Pulse pressure, 139–140
Pupil reflex, 65
Purkinje fibers, 135, 136

Q

QRS-wave (ECG), 136, 136f

R

Range (statistics), 3, 4b
Reaction time, 65–67
Receptors, in feedback, 13
Reciprocal inhibition, 61–62
Red blood cells. *See* erythrocytes
Reference tables
  blood gases, 209t
  blood values, 210t
  cardiac function, 209t
  pulmonary function, 209t
  renal function, 209t
  urine, 209t
Reflex arc
  afferent (sensory) neurons in, 59
  components of, 59
  monosynaptic, 61, 62f
  polysynaptic, 62, 63f
Reflexes, 59–70
  comparative physiology of, 64
  experiments, 65–67
  ipsilateral vs. contralateral, 61
  learned vs. unlearned response, 62
  nervous system function and, 59, 60f, 61
  problem sets, 67, 70
  somatic vs. autonomic, 59, 60f, 61
  terminology of, 59
Refraction (sound), 72, 72f
Refractory period
  action potentials and, 49
  cardiac muscle, 133–134
  experiments, 52–53, 58t
  problem sets, 58
Regression analysis, 7–8, 8b, 8f
Relative frequency, 3
Relative refractory period, 50
Renal corpuscle, 178, 179f
Renal pelvis, 178
Renal physiology, 177–189. *See also* kidney
  experiments, 182, 185t–188t
  normal values, 209t
  problem set, 189
Renal tubules, 178, 179f
Repolarization, 48, 136
Residual volume (RV), 163
Respiratory control center, 161
Respiratory physiology, 159–176
  airway zones, 159
  anatomy of, 159, 160f
  blood supply, 160f
  erythrocytes in, 148
  experiments, 167–169, 173t
  inspiration and expiration, 160
  lung volumes and capacities, 162–165, 163f, 165t
  medulla oblongata and, 161
  muscles of, 161, 162f
  problem sets, 171, 174, 175t–176t
  ventilation rate control, 165–166
Respiratory pigments, 158
Respiratory zone, 159
Respirometry, 193f
Resting membrane potential, 46–47, 46f
  experiments, 51–52, 55t
  problem sets, 56
Restrictive lung diseases, 165
Retina, 74
Rh factor (blood), 149, 150t
Rh-immune globulin (RhoGAM), 149
Rinne's Test, 80, 90
Rods (eye), 74
Rubner's surface rule, 194, 195f
RV (residual volume), 163
Ryanodine, 96

S

SA (sinoatrial) node, 135, 137
Sample (statistics), 3
Sarcolemma, 96, 107
Sarcomere, 93, 95, 95f, 96f
Sarcoplasmic reticulum, 96, 97f
Scalene muscles, 161, 162f
Scientific investigation, 1–12. *See also* statistics
Scientific misconduct, 202
Scientific writing, 201–208
  abstract, 203
  authorship credits, 203
  completeness in, 202
  discussion in, 204
  fundamentals of, 201–202
  introduction in, 203
  literature citations, 204–205
  materials and methods in, 203–204
  peer review process, 205–208
  plagerism and falsified data, 202
  popular writing vs., 201
  results in, 203
  title of manuscript, 202–203
Sclera, 73
Sensorineural deafness, 79–80
Sensors, in feedback, 13
Sensory (afferent) neurons, 59
Sensory physiology, 71–91
  auditory, 72–73, 79–80
  comparative, 86
  experiments, 79–85, 87t–89t
  problem sets, 90–91
  stimulus modality in, 71
  two-point threshold test, 79, 87t–89t, 90
  visual, 73–78, 80–81, 82f–84f, 85
Sensory receptors
  dendritic endings of, 71
  distribution of, 72
  in reflex arc, 59
Set point, 13, 16
Sex change in fishes, 131
SI unit prefixes and symbols, 211
Sickle cell anemia, 148–149, 149f
Sinoatrial (SA) node, 135, 137
Skeletal muscle, 93, 94f–95f, 95–96. *See also* muscle contraction
Sliding-filament mechanism, 96–98
Smooth muscle, 93, 94f
SMR. *See* standard metabolic rate (SMR)
Snakes, sensory receptors in, 86
Snellen eye chart, 76, 78f, 85, 91
Sodium, dietary, 189
Sodium channels, voltage-gated, 47–48, 48f, 50
Sodium (Na^+) ions
  in end-plate potential, 107
  extracellular concentration, 52, 55t, 56
  reabsorption into interstitial fluid, 178
Sodium/potassium ion (Na^+/K^+) pump in, 45

Solute, 23, 24f
Solutions
  concentration and diffusion rate, 27
  definition of, 23
  osmotic pressure, 25, 25t
  tonicity of, 26, 26f
Solvents, 23, 24–25
Sound waves, 72, 73
Specific gravity, 183
Specificity of enzymes, 35
Sphygmomanometry, 138, 138f, 140f
Spirocomp procedure, 168
Spirography, 163, 164f, 165, 165t
Spirotest procedure, 167, 167f
Standard deviation, 4, 4b
Standard error, 4, 4b
Standard metabolic rate (SMR)
  body size and, 197–200, 199t–200t
  measurement of, 192
Stapes, 73, 73f
Statistics, 4–9
  95% confidence intervals, 5–6, 6f
  analysis of variance (ANOVA), 7, 7b
  concepts and tests, 10
  correlation, 8–9, 9b, 9f
  descriptive statistics, 4–5
  measures of central tendency, 4
  measures of dispersion, 4
  normal probability distribution, 5–6, 5f
  problem sets, 11–12
  regression analysis, 7–8, 8b, 8f
  Student's *t*-test, 5f, 6–7
  terms and definitions in, 3–4
Stereocilia, 73
Sternocleidomastoid muscle, 161, 162f
Stethoscope, 138, 140f
Striated muscle, 93. *See also* skeletal muscle
Stroke volume, 141
Student's *t*-test, 6–7, 7b
  paired, 144, 146
Subthreshold stimulus, 49, 108
Suprathreshold stimulus, 49
Suspensory ligaments, 74
Systole, 134, 136
Systolic blood pressure, 138, 138f

T

T-test. *See* Student's *t*-test
T-tubules, 96, 97f
T-wave (ECG), 136, 136f
Taq polymerase, 49
Temperature. *See also* body temperature
  kinetic energy and, 37
Tertiary structure (enzymes), 35, 36, 36f
Testosterone
  hemostatic regulation of, 17, 21
  manipulation in rats, 123–125, 129t
Tetanus, 109, 112, 114
Thermus aquaticus, 49
Threshold test, two-point (touch), 79, 87t–89t, 90
Threshold voltage, 49, 108
Thrombocytes, 148
Thyroid gland, 121, 121f
Thyroid hormones
  metabolic rate and, 193
  secretion of, 120, 121f
Thyroid-stimulating hormone (TSH)
  manipulation in rats (experiment), 123–125, 129t
  secretion of, 120, 121f
Thyrotropin-releasing hormone (TRH)
  manipulation in rats (experiment), 123–125, 129t
  secretion of, 120, 121f
Thyroxine (T_4)
  goldfish ventilation and, 123, 127t–128t, 128–129
  secretion of, 120, 121f
Tidal volume (TV), 162, 168
Tonicity
  definition, 26
  experiments and problem sets, 27–28, 31–32, 31t
  osmotic pressure and, 26, 26f
Total lung capacity (TLC), 163
Total peripheral resistance, 141
Touch, two-point threshold test, 79, 87t–89t, 90
Trachea, 160f
Transduction, sensory, 71
Transfusion reactions, 149
Transport proteins, 24
Transverse tubules (T-tubules), 96, 97f
Treppe, 114, 114f
TRH. *See* thyrotropin-releasing hormone (TRH)
Triiodothyronine (T_3), 120, 121f
Tropomysin, 95
Troponin, 96
Troponin-tropomyosin complex, 96, 109
TSH. *See* thyroid-stimulating hormone (TSH)
Tuning fork, 72, 72f, 79f
TV (tidal volume), 162, 168
Two-point threshold test (touch), 79, 87t–89t, 90
Two-thirds rule (Rubner's surface rule), 194
Tympanic membrane, 73, 73f

U

Unfused tetanus, 109
Unlearned response time, 65–66, 68t–69t, 70
Urea, 180–181, 182
Ureter, 178, 178f
Urethra, 178, 178f
Uric acid, 182
Urinary bladder, 178, 178f
Urinary system, 178f
Urination, 181
Urine
  constituents, normal values, 209t
  ion concentration and, 183, 185t–188t, 189
  production of, 178, 180f

V

Vagus nerve, 137
Variables
  association vs. cause-and-effect, 8–9
  cardiovascular, 140–141, 146
  definition of, 3
Variance, 4b
Variance (statistics), 4
Variate (statistics), 3
Vasopressin, 131, 180
Visual acuity
  clinical investigation of, 80–85
  definition of, 75
  fovea centralis and, 74
  measurement of, 76, 77f, 78f
  problem set, 91
Visual defects, 75–76, 76f
Visual system physiology, 73–78
Vital capacity (VC), 162
  in females, 175t
  in males, 176t
  measurement of, 167, 167f
Vitamin deficiencies, anemia and, 149
Voluntary muscles, 93. *See also* skeletal muscle

W

Water balance, 180
Water channels (kidney), 179
Wave summation, 109, 109f, 114, 114f
Weber's Test, 80, 90
White blood cells. *See* leukocytes
Writing. *See* scientific writing

Z

Z line, 93, 95f, 96f
Zonal fibers, optic, 74

Unit Conversions of Measure

	Quantity	Numerical Value	English Equivalent	Converting English to Metric
Length	kilometer (km)	1000 (10^3) meters	1 km = 0.62 mile	1 mile = 1.609 km
	meter (m)	100 centimeters	1 m = 1.09 yards	1 yard = 0.914 m
			= 3.28 feet	1 foot = 0.305 m
	centimeter (cm)	0.01 (10^{-2}) meter	1 cm = 0.394 inch	= 30.5 cm
	millimeter (mm)	0.001 (10^{-3}) meter	1 mm = 0.039 inch	1 inch = 2.54 cm
	micrometer (μm)	0.000001 (10^{-6}) meter		
	nanometer (nm)	0.000000001 (10^{-9}) meter		
	angstrom (Å)	0.0000000001 (10^{-10}) meter		
Area	square kilometer (km^2)	100 hectares	1 km^2 = 0.3861 square mile	1 square mil = 2.590 km^2
	hectare (ha)	10,000 square meters	1 ha = 2.471 acres	1 acre = 0.4047 ha
	square meter (m^2)	10,000 square centimeters	1 m^2 = 1.1960 square yards	1 square yard = 0.8361 m^2
			= 10.764 square feet	1 square foot = 0.0929 m^2
	square centimeter (cm^2)	100 square millimeters	1 cm^2 = 0.155 square inch	1 square inch = 6.4516 cm^2
Mass	metric ton (t)	1000 kilograms =	1 t = 1.103 ton	1 ton = 0.907 t
		1,000,000 grams		1 pound = 0.4536 kg
	kilogram (kg)	1000 grams	1 kg = 2.205 pounds	1 ounce = 28.35 g
	gram (g)	1000 milligrams	1 g = 0.0353 ounce	
	milligram (mg)	0.001 gram		
	microgram (μg)	0.000001 gram		
Time	second (sec)	1000 milliseconds		
	millisecond	0.001 second		
	microsecond	0.000001 second		
Volume (solids)	1 cubic meter (m^3)	1,000,000 cubic centimeters	1 m^3 = 1.3080 cubic yards	1 cubic yard = 0.7646 m^3
			= 35.315 cubic feet	1 cubic foot = 0.0283 m^3
	1 cubic centimeter (cm^3)	1000 cubic millimeters	1 cm^3 = 0.0610 cubic inch	1 cubic inch = 16.387 cm^3
Volume (liquids)	kiloliter (kl)	1000 liters	1 kl = 264.17 gallons	1 gal = 3.785 l
	liter (l)	1000 milliliters	1 l = 1.06 quarts	1 qt = 0.94 l
	milliliter (ml)	0.001 liter	1 ml = 0.034 fluid ounce	1 pt = 0.47 l
	microliter (μl)	0.000001 liter		1 fluid ounce = 29.57 ml

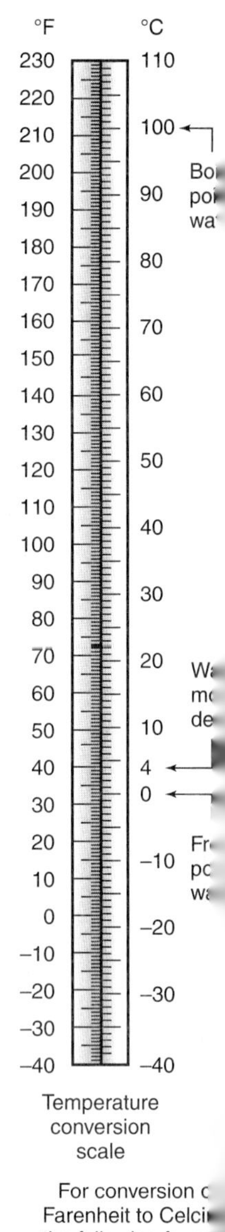

Temperature conversion scale

For conversion c Farenheit to Celci the following form can be used:

$°C = \frac{5}{9}(°F - 32$

For conversion of Ce to Farenheit, the foll formula can be us

$°F = \frac{9}{5}°C + 3$